Sporadic Groups is the first step in a program to provide a uniform, self-contained treatment of the foundational material on the sporadic simple groups necessary for the classification of the finite simple groups. The classification of the finite simple groups is one of the premier achievements of modern mathematics. It demonstrates that each finite simple group is either a finite analogue of a simple Lie group or one of twenty-six pathological sporadic groups. *Sporadic Groups* provides for the first time a self-contained treatment of the foundations of the theory of sporadic groups accessible to mathematicians with a basic background in finite groups such as in the author's text *Finite Group Theory*.

Introductory material useful for studying the sporadics, such as a discussion of large, extraspecial 2-subgroups and Tits's coset geometries, opens the book. A construction of the Mathieu groups as the automorphism groups of Steiner systems follows. The Golay and Todd modules and the 2-local geometry for M_{24} are discussed. This is followed by the standard construction of Conway of the Leech lattice and the Conway group. The Monster is constructed as the automorphism group of the Griess algebra using some of the best features of the approaches of Griess, Conway, and Tits, plus a few new wrinkles. The existence treatment finishes with an application of the theory of large extraspecial subgroups to produce the twenty sporadics involved in the Monster.

The Aschbacher–Segev approach addresses the uniqueness of the sporadics via coverings of graphs and simplicial complexes. The basics of this approach are developed and used to establish the uniqueness of five of the sporadics.

Researchers in finite group theory will find this text invaluable. The subjects treated will interest combinatorists, number theorists, and conformal field theorists.

CAMBRIDGE TRACTS IN
MATHEMATICS

General Editors
B. BOLLOBAS, P. SARNAK, C. T. C. WALL

104   Sporadic groups

MICHAEL ASCHBACHER
Department of Mathematics
California Institute of Technology

Sporadic groups

CAMBRIDGE
UNIVERSITY PRESS

CAMBRIDGE UNIVERSITY PRESS
Cambridge, New York, Melbourne, Madrid, Cape Town, Singapore, São Paulo

Cambridge University Press
The Edinburgh Building, Cambridge CB2 8RU, UK

Published in the United States of America by Cambridge University Press, New York

www.cambridge.org
Information on this title: www.cambridge.org/9780521420495

First published 1994
This digitally printed version 2008

A catalogue record for this publication is available from the British Library

Library of Congress Cataloguing in Publication data
Aschbacher, Michael.
    Sporadic groups / Michael Aschbacher.
      p.   cm. – (Cambridge tracts in mathematics ; 104)
    Includes bibliographical references and indexes.
    ISBN 0-521-42049-0
      1. Sporadic groups (Mathematics)   I. Title.   II. Series.
  QA177.A83   1994
512'.2 – dc20   92–13653
                                                        CIP

ISBN 978-0-521-42049-5 hardback
ISBN 978-0-521-05686-1 paperback

Contents

Preface

The classification of the finite simple groups says that each finite simple group is isomorphic to exactly one of the following:

A group of prime order
An alternating group A_n of degree n
A group of Lie type
One of twenty-six sporadic groups

As a first step in the classification, each of the simple groups must be shown to exist and to be unique subject to suitable hypotheses, and the most basic properties of the group must be established. The existence of the alternating group A_n comes for free, while the representation of A_n on its n-set makes possible a first uniqueness proof and easy proofs of most properties of the group. The situation with the groups of Lie type is more difficult, but while groups of Lie rank 1 and 2 cause some problems, Lie theory provides proofs of the existence, uniqueness, and basic structure of the groups of Lie type in terms of their Lie algebras and buildings.

However, the situation with the sporadic groups is less satisfactory. Much of the existing treatment of the sporadic groups remains unpublished and the mathematics which does appear in print lacks uniformity, is spread over many papers, and often depends upon machine calculation.

Sporadic Groups represents the first step in a program to provide a uniform, self-contained treatment of the foundational material on the sporadic groups. More precisely our eventual aim is to provide complete proofs of the existence and uniqueness of the twenty-six sporadic groups subject to appropriate hypotheses, and to derive the most basic structure of the sporadics, such as the group order and the normalizers of subgroups of prime order.

While much of this program is necessarily technical and specialized, other parts are accessible to mathematicians with only a basic knowledge of finite group theory. Moreover some of the sporadic groups are the automorphism groups of combinatorial objects of independent interest, so it is desirable to make this part of the program available to as large an audience as possible. For example, the Mathieu groups are the automorphism groups of Steiner systems and Golay codes while the largest Conway group is the automorphism group of the Leech lattice.

Sporadic Groups begins the treatment of the foundations of the sporadic groups by concentrating on the most accessible chapters of the subject. It is our hope that large parts of the book can be read by the nonspecialist and provide a good picture of the structure of the sporadics and the methods for studying these groups. At the same time the book provides the basis for a complete treatment of the sporadics.

The book is divided into three parts: Part I, introductory material (Chapters 1-5); Part II, existence theorems (Chapters 6-11); and Part III, uniqueness theorems (Chapters 12-17).

The goal of the existence treatment is to construct the largest sporadic group (the Monster) as the group of automorphisms of the Griess algebra. Twenty of the twenty-six sporadic groups are sections of the Monster. We establish the existence of these groups via these embeddings. To construct the Griess algebra one must first construct the Leech lattice and the Conway groups, and to construct the Leech lattice one must first construct the Mathieu groups, their Steiner systems, and the binary Golay code.

There are many constructions of the Mathieu groups. Our treatment proceeds by constructing the Steiner systems for the Mathieu groups as a tower of extensions of the projective plane of order 4. This method has the advantage of supplying the extremely detailed information about the Mathieu groups, their Steiner systems, and the Golay code module and Todd module necessary both for the construction of the Leech lattice and the Griess algebra, and for the proof of the uniqueness of various sporadics.

The construction given here of the Leech lattice and the subgroups stabilizing various sublattices is the standard one due to Conway in [Co1] and [Co2]. The construction of the Griess algebra combines aspects of the treatments due to Griess [Gr2], Conway [Co3], and Tits [T2], plus a few extra wrinkles. The basis of the construction is Parker's loop and Conway's construction via the Parker loop of the normalizer $\bar{N}$ of a certain 4-subgroup of the Monster. Chapter 4 contains a discussion of a general class of loops which includes the Parker loop. This discussion contains much material not needed to construct the Parker loop or the Griess algebra, but the extra discussion provides a context which hopefully makes the Parker loop and Conway's construction of $\bar{N}$ more natural.

The majority of the sporadic groups contain a *large extraspecial* 2-subgroup. Such subgroups provide one of the unifying features of our treatment. The basic theory of large extraspecial subgroups is developed in Chapter 2. The theory is used to recognize and establish the simplicity of the sporadics contained in the Monster that are not symmetry groups of any nice structure.

The eventual object of the uniqueness treatment is to prove each sporadic is unique subject to suitable hypotheses. Here is a typical hypothesis; let w be a positive integer and L a group. (See Chapter 2 for terminology and notation.)

Hypothesis $\mathcal{H}(w, L)$: *G is a finite group containing an involution z such that $F^*(C_G(z)) = Q$ is an extraspecial 2-subgroup of order 2^{2w+1}, $C_G(z)/Q \cong L$, and z is not weakly closed in Q with respect to G.*

For example, Hypothesis $\mathcal{H}(12, Co_1)$ characterizes the Monster. Hypotheses of this sort are the appropriate ones for characterizing the sporadics for purposes of the classification.

Sporadic Groups lays the foundation for a proof of the uniqueness of each of the sporadics and supplies actual uniqueness proofs for five of the sporadic groups: M_{24}, He, J_2, Suz, and Co_1.

Our approach to the uniqueness problem follows Aschbacher and Segev in [AS1]. Namely given a group theoretic hypothesis $\mathcal{H}$ we associate to each group G satisfying $\mathcal{H}$ a coset graph Δ defined by some family $\mathcal{F}$ of subgroups of G. We prove the amalgam $\mathcal{A}$ of $\mathcal{F}$ is determined up to isomorphism by $\mathcal{H}$ independently of G, and form the free amalgamated product $\tilde{G}$ of $\mathcal{A}$ and its coset graph $\tilde{\Delta}$. Now there exists a covering $d : \tilde{\Delta} \to \Delta$ of graphs. To complete the proof we show Δ is simply connected so d is an isomorphism and hence $G = \tilde{G}$ is determined up to isomorphism by $\mathcal{H}$.

After developing the most basic part of the conceptual base for our treatment of the sporadic groups in Part I, Chapter 5 closes the first part of the book with an overview of the sporadic groups including the hypotheses by which we expect each group to be characterized, the approach for constructing each of the twenty sporadics involved in the Monster, and a number of historical remarks.

While *Sporadic Groups* concentrates on some of the most accessible and least technical aspects of the study of the sporadic groups, a complete treatment of even this material sometimes requires some difficult specialized arguments. The reader wishing to minimize contact with such arguments can do so as follows. As a general rule the book becomes progressively more difficult in the later chapters. Thus most of the material in Part I should cause little difficulty. A possible exception is Chapter 4, containing the discussion of loops. However, much of this material is not needed in the rest of the book, and none is needed outside of Chapter 10, where the Griess algebra is constructed. As Chapter 10 is the most technical part of Part II, some readers may wish to skip both Chapter 4 and Chapter 10.

Part II contains constructions of the Mathieu groups, the Conway group Co_1 and its sporadic sections, and the Monster and its sporadic sections. Two chapters are devoted to the Mathieu groups and two to the Conway groups. In each case the second of the two chapters is the most technical. Thus the reader may wish to read Chapters 6 and 8, while skipping or skimming Chapters 7 and 9. As suggested in the previous paragraph, dilettantes should skip the construction in Chapter 10 of the Griess algebra and the Monster. The existence proofs for the sporadic sections of the Monster not contained in Co_1 appear in the very short Chapter 11.

The Steiner systems and Golay codes associated to the Mathieu groups and the Leech lattice associated to the Conway groups are beautiful and natural objects. Most of the discussion of these objects appears in Chapters 6 and 8. There is some evidence that the Griess algebra is also natural, in that it is the 0-graded submodule of a conformal field theory preserved by the Monster (cf. [FLM]). However, the construction of the Griess algebra in Chapter 10 is not particularly natural or edifying.

The first two chapters of Part III provide the conceptual base for proving the uniqueness of the sporadic groups. These chapters are fairly elementary. Sections 39 through 41 establishing the uniqueness of M_{24} and $L_5(2)$ probably provide the easiest example of how to apply this machinery to establish uniqueness. On the other hand the proofs of the uniqueness of He, J_2, Suz, and Co_1, while more difficult, are also more representative of the complexity involved in proving the uniqueness of the sporadic groups.

The book closes with tables describing the basic structure of the five sporadic groups considered in detail in *Sporadic Groups*: M_{24}, He, J_2, Suz, and Co_1. These tables enumerate the subgroups of prime order of each group G and the normalizers of these subgroups. Much of this information comes out during the proof of the uniqueness of G, but some of the loose ends are tied up in Chapter 17.

PART I

Chapter 1

Preliminary Results

We take as our starting point the text *Finite Group Theory* [FGT], although we need only a fraction of the material in that text. Frequently quoted results from [FGT] will be recorded in this chapter and in other of the introductory chapters.

Chapters 1 and 2 record some of the most basic terminology and notation we will be using plus some elementary results. The reader should consult [FGT] for other basic group theoretic terminology and notation, although we will try to recall such notation when it is first used, or at least give a specific reference to [FGT] at that point. There is a "List of Symbols" at the end of [FGT] which can be used to help hunt down notation.

We begin in Section 1 with a brief discussion of abstract representations of groups. Then in Section 2 we specialize to permutation representations. In Section 3 we consider graphs and in Section 4 geometries (in the sense of J. Tits) and geometric complexes. In the last few sections of the chapter we record a few basic facts about the general linear group and fiber products of groups.

1. Abstract representations

Let C be a category. For X an object in C, we write $Aut(X)$ for the group of automorphisms of X under the operation of composition in C (cf. Section 2 in [FGT]). A *representation* of a group G in the category C is a group homomorphism $\pi; G \to Aut(X)$. For example, a *permutation representation* is a representation in the category of sets and a *linear*

representation is a representation in the category of vector spaces and
linear maps.

If $\alpha : A \to B$ is an isomorphism of objects in $\mathcal{C}$ then α induces a map

$$\alpha^* : Mor(A, A) \to Mor(B, B)$$
$$\beta \mapsto \alpha^{-1}\beta\alpha$$

and α^* restricts to an isomorphism $\alpha^* : Aut(A) \to Aut(B)$. Thus in
particular if $A \cong B$ then $Aut(A) \cong Aut(B)$.

A representation $\pi : G \to Aut(A)$ is *faithful* if π is injective.

Two representations $\pi : G \to Aut(A)$ and $\sigma : G \to Aut(B)$ in $\mathcal{C}$ are
equivalent if there exists an isomorphism $\alpha : A \to B$ such that $\sigma = \pi\alpha^*$ is
the composition of π with α^*. Equivalently for all $g \in G$, $(g\pi)\alpha = \alpha(g\sigma)$.

Similarly if $\pi_i : G_i \to Aut(A_i)$, $i = 1, 2$, are representations of groups
G_i on objects A_i in $\mathcal{C}$, then π_1 is said to be *quasiequivalent* to π_2 if
there exists a group isomorphism $\beta : G_1 \to G_2$ and an isomorphism
$\alpha : A_1 \to A_2$ such that $\pi_2 = \beta^{-1}\pi_1\alpha^*$. Observe that we have a permutation representation of $Aut(G)$ on the equivalence classes of representations of G via $\alpha : \pi \mapsto \alpha\pi$ with the orbits the quasiequivalence classes.
Write $Aut(G)_\pi$ for the stabilizer of the equivalence class of π under this
representation. The following result is Exercise 1.7 in [FGT]:

Lemma 1.1: *Let $\pi, \sigma : G \to Aut(A)$ be faithful representations. Then*

(1) *π is quasiequivalent to σ if and only if $G\pi$ is conjugate to $G\sigma$ in*
     $Aut(A)$.

(2) *$Aut_{Aut(A)}(G\pi) \cong Aut(G)_\pi$.*

If $H \le G$ then write $Aut_G(H) = N_G(H)/C_G(H)$ for the group of
automorphisms of H induced by G. Also

$$C_G(H) = \{c \in G : ch = hc \text{ for all } h \in H\}$$

is the *centralizer* in G of H and $N_G(H)$ is the *normalizer* in G of H,
that is, the largest subgroup of G in which H is normal.

2. Permutation representations

In this section X is a set. We refer the reader to Section 5 of [FGT] for
our notational conventions involving permutation groups, although we
record a few of the most frequently used conventions here. In particular
we write $Sym(X)$ for the symmetric group on X and if X is finite we
write $Alt(X)$ for the alternating group on X. Further S_n, A_n denote the
symmetric and alternating groups of degree n; that is, $S_n = Sym(X)$
and $A_n = Alt(X)$ for X of order n.

Let $\pi : G \to Sym(X)$ be a permutation representation of a group G on X. Usually we suppress π and write xg for the image $x(g\pi)$ of a point $x \in X$ under the permutation $g\pi$, $g \in G$. For $S \subseteq G$, we write $Fix(S) = Fix_X(S)$ for the set of fixed points of S on X. For $Y \subseteq X$,

$$G_Y = \{g \in G : yg = y \text{ for all } y \in Y\}$$

is the *pointwise stabilizer* of Y in G,

$$G(Y) = \{g \in G : Yg = Y\}$$

is the *global stabilizer* of Y in G, and $G^Y = G(Y)/G_Y$ is the image of $G(Y)$ in $Sym(Y)$ under the restriction map. In particular G_y denotes the stabilizer of a point $y \in X$.

Recall the *orbit* of $x \in X$ under G is $xG = \{xg : g \in G\}$ and G is *transitive* on X if G has just one orbit on X. If G is transitive on X then our representation π is equivalent to the representation of G by right multiplication on the coset space G/G_x via the map $G_x g \mapsto xg$ (cf. 5.9 in [FGT]).

A subgroup K of G is a *regular normal subgroup* of G if $K \unlhd G$ and K is *regular* on X; that is, K is transitive on X and $K_x = 1$ for $x \in X$.

Recall a transitive permutation group G is *primitive* on X if G preserves no nontrivial partition on X. Further G is primitive on X if and only if G_x is maximal in G (cf. 5.19 in [FGT]).

Lemma 2.1: *Let G be transitive on X, $x \in X$, and $K \leq G$. Then*

(1) K is transitive on X if and only if $G = G_x K$.

(2) If $1 \neq K \unlhd G$ and G is primitive on X then K is transitive on X.

(3) If K is a regular normal subgroup of G then the representations of G_x on X and on K by conjugation are equivalent.

Proof: These are all well known; see, for example, 5.20, 15.15, and 15.11 in [FGT].

Recall that G is *t-transitive* on X if G is transitive on ordered t-tuples of distinct points of X. In Chapter 6 we will find that the Mathieu group M_{m+t} is t-transitive on $m+t$ points for $m = 19$ and $t = 3, 4, 5$ and $m = 7$ and $t = 4, 5$.

Lemma 2.2: *Let G be t-transitive on a finite set X with $t \geq 2$, $x \in X$, and $1 \neq K \unlhd G$. Then*

(1) G is primitive on X.

(2) K is transitive on X and $G = G_x K$.

(3)   If K is regular on X then $|K| = |X| = p^e$ is a power of some prime p, and if $t > 2$ then $p = 2$.

(4)   If $t = 3 < |X|$ and $|G : K| = 2$ then K is 2-transitive on X.

Proof: Again these are well-known facts. See, for example, 15.14 and 15.13 in [FGT] for (1) and (3), respectively. Part (2) follows from (1) and 1.1. Part (4) is left as Exercise 1.1.

3. Graphs

A *graph* $\Delta = (\Delta, *)$ consists of a set Δ of *vertices* (or objects or points) together with a symmetric relation $*$ called *adjacency* (or incidence or something else). The ordered pairs in the relation are called the *edges* of the graph. We write $u * v$ to indicate two vertices are related via $*$ and say u is *adjacent* to v. Denote by $\Delta(u)$ the set of vertices adjacent to u and distinct from u and define $u^\perp = \Delta(u) \cup \{u\}$.

A *path of length n* from u to v is a sequence of vertices $u = u_0, u_1, \ldots, u_n = v$ such that $u_{i+1} \in u_i^\perp$ for each i. Denote by $d(u, v)$ the minimal length of a path from u to v. If no such path exists set $d(u, v) = \infty$. $d(u, v)$ is the *distance* from u to v.

The relation $\sim$ on Δ defined by $u \sim v$ if and only if $d(u, v) < \infty$ is an equivalence relation on Δ. The equivalence classes of this relation are called the *connected components* of the graph. The graph is *connected* if it has just one connected component. Equivalently there is a path between any pair of vertices.

A *morphism* of graphs is a function $\alpha : \Delta \to \Delta'$ from the vertex set of Δ to the vertex set of Δ' which preserves adjacency; that is, $u^\perp \alpha \subseteq (u\alpha)^\perp$ for each $u \in \Delta$.

A group G of automorphisms of Δ is *edge transitive* on Δ if G is transitive on Δ and on the edges of Δ.

Representations of groups on graphs play a big role in this book. For example, we prove the uniqueness of some of the sporadics G by considering a representation of G on a suitable graph. The following construction supplies us with such graphs.

Let G be a transitive permutation group on a finite set Δ. Recall the *orbitals* of G on Δ are the orbits of G on the set product $\Delta^2 = \Delta \times \Delta$. The *permutation rank* of G is the number of orbitals of G; recall this is also the number of orbits of G_x on Δ for $x \in \Delta$.

Given an orbital Ω of G, the *paired orbital* Ω^p of Ω is

$$\Omega^p = \{(y, x) : (x, y) \in \Omega\}.$$

Evidently Ω^p is an orbital of G with $(\Omega^p)^p = \Omega$. The orbital Ω is said to be *self-paired* if $\Omega^p = \Omega$. For example, the *diagonal orbital* $\{(x,x) : x \in \Delta\}$ is a self-paired orbital.

Lemma 3.1: *(1) A nondiagonal orbital $(x,y)G$ of G is self-paired if and only if (x,y) is a cycle in some $g \in G$.*

(2) If G is finite then G possesses a nondiagonal self-paired orbital if and only if G is of even order.

(3) If G is of even order and permutation rank 3 then all orbitals of G are self-paired.

Proof: See 16.1 in [FGT].

Lemma 3.2: *(1) Let Ω be a self-paired orbital of G. Then Ω is a symmetric relation on Δ, so $\Delta = (\Delta, \Omega)$ is a graph and G is an edge transitive group of automorphisms of Δ.*

(2) Conversely if H is an edge transitive group of automorphisms of a graph $\Delta = (\Delta, *)$ then the set $*$ of edges of Δ is a self-paired orbital of G on Δ, and Δ is the graph determined by this orbital.

Many of the sporadics have representations as rank 3 permutation groups. Indeed some were discovered via such representations; see Chapter 5 for a discussion of the sporadics discovered this way. See also Exercise 16.5, which considers the rank 3 representation of J_2, and Lemmas 24.6, 24.7, and 24.11, which establish the existence of rank 3 representations of Mc, $U_4(3)$, and HS.

In the remainder of this section assume G is of even order and permutation rank 3 on a set X. Hence G has two nondiagonal orbitals Δ and Γ and by 3.1, each is self-paired. Further for $x \in X$, G_x has two orbits $\Delta(x)$ and $\Gamma(x)$ on $X - \{x\}$, where $\Delta(x) = \{y \in X : (x,y) \in \Delta\}$ and $\Gamma(x) = \{z \in X : (x,z) \in \Gamma\}$. By 3.2, $X = (X, \Delta)$ is a graph and G is an edge transitive group of automorphisms of X. Notice $\Delta(x) = X(x)$ in our old notation.

The following notation is standard for rank 3 groups and their graphs: $k = |\Delta(x)|$, $l = |\Gamma(x)|$, $\lambda = |\Delta(x) \cap \Delta(y)|$ for $y \in \Delta(x)$, and $\mu = |\Delta(x) \cap \Delta(z)|$ for $z \in \Gamma(x)$. The integers k, l, λ, μ are the *parameters* of the rank 3 group G. Also let $n = |X|$ be the degree of the representation.

Lemma 3.3: *Let G be a rank 3 permutation group of even order on a finite set of order n with parameters k, l, λ, μ. Then*

(1) $n = k + l + 1$.
(2) $\mu l = k(k - \lambda - 1)$.

(3)   *If* $\mu \neq 0$ *or* k *then* G *is primitive and the graph* $\mathcal{G}$ *of* G *is connected.*

(4)   *Assume* G *is primitive. Then either*

(a) $k = l$ *and* $\mu = \lambda + 1 = k/2$, *or*

(b) $d = (\lambda - \mu)^2 + 4(k - \mu)$ *is a square and setting* $D = 2k + (\lambda - \mu)(k + l)$, $d^{1/2}$ *divides* D *and* $2d^{1/2}$ *divides* D *if and only if* n *is odd.*

Proof: See Section 16 of [FGT].

4.   Geometries and complexes

In this book we adopt a notion of geometry due to J. Tits in [T1].

Let I be a finite set. For $J \subseteq I$, let $J' = I - J$ be the complement of J in I. A *geometry* over I is a triple $(\Gamma, \tau, *)$ where Γ is a set of objects, $\tau : \Gamma \to I$ is a surjective type function, and $*$ is a symmetric incidence relation on Γ such that objects u and v of the same type are incident if and only if $u = v$. We call $\tau(u)$ the *type* of the object u. Notice $(\Gamma, *)$ is a graph. We usually write Γ for the geometry $(\Gamma, \tau, *)$ and Γ_i for the set of objects of Γ of type i.

The *rank* of the geometry Γ is the cardinality of I.

A *flag* of Γ is a subset T of Γ such that each pair of objects in T is incident. Notice our one axiom insures that if T is a flag then the type function $\tau : T \to I$ is injective. Define the *type* of T to be $\tau(T)$ and the *rank* of T to be the cardinality of T. The *chambers* of Γ are the flags of type I.

A *morphism* $\alpha : \Gamma \to \Gamma'$ of geometries is a function $\alpha : \Gamma \to \Gamma'$ of the associated object sets which preserves type and incidence; that is, if $u, v \in \Gamma$ with $u * v$ then $\tau(u) = \tau'(u\alpha)$ and $u\alpha *' v\alpha$. A group G of automorphisms of Γ is *edge transitive* if G is transitive on flags of type J for each subset J of I of order at most 2. Similarly G is *flag transitive* on Γ if G is transitive on flags of type J for all $J \subseteq I$.

Representations of groups on geometries also play an important role in *Sporadic Groups*. For example, the Steiner systems in Chapter 6 are rank 2 geometries whose automorphism groups are the Mathieu groups. Here are some other examples:

Examples   (1) Let V be an n-dimensional vector space over a field F. We associate a geometry $PG(V)$ to V called the *projective geometry* of V. The objects of $PG(V)$ are the proper nonzero subspaces of V, with incidence defined by inclusion. The type of U is $\tau(U) = dim(U)$. Thus

$PG(V)$ is of rank $n - 1$. The projective general linear group on V is a flag transitive group of automorphism of $PG(V)$.

(2) A *projective plane* is a rank 2 geometry Γ whose two types of objects are called points and lines and such that:

(PP1)   Each pair of distinct points is incident with a unique line.

(PP2)   Each pair of distinct lines is incident with a unique point.

(PP3)   There exist four points no three of which are on a common line.

Remarks.   (1) Rank 2 projective geometries are projective planes.

(2) If Γ is a finite projective plane then there exists an integer q such that each point is incident with exactly $q + 1$ lines, each line is incident with exactly $q + 1$ points, and Γ has $q^2 + q + 1$ points and lines.

Examples   (3) If f is a sesquilinear or quadratic form on V then the *totally singular subspaces* of V are the subspaces U such that f is trivial on U. The set of such subspaces forms a subgeometry of the projective geometry. See, for example, page 99 in [FGT].

(4) Let G be a group and $\mathcal{F} = (G_i : i \in I)$ a family of subgroups of G. Define $\Gamma(G, \mathcal{F})$ to be the geometry whose set of objects of type i is the coset space G/G_i and with objects $G_i x$ and $G_j y$ incident if $G_i x \cap G_j y \neq \varnothing$. Observe:

Lemma 4.1: *(1) G is represented as an edge transitive group of automorphisms of $\Gamma(G, \mathcal{F})$ via right multiplication and $\Gamma(G, \mathcal{F})$ possesses a chamber.*

(2) Conversely if H is an edge transitive group of automorphisms of a geometry Γ and Γ possesses a chamber C, then $\Gamma \cong \Gamma(H, \mathcal{F})$, where $\mathcal{F} = (H_c : c \in C)$.

The construction of 4.1 allows us to represent each group G on various geometries. The construction is used in Chapter 13 as part of our machine for establishing the uniqueness of groups. Further the construction associates to each sporadic group G various geometries which can be used to study the subgroup structure of G. The latter point of view is not explored to any extent in *Sporadic Groups*; see instead [A2] or [RS] where such geometries are discussed. We do use the 2-local geometry of M_{24} to study that group in Chapter 7.

Define the *direct sum* of geometries Γ_i on I_i, $i = 1, 2$, to be the geometry $\Gamma_1 \oplus \Gamma_2$ over the disjoint union I of I_1 and I_2 whose object set is the disjoint union of Γ_1 and Γ_2, whose type function is $\tau_1 \cup \tau_2$, and whose incidence is inherited from Γ_1 and Γ_2 with each object in Γ_1 incident with each object in Γ_2.

Example   (5) A *generalized digon* is a rank 2 geometry which is the direct sum of rank 1 geometries. That is, each element of type 1 is incident with each element of type 2.

Lemma 4.2: *Let G be a group and $\mathcal{F} = \{G_1, G_2\}$ a pair of subgroups of G. Then $\Gamma(G, \mathcal{F})$ is a generalized digon if and only if $G = G_1 G_2$.*

Proof: As G is edge transitive on Γ, Γ is a generalized digon if and only if G_2 is transitive on Γ_1 if and only if $G = G_1 G_2$ by 2.1.1.

Given a flag T, let $\Gamma(T)$ consist of all $v \in \Gamma - T$ such that $v * t$ for all $t \in T$. We regard $\Gamma(T)$ as a geometry over $I - \tau(T)$. The geometry $\Gamma(T)$ is called the *residue* of T.

Example   (6) Let $\Gamma = PG(V)$ be the projective geometry of an n-dimensional vector space. Then for $U \in \Gamma$, the residue $\Gamma(U)$ of the object U is isomorphic to $PG(U) \oplus PG(V/U)$.

The category of geometries is not large enough; we must also consider either the category of chamber systems or the category of geometric complexes.

A *chamber system* over I is a set X together with a collection of equivalence relations $\sim_i$, $i \in I$. For $J \subseteq I$ and $x \in X$, let $\sim_J$ be the equivalence relation generated by the relations $\sim_j$, $j \in J$, and $[x]_J$ the equivalence class of $\sim_J$ containing x. Define X to be *nondegenerate* if for each $x \in X$, and $j \in I$, $\{x\} = \bigcap_i [x]_{i'}$ and $[x]_j = \bigcap_{i \in j'} [x]_{i'}$. A morphism of chamber systems over I is a map preserving each equivalence relation.

The notion of "chamber system" was introduced by J. Tits in [T1].

Recall that a *simplicial complex* K consists of a set X of *vertices* together with a distinguished set of nonempty subsets of X called the *simplices* of K such that each nonempty subset of simplex is a simplex. The morphisms of simplicial complexes are the *simplicial maps*; that is, a simplicial map $f : K \to K'$ is a map $f : X \to X'$ of vertices such that $f(s)$ is a simplex of K' for each simplex s of K.

Example   (7) If Δ is a graph then the *clique complex* $K(\Delta)$ is the simplicial complex whose vertices are the vertices of Δ and whose simplices are the finite cliques of Δ. Recall a *clique* of Δ is a set Y of vertices such that $y \in x^\perp$ for each $x, y \in Y$. Conversely if K is a simplicial complex then the *graph* of K is the graph $\Delta = \Delta(K)$ whose vertices are the vertices of K and with $x * y$ if $\{x, y\}$ is a simplex of K. Observe K is a subcomplex of $K(\Delta(K))$.

Given a simplicial complex K and a simplex s of K, define the *star* of s to be the subcomplex $st_K(s)$ consisting of the simplices t of K such that

$s \cup t$ is a simplex of K. Define the *link* $Link_K(s)$ to be the subcomplex of $st_K(s)$ consisting of the simplices t of $st_K(s)$ such that $t \cap s = \varnothing$.

A *geometric complex* over I is a geometry Γ over I together with a collection C of distinguished chambers of Γ such that each flag of rank 1 or 2 is contained in a member of C. The *simplices* of the complex are the subflags of members of C. A morphism $\alpha : C \to C'$ of complexes over I is a morphism of geometries with $C\alpha \subseteq C'$. Notice a geometric complex is just a simplicial complex together with a type function on vertices that is injective on simplices.

Example (8) The *flag complex* of a geometry Γ is the simplicial complex on Γ in which all chambers are distinguished. Notice the flag complex is a geometric complex if and only if each flag of rank at most 2 is contained in a chamber. Further as a simplicial complex, the flag complex is just the clique complex of Γ regarded as a graph.

Many theorems about geometries are best established in the larger categories of geometric complexes or chamber systems. Theorem 4.11 is an example of such a result. We find in a moment in Lemma 4.3 below that the category of nondegenerate chamber systems is isomorphic to the category of geometric complexes. I find the latter category more intuitive and so work with complexes rather than chamber systems. But others prefer chamber systems and there is a growing literature on the subject.

Given a chamber system X define Γ_X to be the geometry whose objects of type i are the equivalence classes of the relation $\sim_{i'}$ with $A * B$ if and only if $A \cap B \neq \varnothing$. For $x \in X$ let C_x be the set of equivalence classes containing x; thus C_x is a chamber in Γ_X. Define C_X to be the set of chambers C_x, $x \in X$, of Γ_X. If $\alpha : X \to X'$ is a morphism of chamber systems define $\alpha_C : C_X \to C_{X'}$ to be the morphism of complexes such that $\alpha_C : A \mapsto A'$ for A a $\sim_{i'}$ equivalence class of X and A' the $\sim_{i'}$ equivalence class containing $A\alpha$.

Conversely given a geometric complex C over I let $\sim_i$ be the equivalence relation on C defined by $A \sim_i B$ if A and B have the same subflag of type i'. Then we have a chamber system X_C with chamber set C and equivalence relations $\sim_i$. Further if $\alpha : C \to C'$ is a morphism of complexes let $\alpha_X : X_C \to X_{C'}$ be the morphism of chamber systems defined by the induced map on chambers.

Lemma 4.3: *The category of nondegenerate chamber systems over I is isomorphic to the category of geometric complexes over I via the maps $X \mapsto C_X$ and $C \mapsto X_C$.*

Example (9) Let G be a group and $\mathcal{F} = (G_i : i \in I)$ a family of subgroups of I. For $J \subseteq I$ and $x \in G$ define

$$S_{J,x} = \{G_j x : j \in J\}.$$

Thus $S_{J,x}$ is a flag of the geometry $\Gamma(G, \mathcal{F})$ of type J. Observe that the stabilizer of the flag $S_J = S_{J,1}$ is the subgroup $G_J = \bigcap_{j \in J} G_j$. Define $\mathcal{C}(G, \mathcal{F})$ to be the geometric complex over I with geometry $\Gamma(G, \mathcal{F})$ and distinguished chambers $S_{I,x}$, $x \in G$. Then $\mathcal{C}(G, \mathcal{F})$ is a geometric complex with simplices $S_{J,x}$, $J \subseteq I$, $x \in G$, and G acts as an edge transitive group of automorphisms of $\mathcal{C}(G, \mathcal{F})$ via right multiplication, and transitively on $\mathcal{C}(G, \mathcal{F})$. Indeed:

Lemma 4.4: *Assume C is a geometric complex over I and G is an edge transitive group of automorphisms with $C = CG$ for some $C \in \mathcal{C}$. Let $G_i = G_{x_i}$, where $x_i \in C$ is of type i, and let $\mathcal{F} = (G_i : i \in I)$. Then the map $x_i g \mapsto G_i g$ is an isomorphism of C with $\mathcal{C}(G, \mathcal{F})$.*

Further we have a chamber system $X(G, \mathcal{F})$ whose chamber set is G/G_I and with $G_I x \sim_i G_I y$ if and only if $xy^{-1} \in G_{i'}$. Observe that the map $G_I x \mapsto S_{I,x}$ defines an isomorphism of the chamber systems $X(G, \mathcal{F})$ and $X_{\mathcal{C}(G, \mathcal{F})}$.

The construction of 4.4 allows us to represent a group G on many complexes. We make use of this construction in Chapter 13 as part of our uniqueness machine.

Let $\mathcal{C} = (\Gamma, \mathcal{C})$ be a geometric complex over I. Given a simplex S of type J, regard the link $Link_\mathcal{C}(S)$ of S to be a geometric complex over J'; thus the objects of $Link_\mathcal{C}(S)$ of type $i \in J'$ are those $v \in \Gamma_i$ such that $S \cup \{v\}$ is a simplex and with $v * u$ if $S \cup \{u, v\}$ is a simplex, and the chamber set $\mathcal{C}(S)$ of $Link_\mathcal{C}(S)$ consists of the simplices $C - S$ with $S \subseteq C \in \mathcal{C}$. For example, $\mathcal{C} = Link_\mathcal{C}(\varnothing)$ is the link of the empty simplex. Notice that if all flags are simplices then the geometry of $Link_\mathcal{C}(S)$ is the residue $\Gamma(S)$ of S in the geometry Γ.

We say $\mathcal{C}$ is *residually connected* if the link of each simplex of corank at least two (including $\varnothing$ if $|I| \geq 2$) is connected. A geometry Γ is residually connected if each flag is contained in a chamber and the flag complex of Γ is residually connected.

Lemma 4.5: *Let $\mathcal{F} = (G_i : i \in I)$ be a family of subgroups of G. Then*

  (1) $\Gamma(G, \mathcal{F})$ is connected if and only if $G = \langle \mathcal{F} \rangle$.

  (2) $Link_\mathcal{C}(S_J) \cong \mathcal{C}(G_J, \mathcal{F}_J)$ for each $J \subseteq I$, where

$$\mathcal{F}_J = (G_{J \cup \{i\}} : i \in J').$$

(3) $C(G, \mathcal{F})$ *is residually connected if and only if* $G_J = \langle \mathcal{F}_J \rangle$ *for all* $J \subseteq I$.

Proof: Notice (1) and (2) imply (3) so it remains to prove (1) and (2).

As $\mathcal{F}$ is a chamber, the connected component Δ of G_i in Γ is the same for each i, and $H = \langle \mathcal{F} \rangle$ acts on Δ. Conversely as G_i is transitive on $\Gamma_j(G_i)$ for each j, $\Delta \subseteq \Delta' = \bigcup_j G_j H$, so $\Delta = \Delta'$ and H is transitive on $\Gamma_i \cap \Delta$ for each i. Thus as G is transitive on Γ_i, Γ is connected if and only if H is transitive on Γ_i for each i, and as $G_i \leq H$ this holds if and only if $G = H$. Thus (1) is established.

In (2) the desired isomorphism is $G_k x \mapsto S_{K,x}$ for $x \in G_J$, $K = J \cup \{k\}$.

Lemma 4.6: *Assume* C *is a residually connected geometric complex over* I, $J \subseteq I$ *with* $|J| \geq 2$, *and* $x, y \in \Gamma$. *Then there exists a path* $x = v_0, \dots, v_m = y$ *in* Γ *with* $\tau(v_i) \in J$ *for all* $0 < i < m$.

Proof: Choose x, y to be a counterexample with $d = d(x, y)$ minimal. As the residue Γ of the simplex $\varnothing$ is connected, d is finite, and clearly $d > 1$. Let $x = v_0 \cdots v_d = y$ be a path. By minimality of d there is a path $v_1 = u_0 \cdots u_m = y$ with $\tau(u_i) \in J$ for $0 < i < m$. Thus if $\tau(v_1) \in J$ then $x u_0 \cdots u_m$ is the desired path, so assume $\tau(v_1) \notin J$.

We also induct on the rank of C; if the rank is 2 the lemma is trivial, so our induction is anchored. Now $Link_C(v_1)$ is a residually connected complex and $x, u_1 \in Link_C(v_1)$, so by induction on the rank of C, there is a path $x = w_0 \cdots w_k = u_1$ with $\tau(w_i) \in J$ for $0 < i < k$. Now $x = w_0 \cdots w_k u_2 \cdots u_m = y$ does the job.

Given geometric complexes C over J and $\bar{C}$ over $\bar{J}$ define $C \oplus \bar{C}$ to be the geometric complex over the disjoint union I of J and $\bar{J}$ whose geometry is $\Gamma \oplus \bar{\Gamma}$ and with chamber set $\{C \cup \bar{C} : C \in \mathcal{C}, \bar{C} \in \bar{\mathcal{C}}\}$.

The *basic diagram* for a geometric complex C over I is the graph on I obtained by joining distinct i, j in I if for some simplex T of type $\{i, j\}'$ (including $\varnothing$ if $|I| = 2$), $Link_C(T)$ is *not* a generalized digon. The basic diagram of a geometry is the basic diagram of its flag complex.

Diagrams containing more information can also be associated to each geometry or geometric complex. The study of such diagrams was initiated by Tits [T1] and Buekenout [Bu].

A graph on I is a *string* if we can order $I = \{1, \dots, n\}$ so that the edges of I are $\{i, i+1\}$, $1 \leq i < n$. Such an ordering will be termed a *string ordering*. A *string geometry* is a geometry whose basic diagram is a string. Most of the geometries considered in *Sporadic Groups* are string geometries; for example:

Example   (10) The basic diagram of projective geometry is a string.

Lemma 4.7: *Assume C is a residually connected geometric complex such that $I = I_1 + I_2$ is a partition of I such that I_1 and I_2 are unions of connected components of the basic diagram of I. Then $C = C^1 \oplus C^2$, where C^i consists of the simplices of type I_i.*

Proof: We may assume $I_i \neq \varnothing$ for $i = 1, 2$. By definition of the basic diagram, the lemma holds if Γ is of rank 2. Thus we may assume I_1 has rank at least 2. Let $x_i \in \Gamma^i$; by 4.6 there exists a path $x_1 = v_0 \cdots v_m = x_2$ with $\tau(v_i) \in I_1$ for $i < m$. Choose this path with m minimal; if $m = 1$ for each choice of x_i we are done, so choose x_i such that m is minimal subject to $m > 1$. Then of course $m = 2$, so $x_i \in Link_C(v_1)$. But by induction on the rank of Γ, x_1 is incident with x_2 in $Link_C(v_1)$, and hence also in Γ.

The proof of the following result is trivial:

Lemma 4.8: *If C is a geometric complex then the following are equivalent:*

     (1)   All flags of Γ are simplices.
     (2) $Link_C(S) = \Gamma(S)$ for each simplex S of C.

Lemma 4.9: *Assume C is a residually connected geometric complex such that the connected components of the basic diagram of C are strings. Then all flags of C are simplices.*

Proof: Assume not and let T be a flag of minimal rank m which is not a simplex. As C is a geometric complex, $m > 2$. Pick a string ordering for I and let $T = \{x_1, \ldots, x_m\}$ with $\tau(x_i) < \tau(x_{i+1})$. Let $x = x_2$. By minimality of m, $\{x_1, x\}$ and $\{x_2, \ldots, x_m\}$ are simplices. Further by 4.7, $Link_C(x) = C^1 \oplus C^2$, where C^i is the subgeometry on I_i, $I_1 = \{1\}$, and $I_2 = \{3, \ldots, n\}$. Thus $\{x_1, x_3, \ldots, x_m\}$ is a simplex in $Link_C(x)$, so T is a simplex of C.

Lemma 4.10: *Let G be a group and $\mathcal{F} = (G_i : i \in I)$ a family of subgroups of G, and assume $C = C(G, \mathcal{F})$ is residually connected. Then the following are equivalent:*

     (1) G is flag transitive on $\Gamma(G, \mathcal{F})$.
     (2)   Each flag of $\Gamma(G, \mathcal{F})$ is a simplex.
     (3) $\Gamma(S_J) = Link_C(S_J) \cong \Gamma(G_J, \mathcal{F}_J)$ for each $J \subseteq I$.

Proof: By 4.5.2 and 4.8, (2) and (3) are equivalent. As G is transitive on simplices of C of type J for each $J \subseteq I$, (1) and (2) are equivalent.

Theorem 4.11: *Let G be a group, $I = \{1, \ldots, n\}$, and $\mathcal{F} = (G_i : i \in I)$ a family of subgroups of G. Assume*

   (a) $C(G, \mathcal{F})$ *is residually connected; that is, $G_J = \langle \mathcal{F}_J \rangle$ for all $J \subseteq I$.*

   (b)   *The diagram of $C(G, \mathcal{F})$ is a union of strings; that is, $\langle G_{i'}, G_{j'} \rangle = G_{i'} G_{j'}$ for all $i, j \in I$ with $|i - j| > 1$.*

Then

   (1)   *G is flag transitive on $\Gamma(G, \mathcal{F})$.*

   (2)   *$\Gamma(S_J) \cong \Gamma(G_J, \mathcal{F}_J)$ for all $J \subseteq I$.*

Proof: This follows from 4.9 and 4.10. Use 4.5 to see that the conditions of (a) are equivalent and 4.2 to see that the conditions of (b) are equivalent.

5. The general linear group and its projective geometry

In this section F is a field, n is a positive integer, and V is an n-dimensional vector space over F. Recall that the group of vector space automorphisms of V is the *general linear group* $GL(V)$. We assume the reader is familiar with basic facts about $GL(V)$, such as can be found in Section 13 of [FGT]. For example, as the isomorphism type of V depends only on n and F, the same is true for $GL(V)$, so we can also write $GL_n(F)$ for $GL(V)$.

Recall that from Section 13 in [FGT] that each ordered basis $X = (x_1, \ldots, x_n)$ of V determines an isomorphism M_X of $GL(V)$ with the group of all nonsingular n-by-n matrices over F defined by $M_X(g) = (g_{ij})$, where for $g \in GL(V)$, $g_{ij} \in F$ is defined by $x_i g = \sum_j g_{ij} x_j$. Thus we will sometimes view $GL(V)$ as this matrix group.

We write $SL(V)$ or $SL_n(F)$ for the subgroup of matrices in $GL(V)$ of determinant 1. Thus $SL_n(F)$ is the *special linear group*. As the kernel of the determinant map, $SL_n(F)$ is a normal subgroup of $GL_n(F)$.

A *semilinear transformation of V* is a bijection $g : V \to V$ that preserves addition and such that there exists $\sigma(g) \in Aut(F)$ such that for each $a \in F$ and $v \in V$, $(av)g = a\sigma(g)v$. Define $\Gamma = \Gamma(V)$ to be the set of all semilinear transformations of V. Notice the map $\sigma : \Gamma \to Aut(F)$ is a surjective group homomorphism with kernel $GL(V)$ and $\Gamma(V)$ is the split extension of $GL(V)$ by the group $\{f_\alpha : \alpha \in Aut(F)\} \cong Aut(F)$ of *field automorphisms* determined by the basis X of V, where

$$f_\alpha : \sum_i a_i x_i \mapsto \sum_i (a_i \alpha) x_i.$$

Notice also that $\Gamma(V)$ permutes the points of the projective geometry $PG(V)$ and this action induces a representation of $\Gamma(V)$ as a group of automorphisms of $PG(V)$ with kernel the scalar matrices. Thus the image $P\Gamma(V)$ is a group of automorphisms of $PG(V)$ which is the split extension of $PGL(V)$ by the group of field automorphisms.

If $F = GF(q)$ is the finite field of order q we write $GL_n(q)$ for $GL_n(F)$, $SL_n(q)$ for $SL_n(F)$, $PGL_n(q)$ for $PGL_n(F)$, and $L_n(q) = PSL_n(q)$ for $PSL_n(F)$.

See Section 13 in [FGT] for the definition of the *transvections* in $GL(V)$ and properties of transvections.

Lemma 5.1: *Let $G = PGL(V)$, $S = PSL(V)$, and H the stabilizer in G of a point p of $PG(V)$. Assume $n \geq 2$. Then*

(1)   *H is the split extension of the group Q of all transvections of V with center p by the stabilizer L of p and a hyperplane U of V complementing p.*

(2)   *$Q \cong U$, $L \cong GL(U)$, and the action of L by conjugation on Q is equivalent to the action of L on U.*

(3)   *Q is the unique minimal normal subgroup of $H \cap L$.*

Proof: Let $\hat{G} = GL(V)$ and regard $\hat{G}$ as a group of matrices relative to a basis X for V such that $p = \langle x_1 \rangle$. Then the preimage $\hat{H}$ of H in $\hat{G}$ consists of all matrices

$$g = \begin{pmatrix} a(g) & 0 \\ \alpha(g) & A(g) \end{pmatrix}$$

with $a(g) \in F^{\#}$, $\alpha(g)$ a row matrix, and $A(g) \in GL(U)$. Moreover Q consists of the matrices g with $a(g) = 1$ and $A(g) = I$, while $\hat{L}$ consists of all matrices h with $\alpha(h) = 0$. Further $g^h \in Q$ with $\alpha(g^h) = a(h)A(h)^{-1}\alpha(g)$. In particular $\hat{H}$ is the split extension of Q by $\hat{L}$, and $Q \cong U$ is abelian. Further $\hat{L} = L_0 \times K$, where K is the group of scalar matrices and L_0 consists of those $h \in \hat{L}$ with $a(h) = 1$. Thus the image L of $\hat{L}$ in G is isomorphic to $L_0 \cong GL(U)$, and the action of L by conjugation on H is equivalent to the action of $L \cong L_0$ on $U \cong Q$.

So (1) and (2) are established. Finally as the action of L on Q is equivalent to its action on U, L (and even $L \cap S$) is faithful and irreducible on Q, so Q is minimal normal in H. Now if M is a second minimal normal subgroup of H, then $\langle M, Q \rangle = M \times Q$, so $M \leq C_H(Q)$ and $M \cap Q = 1$. But as $H = LQ$ with L faithful on Q, $Q = C_H(Q)$, contradicting $M \cap Q = 1$.

The projective plane over the field of order 4 will be the starting point for our construction in Chapter 6 of the Mathieu groups and their Steiner systems. In particular we will need the following result:

Lemma 5.2: *Let F be the field of order 4. Then*

> *(1) The group of automorphisms of the projective plane over F is $P\Gamma_3(F)$.*
>
> *(2) A field automorphism f fixes exactly seven points of $PG(V)$.*

Proof: Let $X = \{x_1, x_2, x_3\}$ be a basis of V. First a proof of (2): A typical point of V is of the form $p = \langle \sum_i a_i x_i \rangle$, with $a_j = 1$ for some j. Then $pf = \langle \sum_i a_i^2 x_i \rangle$, since the automorphism of F of order 2 defining f is $a \mapsto a^2$. Thus $pf = p$ if and only if there exists $b \in F^\#$ with $ba_i = a_i^2$ for each i. It follows that $a_i = 0$ or b for all i. But as $a_j = 1$, also $b = 1$. Hence f fixes p if and only if all coefficients a_i are in $GF(2)$. So there are precisely seven choices for p.

Next let $M = Aut(PG(V))$ and $p = \langle x_1 \rangle$. Then $\Gamma = P\Gamma(V) \le M$. As Γ is transitive on the points of $PG(V)$, $M = \Gamma \cdot M_p$, so it remains to show $M_p \le \Gamma_p = H$. Let Δ be the set of five lines through p. Then $H_\Delta = QB$, where $Q \cong E_{16}$ is the subgroup of 5.1 and $B = \langle \beta \rangle \cong \mathbf{Z}_3$, where $\beta = diag(a, 1, 1)$ and $\langle a \rangle = F^\#$. Further $H^\Delta = Sym(\Delta)$, so $M_p = HM_\Delta$ and it remains to show $M_\Delta = QB$.

Now Q is regular on the sixteen lines not through p, so $M_\Delta = QD$, where D is the subgroup of M_Δ fixing the line $k = \langle x_2, x_3 \rangle$. We must show $D \le \Gamma$.

First D fixes $k \cap m$ for each $m \in \Delta$, so D fixes each point of k. Suppose $d \in D$ fixes a point r on m distinct from p and $k \cap m$. Then for each point t not on m, d fixes $r + ((r + t) \cap k) = r + t$ and then also fixes $t = (r + t) \cap (p + t)$. But then d fixes each point not on m, so $d = 1$.

We have shown D is regular on the three points of m not on k and distinct from p, for each $m \in \Delta$. Hence $D = B \le G$, completing the proof.

6. Fiber products of groups

We will need the notion of the fiber product of groups at several points. For example, the notion is used in the proof of Lemma 8.17 and in the construction in Section 27 of the centralizer of an involution in the Monster.

Let $\alpha_i : A_i \to A_0$, $i = 1, 2$, be group homomorphisms and consider the *fiber product*

$$A = A_1 \times_{A_0} A_2 = \{(a_1, a_2) : a_1 \alpha_1 = a_2 \alpha_2\} \le A_1 \times A_2.$$

Let $p_i : A \to A_i$, $i = 1, 2$, be the ith projection, and observe that we have a commutative diagram:

$$
\begin{array}{ccc}
 & A & \\
p_1 \swarrow & & \searrow p_2 \\
A_1 & & A_2 \\
\alpha_1 \searrow & & \swarrow \alpha_2 \\
 & A_0 & \\
\end{array}
$$

Moreover the fiber product satisfies the following universal property: Whenever we have a commutative diagram

$$
\begin{array}{ccc}
 & B & \\
\beta_1 \swarrow & & \searrow \beta_2 \\
A_1 & & A_2 \\
\alpha_1 \searrow & & \swarrow \alpha_2 \\
 & A_0 & \\
\end{array}
$$

then there exists a unique map $h : B \to A$ such that the following diagram commutes:

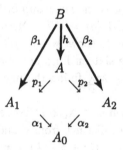

We record this as:

Lemma 6.1: *Let* $\beta_i : B \to A_i$, $i = 1, 2$, *be group homomorphisms with* $\beta_1 \alpha_1 = \beta_2 \alpha_2$. *Then there exists a unique group homomorphism* $h : B \to A = A_1 \times_{A_0} A_2$ *such that*

(1) $h p_i = \beta_i$ *for* $i = 1, 2$.
(2) $\ker(p_1) = \{(1, a_2) : a_2 \in \ker(\alpha_2)\}$.

Lemma 6.2: *Assume* $A_i \leq GL(V_i)$, *for F-spaces* V_i, $i = 1, 2$. *Then*

(1) $p_1 \otimes p_2 : A \to GL(V_1 \otimes V_2)$ *is an FA-representation.*
(2)   *Under the hypotheses of 6.1,* $h(p_1 \otimes p_2) = \beta_1 \otimes \beta_2$ *is an FB-representation on* $V_1 \otimes V_2$.

Proof: First $p_1 \otimes p_2 : A_1 \times A_2 \to GL(V_1 \otimes V_2)$ is a representation of $A_1 \times A_2$ which restricts to a representation of A, so (1) holds. Next under the hypotheses of (2), for $b \in B$, $v_i \in V_i$, we have $(v_1 \otimes v_2)(b(\beta_1 \otimes \beta_2)) = v_1(b\beta_1) \otimes v_2(\beta_2) = v_1(bhp_1) \otimes v_2(bhp_2) = (v_1 \otimes v_2)(bh(p_1 \otimes p_2))$, so (2) holds.

Remarks. The material on rank 3 permutation groups in Section 3 comes from D. Higman [Hi]. Most of the discussion on geometries, complexes, and chamber systems in Section 4 is due to Tits in [T1], with the last few lemmas from Section 4 appearing in [A1].

Our discussion of diagrams associated to geometries and geometric complexes has been restricted to the basic diagram. However, there is a much more extensive theory of diagrams in the literature begun by Tits [T1] and Buekenout [Bu]. See also [A2] and [RS].

Exercises

1. Prove Lemma 2.2.4.
2. Let G be a 4-transitive subgroup of S_6. Prove $G = S_6$ or A_6, and if the stabilizer in G of three points is of order 3 then $G = A_6$.
3. Let Γ be the projective plane over the field of order 4, $L = L_3(4) \leq Aut(\Gamma)$, and $g \in Aut(\Gamma)$ an involution fixing exactly three points on some line of Γ. Let $G = \langle g, L \rangle$ and prove G is L extended by a field automorphism and if x is a point of Γ fixed by g and Δ the set of lines of Γ through x then $G_x^\Delta = S_5$.
4. Let G be a primitive rank 3 group of even order on a set X of finite order n and let $x \in X$. Prove
   (1) If K is a regular normal subgroup of G then $n = p^e$ is the power of a prime and $K \cong E_{p^e}$.
   (2) If n is not a prime power and G_x is simple then G is simple.
5. Let Γ be a string geometry on a string ordered set $I = \{1, \ldots, n\}$. Prove that if $a_i * a_j * a_k$ with $i < j < k$ then $a_i * a_k$.

Chapter 2

2-Structure in Finite Groups

In this chapter we record some facts about the 2-subgroups of finite groups. In particular in Section 7 we recall some standard facts about *involutions*; that is, elements of order 2. Then in Section 8 we consider so-called *large extraspecial 2-subgroups* of a finite group G. Most of the sporadic groups contain such subgroups. They will serve as an important tool both in analyzing the structure of the sporadic groups and as part of the hypotheses under which we characterize many of the sporadics. See Chapter 5 for an idea of how this goes.

7. Involutions

In this section G is a finite group. Recall that an *involution* in G is an element of order 2. The following elementary result appears as 45.2 in [FGT]:

Lemma 7.1: Let x and y be distinct involutions in G, $n = |xy|$, and $D = \langle x, y \rangle$. Then

    (1) D is a dihedral group D_{2n} of order $2n$.

    (2)   Each element in $D - \langle xy \rangle$ is an involution.

    (3)   If n is odd then D is transitive on its involutions, so in particular x is conjugate to y in D.

    (4)   If n is even then each involution in G is conjugate to exactly one of x, y, or z, where z is the unique involution in $\langle xy \rangle$. Further $z \in Z(D)$.

    (5)   If n is even and z is the involution in $\langle xy \rangle$ then xz is conjugate to x in D if and only if $n \equiv 0 \mod 4$.

Given $t \in S \subseteq G$ and $h \in G$, we write $\langle S \rangle$ for the subgroup of G generated by S, $t^h = h^{-1}th$ and $S^h = \{s^h : s \in S\}$ for the conjugates of t and S under h, and $S^G = \{S^g : g \in G\}$ for the conjugacy class of S under G.

Lemma 7.2 (Thompson Order Formula): *Assume G has $k \geq 2$ conjugacy classes of involutions with representatives x_i, $1 \leq i \leq k$, and define n_i to be the number of ordered pairs (u,v) with $u \in x_1^G$, $v \in x_2^G$, and $x_i \in \langle uv \rangle$. Then*

$$|G| = |C_G(x_1)|\,|C_G(x_2)| \left(\sum_{i=1}^{k} n_i / |C_G(x_i)| \right).$$

Proof: This is 45.6 in [FGT]. The proof is an easy counting argument.

Lemma 7.3: *Let I be a G-invariant collection of involutions of G and $H \leq G$. Then the following are equivalent:*

(1) $x^G \cap H = x^H$ *and* $C_G(x) \leq H$ *for each* $x \in I \cap H$.

(2) $H \cap H^g \cap I$ *is empty for* $g \in G - H$.

(3) *The members of* $H \cap I$ *fix a unique point in the permutation representation of G on G/H.*

Proof: See 46.1 in [FGT]. Again the proof is easy.

An involution z is *2-central* in G if z is in the center of a Sylow 2-subgroup of G.

Lemma 7.4: *Assume z is a 2-central involution in G and $H \leq G$ such that H is the unique point of G/H fixed by z. Then*

(1) $|G : H|$ *is odd.*

(2) *If x is a 2-element of H then* $x^G \cap H = x^H$.

Proof: By 7.3, $C_G(z) \leq H$. As z is 2-central, $C_G(z)$ contains a Sylow 2-subgroup T of G. So $T \leq H$ and hence (1) holds.

Let x be a 2-element and $g \in G$ with $x, x^g \in H$. We must show $x^g = x^h$ for some $h \in H$. Conjugating in H we may take $x, x^g \in T$. Then $z \in C(x^g)$, so $z^{g^{-1}} \in K = C_G(x)$. Let R, S be Sylow 2-subgroups of K containing z and $z^{g^{-1}}$, respectively. As H is the unique point of G/H fixed by z and R is nilpotent, H is the unique point of G/H fixed by R (cf. Exercise 2.5). Similarly Hg^{-1} is the unique point fixed by S.

By Sylow's Theorem there is $k \in K$ with $R^k = S$. Then $\{Hg^{-1}\} = Fix(S) = Fix(R)k = \{Hk\}$, so $kg \in H$. Then $x^g = x^{kg}$ with $kg \in H$, completing the proof.

Lemma 7.5: *Let I and J be distinct conjugacy classes of involutions of G and H a subgroup of G such that:*

> (a)   *Each member of $I \cup J$ fixes a unique point of G/H.*
> (b)   *For all $a \in I \cap H$ and $b \in (I \cup J) \cap H$ with $ab = ba$, we have $C_G(ab) \le H$. Then $G = H$.*

Proof: Assume the lemma is false and let $x \in H \cap I$ and $y \in H \cap J$. If $J \subseteq H$ then for all $g \in G$, $\{Hg\} = Fix(y)g = Fix(y^g) = \{H\}$. But then $H = G$, as desired. So let $u \in J - H$ and $D = \langle x, u \rangle$. Then $u \notin x^D$ so by 7.1, xu has even order and $xz \in I \cup J$, where z is the involution in $\langle xu \rangle$. But then $xz \in C_G(x) \le H$, so by hypothesis (b), $u \in C_G(z) \le H$, a contradiction.

A subgroup H of a group G of even order is *strongly embedded* in G if H is a proper subgroup of G and $H \cap H^g$ is of odd order for each $g \in G - H$. Notice that by 7.3, this is equivalent to the assertion that H is proper in G and each nontrivial 2-element in G fixes a unique point of G/H.

Strongly embedded subgroups arise in this book as follows. Let X be some subgroup of G, $M = C_G(X)$, and $H \le M$. We wish to show $H = M$, given that $C_G(\langle X, t \rangle) \le H$ for enough involutions $t \in H$. We use this information to show that if $H \neq M$ then H is strongly embedded in M; then we obtain a contradiction from Lemma 7.6. For example, we may show that some 2-central involution of H fixes a unique point of M/H and then use Exercise 2.10 to show H is strongly embedded in M.

Lemma 7.6: *Let H be a strongly embedded subgroup of G. Then there exists a subgroup of H of odd order transitive on the involutions of H.*

Proof: This is essentially contained in Exercise 16.5 in [FGT]; here are the details. Let I be the set of involutions in G, $t \in I \cap H$, $u \in I - H$, and $K = H \cap H^u$. Then K is the subgroup of G fixing the points H and Hu of G/H and $L = K\langle u \rangle$ is the global stabilizer of $\{H, Hu\}$. Also as each nontrivial 2-element of G fixes a unique point of G/H, we have:

> (a) K is of odd order.

Let $J = \{k \in K : k^u = k^{-1}\}$. By (a), uv is of odd order for each $v \in I \cap L$, so by 7.1, $uJ = I \cap L$ and K is transitive on uJ. We claim:

> (b) $C_H(j)$ is of odd order for each $1 \neq j \in J$.

For if $t \in C_H(j)$ then t is not conjugate to u in $X = C_G(j)\langle u \rangle$ as $C_G(j) \trianglelefteq X$. On the other hand $X \cap H$ is strongly embedded in X, so 7.5 supplies a contradiction. Thus (b) is established. Notice (b) implies:

(c) Distinct involutions in L are in distinct cosets of $C_H(t)$, and $|I \cap L| \leq |H : C_H(t)|$.

Namely the maps $v \mapsto uvC_H(t)$ and $v \mapsto vC_H(t)$ are injections of $I \cap L$ into $H/C_H(t)$ and $G/C_H(t)$, respectively. For if v, w are distinct in $I \cap L$ then $1 \neq wv \in J$ so $wv \notin C_G(t)$. Next we claim:

(d) $|J| = |I \cap H|$.

For let $m = |I \cap H|$ and consider the set S of triples (i, x, y) such that $i \in I$ and (x, y) is a cycle of i on G/H. Observe that $|I| = mn$, where $n = |G : H|$, and i has $(n-1)/2$ cycles of length 2. Hence $|S| = mn(n-1)/2$. But also

$$|S| = \sum_{x,y} M_{x,y} \leq n(n-1)M/2,$$

where $M_{x,y}$ is the number of involutions with cycle (x, y) and M is the maximum of $M_{x,y}$. Observe that by 7.5, G is transitive on I, so by 7.3, H is transitive on $I \cap H$. Thus $m = |H : C_H(t)|$. But by (c), $|H : C_H(t)| \geq M$, so $M \leq m$. It follows that $M_{x,y} = m$ for all x, y; that is, (d) holds. Finally observe:

(e) Distinct elements of J are in distinct cosets of $C_H(t)$.

For if $j, k \in J$ with $k^{-1}j \in C_H(t)$ then $ju, ku \in I \cap L$ with $(uk)^{-1}(uj) = k^{-1}uuj = k^{-1}j \in C_H(t)$, contrary to (c).

It follows from (d) and (e) that $|K : C_K(t)| \geq |J| = |I \cap H|$. But $|K : C_K(t)| = |t^K|$ and $t^K \subseteq I \cap H$, so $t^K = I \cap H$. That is, K is transitive on $I \cap H$, completing the proof.

Let H and S be subgroups of G. We say H *controls fusion* in S if $s^G \cap S = s^H$ for each $s \in S$. We say S is *weakly closed* in H with respect to G if $S^G \cap H = \{S\}$. Part (1) of the following lemma appears as 37.6 in [FGT]; part (2) is easy.

Lemma 7.7: *Let p be a prime and $T \in Syl_p(G)$. Then*

(1)   *If W is a weakly closed subgroup of G then $N_G(W)$ controls fusion in $C_G(W)$.*

(2)   *If $A, B \trianglelefteq T$ and $A \in B^G$ then A and B are conjugate in $N_G(T)$.*

8. Extraspecial groups

The *Frattini subgroup* of a group G is the intersection of all maximal subgroups of G. Write $\Phi(G)$ for the Frattini subgroup of G. Evidently

$\Phi(G)$ is a characteristic subgroup of G. Further

Lemma 8.1: *If G is a group, $X \subseteq G$, and $G = \langle X, \Phi(G) \rangle$ then $G = \langle X \rangle$.*

Lemma 8.2: *Let G be a finite p-group and A a group of automorphisms of G of order prime to p. Then*

  (1) *$\Phi(G)$ is the smallest normal subgroup H of G such that G/H is elementary abelian.*

  (2) *A is faithful on $G/\Phi(G)$.*

Proof: See 23.2 and 24.1 in [FGT].

An *elementary abelian p-group* of order p^m is a direct product of m copies of the group $\mathbf{Z}_p$ of order p and is denoted here by E_{p^m}. We can regard such a group as an m-dimensional vector space over $GF(p)$. Thus if G is a p-group then by 8.2.1 we can regard $G/\Phi(G)$ as a vector space over $GF(p)$, so by 8.2.2, $A \leq GL(G/\Phi(G))$ for each group A of automorphisms of G of order prime to p.

For $x, y \in G$, $[x,y] = x^{-1}y^{-1}xy$ is the *commutator* of x and y, and for $X, Y \leq G$, $[X, Y] = \langle [x, y] : x \in X, y \in Y \rangle$. Recall that $[X, Y] \leq X$ if and only if $Y \leq N_G(X)$ (cf. 8.5.5 in [FGT]).

A p-group E is *extraspecial* if E is finite with $\Phi(E) = Z(E) = [E, E]$ and $Z(E)$ is cyclic. As a consequence, $Z(E)$ is of order p (cf. 23.7 in [FGT]).

Lemma 8.3: *Let E be an extraspecial p-group, $Z = Z(E)$, and $\tilde{E} = E/Z$. Regard Z as $GF(p)$ and $\tilde{E}$ as a vector space over Z. Define $f : \tilde{E} \times \tilde{E} \to Z$ by $f(\tilde{x}, \tilde{y}) = [x, y]$. Then*

  (1) *$(\tilde{E}, f)$ is a symplectic space over Z.*

  (2) *$dim(\tilde{E}) = 2n$ is even.*

  (3) *If $p = 2$ define $Q : \tilde{E} \to Z$ by $Q(\tilde{x}) = x^2$. Then Q is a quadratic form on $\tilde{E}$ associated to f, so $(\tilde{E}, Q)$ is an orthogonal space over Z.*

  (4) *Let $Z \leq U \leq E$. Then U is extraspecial or abelian if and only if U is nondegenerate or totally isotropic, respectively. If $p = 2$ then U is elementary abelian if and only if $\tilde{U}$ is totally singular.*

Proof: See 23.10 in [FGT]. Also see Section 12.

Remark. (1) See Chapter 7 in [FGT] for a discussion of symplectic and quadratic forms. The integer n in (2) is the *width* of E.

We write p^{1+2n} to denote an extraspecial p-group of order p^{1+2n}; if p is odd we also require that the extraspecial group p^{1+2n} be of exponent p.

For odd p this determines p^{1+2n} up to isomorphism; see, for example, 23.13 in [FGT].

In the rest of this section we concentrate on extraspecial 2-groups. Recall D_8 and Q_8 are the dihedral and quaternion groups of order 8, respectively. Notice each is extraspecial of width 1. Given two extraspecial groups E_1 and E_2 define the *central product* $E_1 * E_2$ of E_1 and E_2 to be the group $(E_1 \times E_2)/\langle(z_1, z_2)\rangle$, where $\langle z_i \rangle = Z(E_i)$. Notice $E_1 * E_2$ is also extraspecial. We can extend this construction to form the central product

$$E_1 * \cdots * E_m = (E_1 * \cdots * E_{m-1}) * E_m$$

of an arbitrary number m of extraspecial groups E_i. Thus $E_1 * \cdots * E_m$ is extraspecial. Write $D_8^n Q_8^m$ for the central product of n copies of D_8 with m copies of Q_8.

Lemma 8.4: *Up to isomorphism D_8^n and $D_8^{n-1} Q_8$ are the unique extraspecial groups of width n. The 2-rank of D_8^n is $n + 1$ while $D_8^{n-1} Q_8$ has 2-rank n; hence the groups are not isomorphic.*

Proof: See 23.14 in [FGT].

Remark. (2) Given a prime p, the *p-rank* $m_p(G)$ of a group G is the maximum m such that G has a subgroup isomorphic to E_{p^m}. Notice that if E is an extraspecial 2-group then by 8.3.4, $m_2(E) - 1$ is the Witt index of $\tilde{E}$, so if $E = D_8^n$ then $\tilde{E}$ has Witt index n while if $E = D_8^{n-1} Q_8$ then $\tilde{E}$ has Witt index $n - 1$ (cf. page 78 in [FGT] for the definition of the Witt index).

Lemma 8.5: *Let $E \cong D_8^n$ or $D_8^{n-1} Q_8$, $Z = Z(E)$, $\tilde{E} = E/Z$, and Q the quadratic form on $\tilde{E}$ induced by E. Let $A = Aut(E)$. Then*

(1) $C_A(\tilde{E}) = Inn(E) \cong E_{2^n}$.

(2) $A/C_A(\tilde{E}) = O(\tilde{E}, Q) \cong O_{2n}^\epsilon(2)$, *where $\epsilon = +1$ if $E \cong D_8^n$ and $\epsilon = -1$ if $E \cong D_8^{n-1} Q_8$.*

Proof: See Exercise 8.5 in [FGT]. Also see Section 12.

Remark. (3) Given a group H, $Inn(H)$ is the group of *inner automorphisms* of H: That is, those automorphisms of the form $i_x : h \mapsto h^x$ for some $x \in H$. Notice $Inn(H)$ is a normal subgroup of $Aut(H)$.

Define a subgroup Q of a finite group G to be a *large extraspecial 2-subgroup* of G if:

(L1) Q is an extraspecial 2-subgroup of G.

(L2) $C_G(Q) = Z(Q) = Z$.

(L3) $Q \trianglelefteq C_G(Z)$.

(L4) $I_G(Q, 2') = 1$.

Here $\mathit{M}_G(Q, 2')$ is the set of subgroups K of G of odd order such that $Q \leq N_G(K)$. Given a group A, let $A^{\#} = A - \{1\}$.

Lemma 8.6: *Let $E_4 \cong A \leq G$ and $X \in \mathit{M}_G(A, 2')$. Then*

$$X = \langle C_X(a) : a \in A^{\#} \rangle.$$

Proof: See, for example, Exercise 8.1 in [FGT].

Recall that if π is a set of primes then $O_\pi(G)$ is the largest normal subgroup of G whose order is divisible only by primes in π. Thus if p is a prime then $O_p(G)$ denotes the largest normal p-subgroup of G. Also p' is the set of primes distinct from p and $O_{p'}(G)$ is the largest normal subgroup of order prime to p. $O(G) = O_{2'}(G)$ is the largest normal subgroup of odd order. Similarly $O^\pi(G)$ is the smallest normal subgroup H of G such that G/H has order divisible only by primes in π. So $O^p(G)$ is the subgroup generated by all p'-elements of G.

Lemma 8.7: *Assume Q satisfies (L1)–(L3) in G. Then*

(1) *If $Z \leq O_2(G)$ then Q is large in G.*

(2) *If there exists an involution b in $Q - Z$ with $Z \leq O_2(C_G(b))$ then Q is large in G.*

(3) *Assume the width of Q is at least 2 and $g \in G - N_G(Z)$ with $Z^g \leq Q$. Then $Z \leq Q^g$, $N_G(ZZ^g)/C_G(ZZ^g) \cong S_3$, and Q is large in G.*

Proof: Let $X \in \mathit{M}_G(Q, 2')$. If $Z \leq O_2(G)$ then $[Z, X] \leq O_2(G) \cap X = 1$. Thus $X \leq C_G(Z)$, so as $Q \trianglelefteq C_G(Z)$, $[Q, X] \leq Q \cap X = 1$. Hence $X \leq C_G(Q) = Z$, so $X = 1$. That is, (1) holds.

Assume the hypothesis of (2) and let $Z = \langle z \rangle$. Then $A = \langle z, b \rangle \cong E_4$. Notice $bz = b^x$ for some $x \in Q$, so $z \in O_2(C_G(bz))$. Thus $z \in O_2(C_G(a))$ for all $a \in A^{\#}$. By 8.6, $X = \langle C_X(a) : a \in A^{\#} \rangle$. As $Z \leq O_2(C_G(a))$, $[Z, C_X(a)] \leq O_2(C_G(a)) \cap X = 1$. Thus $X \leq C_G(Z)$, so $X = 1$ by (1) applied to $C_G(Z)$ in place of G.

Finally assume the hypothesis of (3). Here let $b = z^g$ and $A = \langle z, b \rangle$. Again $b^x = bz$ for some $x \in Q$. Observe $x \in M = N_G(A)$ and x induces the transposition (b, bz) on $A^{\#}$. So if $y \in M - C_G(Z)$ then $\langle y, x \rangle \leq M$ induces S_3 on $A^{\#}$. In particular we may choose $g \in M$ to act as (b, z) on $A^{\#}$. Then $Z^g \leq Q$ so $Z = Z^{g^2} \leq Q^g$, and (3) holds.

So assume $M \leq C_G(Z)$. Then $Z \nleq Q^g$. Let $P = Q^g$, $R = C_Q(b)$, and $P^* = P/Z^g$. As Q has width at least 2, $R = Z^g \times S$ with S extraspecial.

In particular $A = Z(R)$. Also $[R \cap P, R] \leq [Q,Q] \cap P = Z \cap P = 1$. So $R \cap P \leq Z(R) \cap P = Z^g$. Thus $[C_P(Z), R] \leq P \cap R = Z^g$. Thus $[C_P(Z)^*, R] = 1$. Also $C_{P^*}(Z) = N_P(A)/Z^g = C_P(Z)/Z^g$ as $M \leq C_G(Z)$. So $C_{P^*}(Z) = C_P(Z)^* = C_{P^*}(R)$.

Let $Y = C_P(Z)R$. Then $Y^* = C_P(Z)^* \times S^*$ with $S^* \cong S$ extraspecial. Thus $\Phi(Y^*) = \Phi(S^*) = Z^*$. Therefore $N_{P^*}(Y^*) \leq C_{P^*}(Z) \leq Y^*$. Thus $P^* \leq Y^*$ (cf. 9.10 in [FGT]) so $P \leq C_G(Z)$, contradicting (L2), and completing the proof.

We now consider some examples. See Section 10 in [FGT] for a discussion of the notions of *extension*, *split extension*, and *complement*.

Lemma 8.8: *Let G be the split extension of $U \cong E_{2^n+1}$, $n \geq 1$, by $M = GL(U)$, and let Z be of order 2 in U. Let $Q = O_2(C_G(Z))$. Then $Q \cong D_8^n$ is large in G and $C_G(Z)$ is the split extension of Q by $L_n(2)$.*

Proof: Observe $C_M(Z)$ is the split extension of the group $W \cong E_{2^n}$ of transvections with center Z by $L \cong L_n(2)$. Let $Q = UW$ and $H = C_G(Z)$. Then $Q = O_2(H)$ so (L3) holds. As M is faithful on U, $C_G(U) \leq U$, so $C_G(Q) \leq Z(Q) \leq U$. Also $Z = C_U(W)$, so $Z = Z(Q)$. Thus (L2) holds. Next $[W,U] = Z$, so $Z = [Q,Q] = \Phi(Q)$. Hence (L1) holds. As $Z \leq O_2(G)$, Q is large in G by 8.7.1.

Now Q is of order 2^{2n+1}, so Q has width n. Then by 8.4, $m(Q) \leq n+1$. But $m(U) = n+1$, so $m(Q) = n+1$ and by 8.4, $Q \cong D_8^n$.

Lemma 8.9: *Let $G = L_{n+2}(2)$, $n \geq 2$, z be a transvection in G, and $Q = O_2(C_G(z))$. Then $Q \cong D_8^n$ is large in G and $C_G(z)$ is the split extension of Q by $L_n(2)$.*

Proof: First $G = GL(V)$ for some $n+2$-dimensional vector space V over $GF(2)$. Let $\langle v \rangle$ be the center of z and $K = C_G(v)$. Then by 5.1, K is the split extension of $U \cong E_{2^n+1}$ by $M = GL(U)$. Further $C_G(z) \leq K$, so $C_G(z) = C_K(z)$. Therefore $Q \cong D_8^n$ satisfies (L1)–(L3) in K, and hence also in G by 8.9. Finally $Z^g \leq U \leq Q$ for $g \in K - N_G(Q)$, so Q is large in G by 8.7.3.

Lemma 8.10: *Let $G = M_{24}$ and $(X, \mathcal{C})$ the Steiner system for G (cf. Section 18). Let z be an involution in G with $C_X(z) \in \mathcal{C}$. Then $Q = O_2(C_G(z)) \cong D_8^3$ is large in G and $C_G(z)$ is the split extension of Q by $L_3(2)$.*

Proof: Let $B \in \mathcal{C}$. By 19.1, $N_G(B)$ is the split extension of $U \cong E_{16}$ by $M \cong L_4(2)$. Further for $z \in U^\#$, $B = Fix_X(z)$; for example, if T is a 3-set in B then $B - T$ is a line in the projective plane on $X - T$ and z is a transvection in $L_3(4)$ with axis $B - T$.

So $C_G(z) \le N_G(B)$. Now complete the proof as in 8.9.

Remark. (4) Notice by 8.9 and 8.10, $G_1 = M_{24}$ and $G_2 = L_5(2)$ are simple groups such that there exist involutions $z_i \in G_i$ with $C_{G_1}(z_1) \cong C_{G_2}(z_2)$. There is one further simple group G_3 possessing an involution z_3 with $C_{G_3}(z_3) \cong C_{G_1}(z_1)$: Namely the sporadic group He of Held. This is the only example of three nonisomorphic simple groups G_i, $1 \le i \le 3$, possessing involutions z_i with $C_{G_i}(z_i) \cong C_{G_j}(z_j)$ for all i, j. However, the classification of the finite simple groups is required to verify this fact. On the other hand the following result has an elementary proof:

Brauer–Fowler Theorem: *Let H be a finite group. Then there exist at most a finite number of finite simple groups G with an involution t such that $C_G(t) \cong H$.*

See, for example, 45.5 in [FGT] for a proof of the Brauer–Fowler Theorem, or the original proof in [BF]. The Brauer–Fowler Theorem supplies the philosophical base for the classification. We find in Chapter 14 that M_{24}, $L_5(2)$, and He are the only simple groups G possessing an involution z with $C_G(z)$ isomorphic to the centralizer of a transvection in $L_5(2)$.

The next two lemmas show that a group with a large extraspecial 2-subgroup is close to being simple.

Lemma 8.11: *Let Q be a large extraspecial 2-subgroup of G, $Z(Q) = Z = \langle z \rangle$, and $M = \langle Z^G \rangle$. Then one of the following holds:*

(1) $Q \trianglelefteq G$.

(2) M is abelian and $F^*(G) = O_2(G)$.

(3) $M = F^*(G)$ is a nonabelian simple group.

(4) $M = L \times L^u = F^*(G)$ for some nonabelian simple group L and $u \in Q$. Further $z = z_1 z_1^u$ for some $z_1 \in L$ with $C_L(z_1) = R \in Syl_2(L)$ and $\Phi(R) \le \langle z_1 \rangle$. Moreover $Q \ne O_2(C_G(z))$.

Remarks. (5) See Section 31 in [FGT] for the definition of the *generalized Fitting subgroup* $F^*(G)$ of G and a discussion of the properties of this subgroup. In particular $F^*(G)$ is the product of the *Fitting subgroup* $F(G)$ of G with the product $E(G)$ of the components of G. Further the *components* of G are the subnormal quasisimple subgroups of G, while a group L is *quasisimple* if $L = [L, L]$ and $L/Z(L)$ is simple.

(6) With some extra work it can be shown that in case (4), $G \cong Z_2 wr A_6$, $G \cong Z_2 wr L_3(2)$, or G is of index 2 in $Z_2 wr S_5$. For example, Exercise 2.3 is a beginning in that direction. See Section 11 in [FGT] for the definition of the *wreath product* $A wr B$ of groups A and B.

Now the proof of 8.11. First $O(G) \in \mathcal{U}_G(Q, 2') = 1$ by (L4).

Suppose $P = O_2(G) \neq 1$. Then $1 \neq C_P(PQ)$, so $Z = C_P(QP)$ by (L2). Then $Z \leq Z(P) \trianglelefteq G$, so $M = \langle Z^G \rangle \leq Z(P)$. In particular M is abelian. Also by (L2), $O^2(F^*(C_G(Z))) = 1$. But as $Z \leq O_2(G)$, $O^2(F^*(G)) = O^2(F^*(C_G(Z)))$ (cf. 31.14.2 in [FGT]). Thus $F^*(G) = O_2(G)$ and (2) holds in this case.

So we may assume G has no nontrivial solvable normal subgroup. Hence $F^*(G) = E(G)$. Let L be a component of G and $Y = \langle L^Q \rangle$. As $O(G) = 1$, L has even order, so by (L2), $Z \leq Y$. Now if K is a component of G not in L^Q then $K \leq C_G(Y) \leq C_G(Z)$, so $K \leq E(C_G(Z)) = 1$. Thus $Y = F^*(G)$.

It remains to show (3) or (4) holds, so we may assume $G = YQ$ but $Y \neq L$. Hence as Q is generated by involutions, there is an involution $u \in Q - N_G(L)$. Let $R = C_L(z)$ and z_1 the projection of z on L. Define a map $\pi : R \to Y$ by $x\pi = [x, u] = x^{-1}x^u$. As $\langle L, L^u \rangle = L \times L^u$, π is an injection and the restriction of π to any abelian subgroup of R is a group homomorphism. Further $[x, u] \in Q$ by (L3), so $R\pi \subseteq Q$. Then for $x \in R$, $\pi : \langle x \rangle \to Q$ is an injective group homomorphism, so $|x|$ divides 4 with $(x\pi)^2 = z$ in case $|x| = 4$. Therefore $x^2 = z_1$ if $|x| = 4$. We conclude R is a 2-group with $\Phi(R) \leq \langle z_1 \rangle$. By (L2), Z is in the center of a Sylow 2-group of G and hence z_1 is in the center of a Sylow 2-group of L. So $R = C_L(z) = C_L(z_1) \in Syl_2(L)$. As L is simple, $|R| > 2$ (cf. 39.2 in [FGT]), so there is $r \in R - \langle z_1 \rangle$ and as $[r, u] \notin Z$, $r \notin Q$. Thus as $R \leq O_2(C_G(Z)))$, $Q \neq O_2(C_G(Z))$.

It remains to show $Y = LL^u$. As $|R| > 2$, $R\pi$ contains a subgroup of Q of order at least 4, so for each $v \in Q$, $1 \neq C_{R\pi}(v)$. However, $r \in C_{R\pi}(v)$ projects only on L and L^u, so v acts on $\{L, L^u\}$. Thus as $Y = \langle L^Q \rangle$, $Y = LL^u$.

Lemma 8.12: *Let z be an involution in a finite group G such that*

(1) $F^*(C_G(z)) = Q$ *is an extraspecial 2-group of width at least 2.*

(2) $K \leq G$ *with* $U = \langle z^K \rangle$ *abelian and* $C_G(z) = \langle U^{C_G(z)} \rangle$.

(3) *Either* $z^K \cap Q \neq \{z\}$ *or there exists* $u \in U \cap Q - \langle z \rangle$ *with* $z \in O_2(C_G(u))$.

Then G is simple and Q is large in G.

Proof: Let $H = C_G(z)$. As $Q = F^*(H)$, $C_G(Q) = C_H(Q) \leq Q$ (cf. 31.13 in [FGT]). So Q satisfies (L1)–(L3) in G.

Next if $z^K \cap Q \neq \{z\}$ then Q is large in G by 8.7.3. Similarly if $u \in U \cap Q - \langle z \rangle$ with $z \in O_2(C_G(u))$ then Q is large in G by 8.7.2.

So in any event Q is large in G. Let $M = \langle z^G \rangle$. Then $U \leq M$, so

$H = \langle U^H \rangle \leq M$. In particular as $Q = O_2(H)$, $M \not\leq O_2(G)$. Also as $U = \langle z^K \rangle \neq \langle z \rangle$ and $\langle z \rangle = Z(Q)$, Q is not normal in G. Hence $M = F^*(G)$ is simple by 8.11.

Finally as $H \leq M$, Exercise 2.1 says $G = M$. That is, G is simple.

The remaining lemmas in this section, while crucial to the analysis in Part III, are more difficult and technical. The reader may wish to skip or postpone these lemmas.

Lemma 8.13: *Let z be an involution in G with $Q = F^*(C_G(z))$ extraspecial and A a subgroup of odd order in $C_G(z)$. Let $R = C_Q(A)$ and assume $|R| \geq 32$, $O(C_G(AR)) \leq A$, and $g \in G$ with $z^g \in R - Z$. Let $M = C_G(A)A$, and $M^* = M/A$. Then $AZ = C_M(RA)A$, R^* is a large extraspecial subgroup of M^*, and z^g is conjugate to z in $\langle R, C_{Q^g}(A) \rangle$.*

Proof: By Exercise 2.2, R is extraspecial and $Q = R * [Q, A]$. Thus if we let $D = C_M(RA)$ and $P \in Syl_2(D)$, we have $P \cap Q = Z$. Therefore P/Z is faithful on Q/Z and hence by the Thompson $A \times B$ Lemma (cf. 24.2 in [FGT]), P/Z is faithful on R/Z. That is, $P = Z$. So $DA = A \times Z$.

Next by 8.7, $z \in Q^g$. So $Z^g \neq R_1 = C_{Q^g}(A)$. Then by Exercise 2.2, R_1 is extraspecial, so z is fused to zz^g under R_1. Hence $\langle R, R_1 \rangle$ is transitive on $\langle z, z^g \rangle^\#$, so we may choose $g \in \langle R, R_1 \rangle$.

As $AZ = C_M(RA)A$, $Z^* = C_{M^*}(R^*)$, so R^* satisfies (L1)–(L3) in M^*. As $g \in M$ with $z^g \in R - Z$, R^* is large in M^* by 8.7.

Lemma 8.14: *Let z be an involution in G, $H = C_G(z)$, $Q = F^*(H)$ extraspecial of width at least 3, and t an involution in $Q \cap Q^g - z^G$ for some $g \in G$ with $z^g \in Q - Z$. Assume $C_H(C_Q(t)/\langle t, z \rangle) = Q$ and let $M = C_G(t)$, and $M^* = M/\langle t \rangle$. Then $C_Q(t)^*$ is a large extraspecial subgroup of M^* and z^g is conjugate to z in $\langle C_Q(t), C_{Q^g}(t) \rangle$.*

Proof: Let $P = C_Q(t)$ and observe that P^* is extraspecial. As $t \notin z^G$ but $tz \in t^Q$, z is weakly closed in $\langle z, t \rangle$ with respect to G, so $C_{M^*}(z^*) = C_H(\langle z, t \rangle)^*$. In particular $P^* \trianglelefteq C_{M^*}(z^*)$ and as $C_H(P/\langle t, z \rangle) = Q$, $C_{M^*}(P^*) = Z^*$. That is, P^* satisfies (L1)–(L3) in M^*.

Next $t \in Q \cap Q^g$ and as $t \notin z^G$, $t \notin \langle z, z^g \rangle = A$, so there exist $r \in P - C(z^g)$ and $s \in C_{Q^g}(t) - H$. Then $\langle r, s \rangle$ induces S_3 on A so we may take $g \in \langle P, C_{Q^g}(t) \rangle$. Therefore P^* is large in M^* by 8.7.3.

Lemma 8.15: *Let Q be a large extraspecial 2-subgroup of a finite group G, $|Q| = 2^{1+2w}$ with $w \geq 2$, $Z = \langle z \rangle = Z(Q)$, $H = C_G(z)$, $g \in G - H$ with $z^g \in Q$, $V = \langle z, z^g \rangle$, $X = \langle Q, Q^g \rangle$, $R = (Q^g \cap H)(Q \cap H^g)$, and*

$E = Q \cap Q^g$. Then

(1) $R = C_X(V) = O_2(X)$ and $X/R \cong S_3$.

(2) $V = Z(R)$, $E \le Z_2(R)$, and R has nilpotence class at most 3.

(3) E and R/E are elementary abelian with $R/E = (Q \cap H^g)/E \times (Q^g \cap H)/E$.

(4) $[X, E] \le V$.

(5) $N_G(V) = XC_H(V)$ with $X \trianglelefteq N_G(V)$ and $N_G(V)/R = X/R \times C_H(V)/R$.

(6) R/E is the tensor product of the 2-dimensional irreducible for $X/R \cong S_3$ with the module $(Q \cap H^g)/E$ for $C_H(V)/R$.

(7) $m_2(E) \le w + 1$ and in case of equality $Q \cong D_8^w$, E/V is dual to $(Q \cap H^g)/E$ as a $C_H(V)/R$-module, and $(Q^g \cap H)/E$ induces the full group of transvections on E/Z with center V/Z.

(8) Let $m + 1 = m_2(E)$. Then $E_{2^{2w-m-1}} \cong RQ/Q \trianglelefteq C_{H/Q}(V/Z)$. In particular if $E_{2^{w-1}} \cong O_2(C_{H/Q}(V/Z))$ then $m(E) = w + 1$.

Proof: As Q is a large extraspecial 2-subgroup of G, $Q \trianglelefteq H$, so $Q^g \cap H$ and $Q \cap H^g$ are normal in R. Further $R = C_{QR}(V) \trianglelefteq QR$. Also by 8.7, $X/C_X(V) \cong S_3$ is transitive on $V^\#$, so we have symmetry between z and z^g, and hence $Q^g \le N_G(R)$, so that $R \trianglelefteq X$.

As $|Q : Q \cap R| = 2$, X/R is dihedral. Indeed as Q is conjugate to Q^g in X, $X/R \cong D_{2n}$ with n odd by 7.1. Then QR/R inverts $O(X/R)$ while $[Q, C_X(V)] \le C_Q(V) \le R$, so $C_X(V) = R$. Then as $X/C_X(V) \cong S_3$, (1) is established.

Next $[Q, E] = Z \le V$, so (4) holds and $E \le Z_2(R)$. As $[Q^g \cap H, Q \cap H^g] \le Q \cap Q^g = E$, we conclude $R/E = (Q \cap H^g)/E \times (Q^g \cap H)/E$ and as $\Phi(Q) = Z$, R/E is elementary abelian. Similarly $\Phi(E) \le Z \cap Z^g = 1$, so E is elementary abelian and (2) and (3) are established, except for the statement $V = Z(R)$. But suppose $y \in Z(R) - V$. As $V = Z(Q \cap H^g)$, $y \notin Q$. Similarly $y \notin Q^g$, so $y = ab$, $a \in Q \cap R - Q^g$ and $b \in Q^g \cap R - Q$. Now $[Q \cap H^g, b] = [Q \cap H^g, a] = Z$, so $y^* = b^*$ induces a transvection on $\tilde{Q}$ with axis $\tilde{V}^\perp = (Q \cap H^g)/Z$. This is impossible as the axis of a transvection in $O(\tilde{Q})$ is the subspace orthogonal to a nonsingular point, whereas $\tilde{V}$ is singular. So the proof of (2) is complete.

As X induces $S_3 = Aut(V)$ on V, $N_G(V) = XC_H(V)$. As $X = \langle Q^x : x \in N_G(V) \rangle$, $X \trianglelefteq N_G(V)$. Then as $R = C_X(V)$, $N_G(V)/R = X/R \times C_H(V)/R$.

Now Q/R interchanges $(Q^g \cap H)/E$ and $(Q^{gt} \cap H)/E$ for $t \in Q - R$, so

$$m(C_{R/E}(t)) = m(R/E)/2 = m((Q \cap H^g)/E)$$

and as $[Q \cap H^g, t] \leq Z$, $(Q \cap H^g)/E = C_{R/E}(t)$. Hence R/E is the sum of $m((Q \cap H^g)/E)$ 2-dimensional irreducibles for $X/R \cong S_3$, and thus (6) holds.

Finally as $\Phi(E) = 1$, $m(E) \leq w + 1$ by 8.3. By (5), $RQ/Q \trianglelefteq C_{H/Q}(V/Z)$. Also $RQ/Q \cong R/(R \cap Q) = (Q^g \cap H)(R \cap Q)/(R \cap Q) \cong (Q \cap H^g)/E \cong E_{2^{2w-m-1}}$. Thus (8) holds.

Assume $m(E) = w+1$. Then by 8.4, $Q \cong D_8^w$. Further $\tilde{E}$ is a maximal totally singular subspace of $\tilde{Q}$, so $\tilde{Q}/\tilde{E}$ is dual to $\tilde{E}$ as an $N_H(E)$-module, and hence as $\tilde{V}^{\perp} = (Q \cap H^g)/Z$, also $(Q \cap H^g)/E$ is dual to E/V as an $N_H(E)$-module.

Lemma 8.16: *Let Q be a large extraspecial 2-subgroup of a finite group G of width $w \geq 2$, $Z = \langle z \rangle$, $H = C_G(z)$, $g, k \in G$ with $E_8 \cong V = \langle z, z^g, z^k \rangle \leq Q \cap Q^g \cap Q^k = U$. Let $X = \langle Q, Q^g, Q^k \rangle$, $R = C_Q(V)C_{Q^g}(V)C_{Q^k}(V)$, $S = (Q \cap Q^g)(Q \cap Q^k)(Q^g \cap Q^k)$, $m + 1 = m(Q \cap Q^g)$, and $u = m(U)$. Then*

(1) $R = C_X(V) = O_2(X)$ *and* $X/R = GL(V) \cong L_3(2)$.

(2) $V \leq Z(R)$ *and* $[X, U] \leq V$ *with* $\Phi(U) = 1$.

(3) $N_G(V) = XC_H(V)$ *with* $X \trianglelefteq N_G(V)$ *and* $N_G(V)/R = X/R \times C_H(V)/R$.

(4) S/U *is the sum of* $m - u + 1$ *copies of the dual of* V^* *and* R/S *is the sum of* $2(w - m) + u - 3$ *copies of* V *as an* X/R-module.

Proof: The proof is much like that of 8.15. As in 8.15, $R = C_{QR}(V) \trianglelefteq QR$, and then $R \trianglelefteq X$. Moreover Q, Q^g, Q^k induce the group of transvections on V with center Z, Z^g, Z^k, respectively, so $V \trianglelefteq X$ and $X/C_X(V) = GL(V) \cong L_3(2)$.

Next $[Q, C_X(V)] \leq C_Q(V) \leq R$, so $C_X(V)/R \leq Z(X/R)$. Let $Y = \langle Q, Q^g \rangle$. By 8.15, $YR/R \cong S_4$, so by Gaushutz's Theorem (cf. 10.4 in [FGT]) X/R splits over $C_X(V)/R$. Further $Q \cap H^g \leq O^2(Y)R \leq X^{\infty}R$ and similarly $Q \cap H^k \leq X^{\infty}R$, so $Q = (Q \cap H^g)(Q \cap H^k) \leq X^{\infty}R$ and hence $X = X^{\infty}R$ and $R = C_X(V)$. Thus (1) is established.

The proof of (2) and (3) are as in 8.15. Next $[Y, Q \cap Q^g] \leq ZZ^g \leq V$ by 8.15, and $S/(Q \cap Q^g) = (S \cap Q)(S \cap Q^g)/(Q \cap Q^g)$ is the sum of natural modules for $Y/O_2(Y)$ by 8.15, with $[S, O_2(Y)] \leq Q \cap Q^g$. Similarly $\langle Q, Q^k \rangle$ acts on S, so $S \trianglelefteq X$. Then S/U is the sum of $m((Q \cap Q^g)/U)$ copies of V^* by Exercise 2.9. Similarly $Q^g \cap H$ and $Q^k \cap H$ centralize $C_Q(V)S/S$, so $H \cap X = \langle Q^g \cap H, Q^k \cap H \rangle$ centralizes $C_Q(V) = Q \cap R$ modulo S. Further by 8.15, $R/(Q \cap R)S \cong (Q^g \cap R)(Q^k \cap R)/(Q^g \cap Q^k)$ is the sum of natural modules for $(H \cap X)/O_2(H \cap X)$, and then (4) holds by Exercise 2.9.

Lemma 8.17: *Let G be a perfect group with $F^*(G) = Q$ an extraspecial 2-group and G/Q quasisimple. Let $\rho : \hat{G} \to G$ be the universal covering of G, $\hat{Z} = \ker(\rho)$, and $\hat{Q} = \rho^{-1}(Q)$. Assume*

(a) *$\tilde{Q} = Q/Z(G)$ is an absolutely irreducible $GF(2)G$-module, and*
(b) *$H^1(G, \tilde{Q}) = 0$.*

Then

(1) *$\hat{Q} = [\hat{Q}, G] \times \hat{Z}$ with $[\hat{Q}, G] \cong Q$.*
(2) *The natural map $\sigma : \hat{G}/\hat{Q} \to G/Q$ is the universal covering of G/Q with $\hat{Z} \cong \ker(\sigma)$.*
(3) *H is a finite group with $F^*(H) \cong Q$ and $H/Z(H) \cong G/Z(G)$ if and only if $H \cong \hat{G}/V$ for some complement V to $[\hat{Q}, G]$ in $\hat{Q}$.*
(4) *If $G/Q = \Omega(\tilde{Q})$ and α is a transvection in $O(\tilde{Q})$ then α lifts to an automorphism of $\hat{G}$.*

Remark. (7) See Section 33 in [FGT] for a discussion of coverings. Recall $O(\tilde{Q})$ and $\Omega(\tilde{Q})$ are the isometry group of the orthogonal space $\tilde{Q}$ and the commutator group of that isometry group, respectively.

Now the proof of lemma 8.17. Let $R = O_2(\hat{Q})$. As $\hat{G}/\hat{Z} \cong G$ is perfect with $\hat{Z} \leq Z(\hat{G})$ and $Z(Q) = Z(G)$, $Z(\hat{G}) = \rho^{-1}(Z(Q))$. Let $Z = Z(\hat{G})$. Observe $\hat{Q} = R\hat{Z}$. As Q is extraspecial, R is of class 2. Let $P = [R, \hat{G}]$.

As $H^1(G, \tilde{Q}) = 0$, $R/\Phi(R)$ splits over $(Z \cap R)/\Phi(R)$ by 17.12 in [FGT]. Thus $P/\Phi(P) \cong \tilde{Q}$ and $\hat{Q} = P\hat{Z}$ with $\Phi(P) = P \cap Z$. As $\tilde{Q}$ is of exponent 2 and P is of class 2, $\Phi(P)$ is of exponent 2 (cf. 23.7 in [FGT]).

Let U be a hyperplane of $\Phi(P)$. Then P/U is extraspecial so the commutator map induces a nondegenerate bilinear form $(,)$ on $P/\Phi(P) \cong \tilde{Q}$ preserved by $\hat{G}/\hat{Q} \cong G/Q$ as described in 8.3. As G is absolutely irreducible on $\tilde{Q}$, $(,)$ is the unique G-invariant nondegenerate bilinear form on $\tilde{Q}$ (cf. Exercise 9.1 in [FGT]). Pick $x, y \in P$ with $(\bar{x}, \bar{y}) \neq 0$ and let $u = [x, y]$. If $\Phi(P) \neq \langle u \rangle$ we can pick U with $u \in U$, so $0 = [x, y] \mod U$, contradicting $(\bar{x}, \bar{y}) \neq 0$.

Thus $\Phi(P) = \langle u \rangle$, so $P \cong Q$ and $\hat{Q} = P \times \hat{Z}$. That is, (1) holds.

As G is perfect, so is $\hat{G}$ and hence also $\hat{G}/P$. Thus as G/Q is quasisimple, so is $\hat{G}/P$. Let $\alpha_1 : L \to \hat{G}/P$ be the universal covering of $\hat{G}/P$ and $\alpha_2 : \hat{G} \to \hat{G}/P$ the natural map. Form the fiber product $A = L \times_{\hat{G}/P} \hat{G}$ with respect to the maps α_1 and α_2 as in Section 6, and let $p_1 : A \to L$ be the projection. By 6.1, $\ker(p_1) = \{(1, a) : a \in \ker(\alpha_2)\} \cong P$.

As $P = [P, \hat{G}]$, P is generated by commutators $a = [x, b]$, $b \in P$, $x \in \hat{G}$. As α_1 is surjective there is $y \in L$ with $y\alpha_1 = x\alpha_2$. Then $(y, x) \in A$ and

$[(y, x), (1, b)] = ([y, 1], [x, b]) = (1, a)$. Therefore $ker(p_1) \leq [A, A]$. But as L is perfect, $A = [A, A]ker(p_1)$, so A is perfect.

Now A is perfect and $p_2 : A \to \hat{G}$ is a surjection with $ker(p_2) = \{(c, 1) : c \in ker(\alpha_1)\}$. But $ker(\alpha_1) \leq Z(L)$, so $ker(p_2) \leq Z(A)$, and hence p_2 is a covering. Therefore as $\rho : \hat{G} \to G$ is universal, p_2 is an isomorphism (cf. 33.7 in [FGT]). Thus $1 = ker(\alpha_1)$ and $\hat{G}/P \cong L$. Thus (2) is established.

Assume the hypotheses of (3). Then H is a perfect central extension of $G/Z(G)$. But as $\hat{G}$ is the universal covering group of G, it is also the universal covering group of $G/Z(G)$ (cf. 33.7 in [FGT]), so we have a surjection $\phi : \hat{G} \to H$ with $ker(\phi) \leq Z$. Then as $\phi : P \to O_2(H)$ is an isomorphism, $ker(\phi)$ is a complement to P in $\hat{Q}$, establishing (3).

Finally assume the hypotheses of (4). Then α lifts to an automorphism β of $G/Z(G)$ as $G/Z(G) = Aut(Q)$ (cf. Exercise 8.5.3 in [FGT] or 12.16). Then as $\hat{G}$ is the covering group of $G/Z(G)$ and β is a covering of G, β lifts to an automorphism of $\hat{G}$.

Remarks. The Brauer–Fowler Theorem appears in [BF]. See Chapter 5 for more discussion of its importance in the study of simple groups. I believe 7.6 is due to Feit.

Phillip Hall introduced the notion of an extraspecial group, and Thompson and Janko did the early work on groups with a large extraspecial 2-subgroup. Eventually such groups were classified via the efforts of a number of mathematicians, most notably Timmesfeld in [Tm]. Many of the later results in Section 8 come from Section 17 of [A2].

Exercises

1. Let Q be a large extraspecial subgroup of G and $Q \leq L \trianglelefteq G$. Prove $G = LN_G(Q)$.

2. Let Q be a large extraspecial 2-subgroup of G and A a subgroup $N_G(Q)$ of odd order. Prove $Q = C_Q(A) * [Q, A]$, $[Q, A]$ is extraspecial, and either $C_Q(A) = Z(Q)$ or $C_Q(A)$ is extraspecial.

3. Let L be a nonabelian simple group and z an involution in L such that $R = C_L(z) \in Syl_2(L)$ and $\Phi(R) = \langle z \rangle$. Prove $R \cong D_8$.
   (Hint: Use transfer and fusion arguments such as in Section 37 and Exercise 13.1 in [FGT]. In particular prove z is not weakly closed in R with respect to G and exploit this fact.)

4. Assume z is an involution in a finite group G, $H = C_G(z)$, $Q = F^*(H)$ is extraspecial of width at least 2, z is not weakly closed in Q with

respect to G, H/Q is simple, H is irreducible on $Q/\langle z \rangle$, and Q is not the weak closure of z in H. Prove G is simple.

5. Assume G is transitive on X and $z \in G$ fixes a unique point of X. Prove each nilpotent subgroup of G containing z fixes a unique point of X.

6. Let G be a finite group with $F^*(G) = Q \cong D_8^4$ and $G/Q \cong \Omega_8^+(2)$. Prove there exists $\mathbf{Z}_3 \cong A \leq G$ with $C_Q(A) = R \cong Q_8^3$, and that $C_G(A) = A \times L$ with $R = F^*(L)$ and $L/R \cong \Omega_6^-(2)$.

7. Let A be a nontrivial elementary abelian 2-subgroup of G, $M = N_G(A)$, and assume each element of $A^{\#}$ fixes a unique point of G/M. Prove either $G = M$ or $N_M(\langle t, A \rangle)$ is transitive on $C_A(t)^{\#}$ for each involution $t \in M$.

8. Let G be a 2-group, H be of index 2 in G, $t \in G - H$, and assume

$$1 = H_0 \leq H_1 \leq \cdots \leq H_n = G$$

is a series of normal subgroups of G such that H_{i+1}/H_i is elementary abelian and $|C_{H_{i+1}/H_i}(t)|^2 = |H_{i+1}/H_i|$ for all $0 \leq i < n$. Prove
   (1) G is transitive on the involutions in tH.
   (2) $|C_H(t)|^2 = |H|$.

9. Let $G = L_3(2)$, $S_4 \cong H \leq G$, F the field of order 2, X the permutation module for G on G/H over F, and Y the natural 3-dimensional module FG-module in which H fixes a point. Prove
   (1) $dim(C_X(H)) = 2$.
   (2) $X = [X, G] \oplus C_X(G)$ with $dim(C_X(G)) = 1$, $Soc([X, G]) \cong Y^*$, and $X/Soc(X) \cong Y$.
   (3) If V is an FG-module in which $[V, O_2(H)] \leq C_V(H)$ and $V/C_V(H)$ is the sum of natural modules for $H/O_2(H) \cong L_2(2)$, then V is the sum of $dim(C_V(H))$ copies of Y as an FG-module.

10. Let G be a finite group, H a proper subgroup of G, and z a 2-central involution of H such that z fixes a unique point of G/H. Let $\mathcal{U}$ be the set of 2-subgroups U of H such that $C_G(U)$ is not contained in H. Then
   (1) G has a normal subgroup M such that $M \cap H = O^2(H)$. In particular if $z \in O^2(H)$ then $M, O^2(H)$ satisfy our hypotheses.
   (2) $\mathcal{U} \neq \varnothing$ and if U is maximal in $\mathcal{U}$, $K = C_G(U)$, and $K^* = K/U$ then
      (a) $(H \cap K)^*$ is strongly embedded in K^*.
      (b) $z^G \cap U = \varnothing$.

   (c) $(H \cap K)^*$ has a subgroup of odd order transitive on the involutions of $(H \cap K)^*$.

   (Hint: Use 37.5 in [FGT]) for (1).)

11. Let H be a finite group with $F^*(H) = Q$ an extraspecial 2-group, $\langle z \rangle = Z(Q)$, $\tilde{H} = H/\langle z \rangle$, and $H^* = H/Q$. Recall by 8.3 that $\tilde{Q}$ is an orthogonal space over $GF(2)$ and $H^* \leq O(\tilde{Q})$. Prove

   (1) Each involution t^* in H^* is of type a_m, b_m, or c_m, where $m = m([\tilde{Q}, t])$, m is even for type a and c but odd for type b, $[\tilde{Q}, t]$ is totally isotropic but $[\tilde{Q}, t]$ is totally singular if and only if t^* is of type a, and finally $t^* \notin \Omega(\tilde{Q})$ if and only if t^* is of type b.

   (2) $tz \in t^Q$ if and only if t is of type b or c.

   (3) $\tilde{t}^Q = \tilde{t}\tilde{Q}_t^-$, where $Q_t^-/\langle z \rangle = [\tilde{Q}, t]$.

   (4) Each involution in tQ is in tQ_t^+, where $Q_t^+/\langle z \rangle = C_{\tilde{Q}}(t)$.

   (5) $O(\tilde{Q})$ is transitive on involutions of type a_m, b_m, and c_m for each m, and $\Omega(\tilde{Q})$ is also transitive except that if $m = dim(\tilde{Q})/2$ then $\Omega(\tilde{Q})$ has two orbits a_m^1 and a_m^2, where for $t_i^* \in a_m^i$, $[\tilde{Q}, t_i]$, $i = 1, 2$, are the two classes of maximal totally singular subspaces of $\tilde{Q}$ under $\Omega(\tilde{Q})$.

   (Hint: See [ASe].)

Chapter 3

Algebras, Codes, and Forms

The *Griess Algebra* is a certain nonassociative, commutative algebra on 196,884 dimensional Euclidean space constructed by R. Griess. Its automorphism group is the largest sporadic group, the Monster. In Section 9 we record a few elementary facts about algebras; in particular we find that a nondegenerate bilinear form γ on a space V determines an isomorphism between the space of algebras on V and the space of trilinear forms on V. Thus the Griess algebra corresponds to a trilinear form and in Chapter 10 we use both the algebra and the form to construct the Monster.

In Section 11 we find that each map $P : V \to F$ of a vector space V over the field F of order 2 into F with $P(0) = 0$ determines a sequence of *derived forms* $P_m : V^m \to F$. This sequence is used in Chapter 4 to study a certain class of loops we call *symplectic 2-loops* which are generalizations of 2-groups of symplectic type. Following Conway, we use a certain symplectic 2-loop discovered by Parker to construct a 2-local in the Monster and then in Chapter 10 use this 2-local to assist in the construction of the Griess algebra.

In Section 10 we briefly recall a few elementary facts about error correcting codes. In Chapter 6 the Steiner system for the Mathieu group M_{24} is used to construct the *Golay code*, a doubly even binary code preserved by M_{24}. In Section 11 we find that each binary code V naturally determines a map P and its derived sequence. If the code is doubly even then P_3 is a trilinear form. In Chapter 7 we use the trilinear form defined by the Golay code to study the Mathieu groups.

9. Forms and algebras

In this section F is a field and V an F-space. Given vector spaces V_i, $1 \le i \le n$, denote by $M(V_1, \dots, V_n; V)$ the F-space of all functions $f : V_1 \times \cdots \times V_n \to V$ where addition and scalar multiplication are defined by

$$(\alpha + \beta)(x_1, \dots, x_n) = \alpha(x_1, \dots, x_n) + \beta(x_1, \dots, x_n)$$
$$(a\alpha)(x_1, \dots, x_n) = a \cdot \alpha(x_1, \dots, x_n)$$

for $\alpha, \beta \in M(V_1, \dots, V_n; V)$, $x_i \in V_i$, and $a \in F$. Write $L(V_1, \dots, V_n; V)$ for the subspace of all n-linear maps α; that is, for all i and each choice of $x_j \in V_j$, $j \ne i$, the map $x_i \mapsto \alpha(x_1, \dots, x_n)$ is a linear map from V_i to V. We write $L^n(V)$ for $L(V_1, \dots, V_n; F)$ and $M^n(V)$ for $M(V_1, \dots, V_n; F)$, when $V_i = V$ for each i. Thus $L^n(V)$ is the space of n-*linear forms* on V.

Given a group G represented on each V_i, we obtain a representation of G on $M(V_1, \dots, V_n; V)$ via $(\alpha g)(x_1, \dots, x_n) = \alpha(x_1 g^{-1}, \dots, x_n g^{-1})$. In particular we have that $GL(V)$ acts on $L^n(V)$ in this manner.

We also have a representation of the symmetric group

$$S_n = Sym(\{1, \dots, n\})$$

on $L^n(V)$ via $(\alpha s)(x_1, \dots, x_n) = \alpha(x_{1s^{-1}}, \dots, x_{ns^{-1}})$. The form α is *symmetric* if α is fixed by each element of S_n and α is *alternating* if $\alpha s = sgn(s)\alpha$ for each $s \in S_n$, where $sgn(s) = 1$ if $s \in A_n$ and $sgn(s) = -1$ if $s \in S_n - A_n$.

Let $\alpha \in M^n(V)$. An *isometry* of (V, α) is some $g \in GL(V)$ such that $\alpha g = \alpha$, or equivalently

$$\alpha(x_1 g, \dots, x_n g) = \alpha(x_1, \dots, x_n)$$

for all $x_i \in V$. Write $O(V, \alpha)$ for the *isometry group* of the form.

Let $X = \{x_1, \dots, x_m\}$ be a basis for V and write X^n for the set of all ordered n-tuples from X. Given $\alpha \in L^n(V)$, we write

$$\alpha = \sum_{(y_1, \dots, y_n) \in X^n} a_{y_1, \dots, y_n} y_1 \cdots y_n$$

to indicate that $\alpha(y_1, \dots, y_n) = a_{y_1, \dots, y_n}$. The term $a_{y_1, \dots, y_n} y_1 \cdots y_n$ is called a *monomial* of α. Observe that as α is n-linear, α is determined by its monomials. Further if α is symmetric then $a_{y_1, \dots, y_n} = a_{y_{1s}, \dots, y_{ns}}$ for all $s \in S_n$, so our convention is to write down just one monomial from each orbit of S_n.

Lemma 9.1: *Let* $V = \bigoplus_{i \in I} V_i$ *and* $\alpha \in L^3(V)$. *Then* $\alpha = \sum_{J \in I^3} \alpha_J$, *where for* $J = (i, j, k) \in I^3$, $\alpha_J \in L(V_i, V_j, V_k; F)$ *is defined by* $\alpha_J(y_i, y_j, y_k) = \alpha(y_i, y_j, y_k)$, *for* $y_r \in V_r$.

Remark 9.2. If α is symmetric then $\alpha = \sum_{J \in I^3/S_3} \alpha_J$, where I^3/S_3 is some set of representatives for the orbits of S_3 on I^3 acting via

$$s : (i_1, i_2, i_3) \mapsto (i_{1s^{-1}}, i_{2s^{-1}}, i_{3s^{-1}}),$$

subject to our convention of displaying only one monomial of α in an orbit of S_3.

Lemma 9.3: *Assume α is a symmetric trilinear form on V, $G \leq O(V, \alpha)$ is a 2-group of exponent 2 acting on Fx for each x in some basis X for V, and $a_{xyz}xyz$ is a monomial of α with $a_{xyz} \neq 0$. Then*

(1) $C_G(\langle x, y \rangle) = C_G(z)$.
(2) If $g \in G$ inverts x and y then $zg = z$.

Proof: Let $g \in G$; by hypothesis $wg = b_w w$ for each $w \in X$ and some $b_w \in F$. As G is of exponent 2, $b_w = \pm 1$. Now $a_{xyz} = \alpha(x, y, z) = \alpha(xg, yg, zg) = b_x b_y b_z a_{xyz}$, so as $a_{xyz} \neq 0$, $b_x b_y b_z = 1$.

An *algebra* on V is some $\tau \in L^2(V; V) = L(V, V; V)$. We write $u * v$ for the image $\tau(u, v)$ of the pair $(u, v) \in V \times V$ under τ and call $u * v$ the *product* of u and v. Thus an algebra is a vector space together with a bilinear product.

Lemma 9.4: *Assume γ is a nondegenerate bilinear form on V. Then*

(1) *There exists an isomorphism $b_\gamma : \tau \mapsto b_\gamma(\tau)$ of the space $L^2(V; V)$ of all algebras on V with the space $L^3(V)$ of all trilinear forms on V, where $b_\gamma(\tau)(x, y, z) = \gamma(x * y, z)$.*

(2) *The inverse of this isomorphism is $t_\gamma : \alpha \to t_\gamma(\alpha)$, where $t_\gamma(\alpha)(x, y)$ is the unique element of V such that $\gamma(t_\alpha(x, y), z) = \alpha(x, y, z)$ for all $z \in V$.*

(3) *The stabilizers in $O(V, \gamma)$ of τ and $b_\gamma(\tau)$ are the same.*

We have the following analogue of 9.1:

Lemma 9.5: *Let $V = \bigoplus_{i \in I} V_i$ and τ an algebra on V. Then $\tau = \sum_{J \in I^3} \tau_J$, where for $J = (i, j, k) \in I^3$, $\tau_J \in L(V_i, V_j; V_k)$ is defined by $\tau_J(y_i, y_j) = p_k(\tau(y_i, y_j))$, for $y_r \in V_r$ and $p_k : V \to V_k$ the kth projection.*

Remark 9.6. Assume γ is a nondegenerate bilinear form on V. Define an algebra τ to be *symmetric* with respect to γ if the trilinear form $b = b_\gamma(\tau)$ of lemma 9.4 is symmetric. Notice that this forces τ to be *commutative*; that is, $\tau(x, y) = \tau(y, x)$ for all $x, y \in V$.

Assume in addition that V is the orthogonal direct sum of subspaces V_i, $i \in I$; that is, $(V_i, V_j) = 0$ for $i \neq j$. (We write $V = V_1 \perp \cdots \perp V_k$ to indicate that V is the orthogonal direct sum of $V_1, \ldots, V_k$.) By Remark 9.2, $b = \sum_{J \in I^3/S_3} b_J$, where I^3/S_3 is some set of representatives for the orbits of S_3 on I^3. We claim:

Lemma 9.7: $b_J = b_\gamma(\tau_J)$.

For let $J = (i, j, k)$. Then for $x_r \in V_r$,

$$b_J(x_i, x_j, x_k) = b(x_i, x_j, x_k) = (\tau(x_i, x_j), x_k).$$

Now $\tau(x_i, x_j) = \sum_r p_r(\tau(x_i, x_j)) = \sum_r \tau_{ijr}(x_i, x_j)$ and then as our sum is orthogonal, $(\tau(x_i, x_j), x_k) = (\tau_{ijk}(x_i, x_j), x_k) = b_\gamma(\tau_J)(x_i, x_j, x_k)$, establishing the claim.

It follows from 9.2, 9.4, and 9.7 that τ is determined by the maps τ_J, $J \in I^3/S_3$, so we abuse notation and write

$$\tau = \sum_{J \in I^3/S_3} \tau_J.$$

Notice that for $i, j \in I$, $\tau_i = \tau_{iii}$ is symmetric with respect to the restriction of γ to V_i and τ_{iij} is commutative.

Conversely if for each $J = \{i, j, k\} \in I^3/S_3$ we are given maps $t_J \in L(V_i, V_j; V_k)$ such that t_J is symmetric with respect to the restriction of γ to V_i when $i = j = k$ and t_J is commutative when $i = j$, then $t = \sum_{J \in I^3/S_3} t_J$ is a symmetric algebra (subject to our notational conventions) with $b_t = \sum_{J \in I^3/S_3} b_{t_J}$.

Lemma 9.8: *Assume X is a basis for V and $G \leq GL(V)$ permutes $Y = \pm X$. Let β be a symmetric trilinear form on V and*

$$\mathcal{M} = \{a_{xyz}xyz : (x, y, z) \in Y^3, \ a_{xyz} \neq 0\}$$

the set of nonzero monomials of β on V. Then

(1) *$G \leq O(V, \beta)$ if and only if G permutes $\mathcal{M}$. That is, for all $g \in G$ and $(x, y, z) \in Y^3$, $a_{xyz} = a_{xg, yg, zg}$.*

(2) *Assume $H \trianglelefteq G$, $K \leq G$ with $G = HK$, and $\mathcal{M}_0 \subseteq \mathcal{M}$ such that K permutes $\mathcal{M}_0$, $H \leq O(V, \beta)$, and each member of $\mathcal{M}$ is H-conjugate to some member of $\mathcal{M}_0$. Then $G \leq O(V, \beta)$.*

Proof: Part (1) follows as β is determined by its monomials. Assume the hypothesis of (2). As $H \leq O(V, \beta)$ and $G = HK$, it suffices to show $K \leq O(V, \beta)$. Hence by (1) it suffices to show K permutes $\mathcal{M}$. But if $a_{xyz}xyz \in \mathcal{M}$ then by hypothesis there are $h \in H$ and $a_{\bar{x}, \bar{y}, \bar{z}}\bar{x}\bar{y}\bar{z} \in \mathcal{M}_0$ with $(\bar{x}, \bar{y}, \bar{z})h = (x, y, z)$. As K acts on $\mathcal{M}_0$, $a_{\bar{x}k, \bar{y}k, \bar{z}k} = a_{\bar{x}, \bar{y}, \bar{z}}$ for all $k \in K$, and as $H \leq O(V, \beta)$, $a_{uvw} = a_{ug, vg, wg}$ for all $g \in H$ and

$(u, v, w) \in Y^3$. Hence $a_{xyz} = a_{\bar{x}\bar{y}\bar{z}} = a_{\bar{x}k,\bar{y}k,\bar{z}k} = a_{\bar{x}kh^k,\bar{y}kh^k,\bar{z}kh^k} = a_{\bar{x}hk,\bar{y}hk,\bar{z}hk} = a_{xk,yk,zk}$, completing the proof.

Lemma 9.9: *Let $char(F) \neq 2$, γ a nondegenerate symmetric bilinear form on V, τ a symmetric algebra map on V, $b = b_\gamma(\tau)$, $G \leq O(V, \gamma, b)$, and z an involution in the center of G. Let $V_+ = C_V(z)$, $V_- = [V, z]$, $Q \leq G$ a 2-group with $\Phi(Q) \leq \langle z \rangle$, $V_s = C_V(Q)$, and $V_r = [V_+, Q]$. For $a \in V$ define*

$$\lambda_a : V \to V$$

$$v \mapsto v * a.$$

Then

(1) $V = V_- \perp V_s \perp V_r$.
(2) $\lambda \in Hom(V, End(V))$, where $\lambda : a \mapsto \lambda_a$.
(3) V_+ is a subalgebra of V.
(4) If U is a G-invariant subspace of V then G acts on $End(U)$ via $g : \rho \mapsto \rho^g$, where $\rho^g(u) = \rho(ug^{-1})g$. Further $\lambda_a^g = \lambda_{ag}$ for each $a \in V$ and $g \in G$.
(5) λ induces $\lambda^{U,W} \in Hom(U, End(W))$ for $(U, W) = (V_+, V_-)$, (V_s, V_-), (V_r, V_-), and (V_s, V_r), where $\lambda_u^{U,W}(w) = \lambda_u(w)$ for $u \in U$ and $w \in W$.

Proof: As $char(F) \neq 2$ and $\Phi(Q) \leq \langle z \rangle$, we can diagonalize Q on V_+. Pick a basis X for V that is the union of bases for V_s, V_r, and V_-, and such that Q is diagonal with respect to X on V_+. Let $X_i = X \cap V_i$.

Now for $x \in X_+$, $y \in X_-$, $\gamma(x, y) = \gamma(xz, yz) = \gamma(x, -y) = -\gamma(x, y)$, so $\gamma(x, y) = 0$ and $V = V_- \perp V_+$. Similarly as Q is diagonal on V_+, $V_+ = V_s \perp V_r$.

As $\tau : (u, v) \mapsto u * v$ is bilinear, (2) holds. As z preserves τ, (3) holds. The first statement in (4) is easy. For the second, $\lambda_a^g(v) = \lambda_a(vg^{-1})g = (a * vg^{-1})g = ag * v = \lambda_{ag}(v)$, so $\lambda_a^g = \lambda_{ag}$.

Suppose $V = U \perp W$ and $b(u, u', w) = 0$ for all $u, u' \in U$, $w \in W$. Then $0 = b(u, u', w) = (u * w, u') - (\lambda_u(w), u')$, so $\lambda_u(W) \subseteq U^\perp = W$, and hence $\lambda^{U,W} \in Hom(U, End(W))$ by restricting (2). In particular $b(V_+, V_+, V_-) = 0$ by 9.3 applied to $G = \langle z \rangle$, so (5) holds for $U = V_+$ and $W = V_-$. Then (5) also holds for $U = V_s$ and V_r by restriction. Similarly $b(V_s, V_s, V_r) = 0$ by 9.3 applied to $G = Q/\langle z \rangle$ acting on V_+, so (5) holds for $(U, W) = (V_s, V_r)$.

10. Codes

A *linear binary error correcting code* is a triple $C = (V, X, U)$, where V is a finite-dimensional vector space over $GF(2)$, X is a basis for V, and U is a subspace of V. The elements of U are the *code words*. If $dim(V) = n$ and $dim(U) = m$ then the code is an *(m,n)-code*.

We can view V as the power set of X with addition of two subsets u, v of X defined to be the symmetric difference of u and v. From this point of view the basis X consists of the one element subsets and $v = \sum_{x \in X} a_x x$ is identified with its support $\{x \in X : a_x = 1\}$. The *weight* of a code word u is the order of its support in the basis X, or equivalently if we regard u as a subset of X by identifying u with its support, then the weight of u is $|u|$. The *Hamming distance* d on U is defined by

$$d(u, v) = |u + v|.$$

Thus d counts the number of places in which u and v differ. The *minimum weight* of the code is the minimum weight of a nonzero code word. Notice this is also the minimum distance between distinct code words.

The code words can be regarded as the admissible words sent as a string of n zeros and ones. If errors are introduced in transmission we receive a different word, hopefully in $V - U$. If M is the minimum weight of the code, then the code can correct e errors, where $e = [(M - 1)/2]$. Namely if we receive a word v then we decode v as the member of U of minimum distance from v. As long as at most e errors have been introduced, there will be a *unique* code word u at minimum distance from v, and u will have been the word sent. Thus we will have corrected the error introduced in transmission.

Given any $u \in U$ and positive integer r, we can consider the ball

$$B_r(u) = \{v \in V : d(u, v) \le r\}$$

of radius r around u. A code C is *perfect* if C has minimum weight $2e+1$ and for each $v \in V$, there is a unique $u \in U$ with $v \in B_r(u)$. That is, the code is e-error correcting and every member of V can be decoded uniquely.

The group of automorphisms of the code (V, X, U) is the subgroup of $Sym(X)$ acting on U. Equivalently it is the subgroup of $GL(V)$ acting on X and U.

The *level* of the code is the greatest integer k such that $|u| \equiv 0$ mod 2^k for all $u \in U$. If the level of the code is at least 1 the code is said to be *even* and if the level is at least 2 the code is *doubly even*. Define the code to be *strictly doubly even* if the code has level 2.

In Section 19 we will encounter codes preserved by the Mathieu groups including the *binary Golay code* which is a perfect $(23, 12)$-code with minimum weight 7 and correcting 3 errors.

11. Derived Forms

In this section V is a finite-dimensional vector space over $F = GF(2)$ and $P : V \to F$ is a function with $P(0) = 0$. For m a positive integer and $x = (x_1, \dots, x_m) \in V^m$, write 2^m for the power set of $\{1, \dots, m\}$ and for $I \in 2^m$ let $I(x) = \sum_{i \in I} x_i$. Define

$$P_m : V^m \to F$$

$$x \mapsto \sum_{I \in 2^m} P(I(x)).$$

The form P_m is the *mth derived form* of P.

Lemma 11.1: *Let $P \in M(V; F)$ with $P(0) = 0$ and P_m the mth derived form of P. Then*

(1) P_m *is symmetric.*

(2)   *For all $x, y, x_i \in V$,*

$$P_m(x + y, x_2, \dots, x_m) = P_m(x, x_2, \dots, x_m)$$
$$+ P_m(y, x_2, \dots, x_m) + P_{m+1}(x, y, x_2, \dots, x_m).$$

(3)   *If $v = \{x_1, \dots, x_m\}$ is linearly dependent then $P_m(v) = 0$.*

(4)   *If $v = \{x_1, \dots, x_m\}$ is linearly independent then $P_m(v) = \sum_{u \in \langle v \rangle} P(u)$.*

Proof: Parts (1) and (4) are immediate from the definition of P_m. In part (2) let $z = (x_2, \dots, x_m)$. Then by definition of P_m,

$$P_{m+1}(x, y, z) = P_m(x, z) + P_m(y, z) + P_m(x + y, z) - 2P_{m-1}(z),$$

so (2) holds.

We prove (3) by induction on m. If $m = 1$ then (3) holds as $P(0) = 0$. Assume the result for m. Let $y = x_1$ and $x = x_{m+1}$. We may assume $x \in \langle y, x_2, \dots, x_m \rangle$ and $x + y \in \langle x_2, \dots, x_m \rangle$. So by induction on m, $P_m(x + y, z) = 0$. Further if $\{x_1, \dots, x_m\}$ is linearly dependent then $P(x, z) = P(y, z) = 0$ by induction, while if $\{x_1, \dots, x_m\}$ is linearly independent then $P(x, z) = P(y, z)$ by (4). Thus $P_{m+1}(x, y, z) = P_m(x, z) + P_m(y, z) = 0$, so that (3) holds.

Define $deg(P)$ to be the maximum d such that $P_d \neq 0$. By 11.1.2, $P_m = 0$ for $m > dim(V)$, so $deg(P)$ is well defined and $deg(P) \leq dim(V)$. Observe also that by 11.1:

Lemma 11.2: *Let $P \in M(V; F)$ with $P(0) = 0$ and $d = deg(P)$. Then P_d is a symmetric d-linear form on V.*

Examples (1) The form P is of degree 1 if and only if P is linear.

(2) The form P is of degree 2 if and only if P is a quadratic form on V (cf. Chapter 7 in [FGT]). Moreover in that event P_1 is the bilinear form associated with P and $O(V, P)$ is an orthogonal group.

Lemma 11.3: *If X is a set, V the $GF(2)$ space of all subsets of X under symmetric difference, and $u, v, w \in V$ then*

(1) $(u + v) \cap w = (u \cap w) + (v \cap w)$.
(2) $|(u + v) \cap w| = |u \cap w| + |v \cap w| - 2|u \cap v \cap w|$.

Lemma 11.4: *Assume (V, X, U) is a binary error correcting code of level l and identify each $v = \sum_{x \in X} a_x x \in V$ with its support $\{x \in X : a_x = 1\}$. Define $P \in M(U; F)$ by $P(u) = |u|/2^l \mod 2$. Let $(x_1, \dots, x_m) \in U^m$. Then*

(1) $|\bigcap_{i=1}^m x_i| \equiv 0 \mod 2^{l-m+1}$ *for* $m \leq l + 1$.
(2) $P_m(x_1, \dots, x_m) = |x_1 \cap \cdots \cap x_m|/2^{l+1-m} \mod 2$.
(3) $deg(P) \leq l + 1$.

Proof: We first prove (1) and (2) by induction on m. Notice when $m = 1$, (1) holds as our code is of level l while (2) holds by definition of P. So the induction is anchored. Next let $x = x_1$, $y = x_2$, $z = (x_3, \dots, x_m)$, and $v = \bigcap_{i=3}^m x_i$. Then by induction on m and 11.3,

$$P_{m-1}(x + y, z) \equiv |(x + y) \cap v|/2^{l-m+2}$$

$$= (|x \cap v| + |y \cap v| - 2|x \cap y \cap v|)/2^{l-m+2}$$

$$\equiv P_{m-1}(x, z) + P_{m-1}(y, z) - |x \cap y \cap v|/2^{l-m+1} \mod 2.$$

Thus as $P_{m-1}(u, z) \in \mathbf{Z}$ for $u = x, y, x + y$, we conclude $|x \cap y \cap v| \equiv 0 \mod 2^{l-m+1}$ and by 11.1.2, $P_m(x, y, z) = |x \cap y \cap v|/2^{l-m+1} \mod 2$. That is, (1) and (2) are established. Finally when $m > l + 1$, (2) says $P_m(x, y, z) = 0$, so (3) holds.

Example (3) Let X be a basis for the $GF(2)$-space V and U the *core* of V with respect to X; that is, U consists of the vectors of even weight. Then (V, X, U) is of level 1, so the form P of 11.4 is of degree 2. That is, P is a quadratic form on U.

We now specialize to the case where P is a form of degree 3, and let $C = P_2$ and $f = P_3$. Thus f is a symmetric trilinear form by 11.2. The triple (P, C, f) induces extra geometric structure on the projective geometry $PG(V)$ of V (cf. Example 1 in Section 4) by allowing us to distinguish subspaces of the same dimension via the restriction of our forms to such subspaces. The next few lemmas introduce some geometric concepts useful in studying this geometry.

Recall the *radical* of a symmetric bilinear form b on V is $Rad(b) = V^{\perp} = \{u \in V : (u, v) = 0 \text{ for all } v \in V\}$. Similarly a vector $v \in V$ is *singular* with respect to a quadratic form q on V if $q(v) = 0$. More generally see Chapter 7 in [FGT] for a discussion of bilinear and quadratic forms. In particular the proof of the following lemma follows trivially from the definition of a quadratic form.

Lemma 11.5: *For $x \in V$ let $f_x \in L^2(V)$ and $C_x \in M^1(V)$ be defined by $f_x(y, z) = f(x, y, z)$ and $C_x(y) = C(x, y)$. Then C_x is a quadratic form on V with associated bilinear form f_x.*

Define $R(x) = Rad(f_x)$ and for $U \leq V$ define $R(U) = \bigcap_{u \in U} R(u)$ and $U\theta = \{x \in V : f_x = 0 \text{ on } U\}$. Observe that as f is trilinear, $U\theta$ is a subspace of V. Define U to be *subhyperbolic* if $U\theta$ is a hyperplane of V.

Lemma 11.6: *Let U be a subhyperbolic subspace and define $f_U \in M^2(U)$ by $f_U(x, y) = f(z, x, y)$ for $z \in V - U\theta$. Then*

(1)   *f_U is independent of the choice of $z \in V - U\theta$.*

(2)   *f_U is a symmetric bilinear form on U.*

(3)   *For $A \leq U$, $R(A) \cap U$ is the subspace of U orthogonal to A with respect to f_U.*

Proof: Parts (1) and (2) are trivial. For $u \in U$ and $v \in U\theta$, $f(A, u, v) = 0$, so $u \in R(A)$ if and only if $f_U(A, u) = 0$. Thus (3) holds.

Define a $U \leq V$ to be *singular* if P is trivial on U and $V = U\theta$. Define a line l of V to be *hyperbolic* if all points of l are singular but l is not singular.

Lemma 11.7: *Let $S \subseteq V$ with $P(s) = 0$ for all $s \in S$. Then $\langle S \rangle$ is singular if and only if $S \subseteq R(s)$ for each $s \in S$.*

Proof: Let $U = \langle S \rangle$ and $x \in V$. Certainly the condition on S is necessary for f_x to be trivial on U for all $x \in V$. Conversely given the condition, U is generated by a set of pairwise orthogonal singular points with respect to the quadratic form C_x, so C_x is trivial on U. Finally let $u = \sum_{t \in T} t$ for some $T \subseteq S$; we prove $P(u) = 0$ by induction on $|T|$. Namely $u = s + v$, where $v = \sum_{t \neq s} t$ and by induction $P(v) = 0$, so $P(u) = P(v) + P(s) + C(s, v) = 0$.

Lemma 11.8: *Assume (V, X, U) is a strictly doubly even code and define $P \in M^1(V)$, $C \in M^2(V)$, and $f \in M^3(V)$ by*

$$P(x) = |x|/4 \bmod 2;$$

$$C(x, y) = |x \cap y|/2 \bmod 2;$$

$$f(x, y, z) = |x \cap y \cap z| \bmod 2.$$

Then

(1) $C = P_2$ *and* $f = P_3$ *are the derived forms of* P.
(2) $\deg(P) \leq 3$ *so* f *is a symmetric trilinear form on* V *with* $f(x, x, y) = 0$ *for all* $x, y \in V$.
(3) *If* $\dim(V) \equiv 0 \bmod 8$ *then there are induced forms on* $V/\langle X \rangle$.

Proof: This follows from earlier lemmas.

Define a 3-*form* on V to be a triple $\mathcal{F} = (T, Q, f)$ such that $T : V \to F$, $Q : V^2 \to F$, Q is linear in the first variable, and for all $x, y, z \in F$:

(F1) $T(x + y) = T(x) + T(y) + Q(x, y) + Q(y, x)$.
(F2) f *is a symmetric trilinear form on* V.
(F3) Q_x *is a quadratic form with associated bilinear form* f_x; *that is,*

$$Q_x(y + z) = Q_x(y) + Q_x(z) + f_x(y, z),$$

where $Q_x : y \mapsto Q(x, y)$ *and* $f_x : (y, z) \mapsto f(x, y, z)$.

Thus T is a cubic form and Q is a linear family of quadratic forms.

Lemma 11.9: *Let X be an ordered basis for V, and f a symmetric trilinear form on V with $f(x, x, y) = 0$ for all $x, y \in X$. Then*

(1) $\mathcal{F} = (T, Q, f)$ *is a 3-form, where*

$$Q\left(v, \sum_{x \in X} a_x x\right) = \sum_{x < y} a_x a_y f(v, x, y),$$

$$T\left(\sum_x a_x x\right) = \sum_{x < y < z} a_x a_y a_z f(x, y, z).$$

(2) $\mathcal{F}$ *is unique subject to* $T(x) = Q(v, x) = 0$ *for all* $v \in V$, $x \in X$.
(3) *If* $G \leq O(V, f)$ *and for all* $x \in X$, $u \in xG$, *and* $v \in V$, $T(u) = Q(v, u) = 0$, *then* $G \leq O(V, \mathcal{F})$.

Proof: Exercise 3.1.

Remarks. The material in Section 11 comes from [A4] but part of that material is a rediscovery of earlier work of H. Ward. In particular the notion of a derived form is a special case of what Ward calls *combinatorial polarization* in [Wa2], while the coding theoretic construction of Lemma 11.4 appears first in [Wa3].

Exercises

1. Prove Lemma 11.9.
2. Assume the hypotheses of Lemma 11.8 and define

$$\epsilon : U \to V^*$$

$$J \mapsto \epsilon_J$$

by $\epsilon_J(v) = |J \cap v| \mod 2$. Prove:
   (1) ϵ is a well-defined surjective linear map with $V \le ker(\epsilon)$.
   (2) If $dim(V) = dim(U)/2$ then ϵ induces an isomorphism of U/V and V^*.
   (3) For $u, v, w \in V$, $\epsilon_{u \cap w}(v) = f(u, w, v)$.
3. Let V be a $GF(2)$-space and $P : V \to GF(2)$ a form. Suppose $W \le V$ with $P(w) = 1$ for all $w \in W^{\#}$. Prove:
   (1) If $dim(W) = 3$ then $P_3(x, y, z) = 1$ for each basis $\{x, y, z\}$ of V.
   (2) If P has degree at most 3 then $dim(W) \le 3$.

Chapter 4

Symplectic 2-Loops

Recall that a *loop* is a set L together with a binary operation possessing an identity, and such that for each $x \in L$, the left and right translation maps $a \mapsto xa$ and $a \mapsto ax$ are permutations of L. Notice that the associative loops are precisely the groups. A loop is *Moufang* if it satisfies a certain weak form of associativity (cf. Section 12). For example, by Moufang's Theorem (cf. 12.2) each 2-generator subloop of a Moufang loop is a group.

In Section 12 we study loops L which are extensions of the additive group of the field F of order 2 by a finite-dimensional vector space V over F. We term such loops *symplectic 2-loops*, as the groups of this form are the direct product of an elementary abelian 2-group with a 2-group of symplectic type (cf. page 109 in [FGT]). In particular the extraspecial 2-groups of Section 8 are examples of symplectic 2-loops.

Each symplectic 2-loop L is determined by a cocycle $\theta : V \times V \to F$ and comes equipped with a power map P, a commutator C, and an associator A. We view P, C, A as forms $P : V \to F$, $C : V^2 \to F$, and $A : V^3 \to F$. In Section 12 we find that the central isomorphism type of L is determined by the equivalence class of θ modulo the coboundaries, and that this class is determined in turn by the parameters $par(L) = (P, C, A)$.

Then in Section 13 we find that L is Moufang if and only if (in the language of Section 11) P has degree at most 3 and $C = P_2$ and $A = P_3$ are the derived forms of P. Moreover if L is Moufang then there exists a cocycle θ defining L such that (P, θ, A) is a 3-form in the sense of Section 11.

Finally in Section 14 we see that if L is a Moufang symplectic 2-loop of order 2^{n+1} such that the radical of the associator A of L is 1-dimensional, then a construction of Conway supplies a group N containing a large extraspecial 2-group $Q \cong D_8^n$ such that $|N : N_N(Q)| = 3$.

In Chapter 7 we will construct a form P of degree 3 on a 12-dimensional F-space preserved by the Mathieu group M_{24}. The corresponding symplectic 2-loop is the *Parker loop*. When Conway's construction is applied to the Parker loop we obtain the normalizer N of a 4-group in the Monster. This group is used in Chapter 10 to construct the Griess algebra and the Monster.

This chapter is used elsewhere in the book only in Chapter 10 in the construction of the Griess algebra and the Monster. Thus readers skipping that chapter can skip this one too. Also much of the discussion in Sections 12 and 13 is not necessary for the construction of the Monster. The Remarks at the end of this chapter indicate which lemmas are necessary. On the other hand the material in these sections places the Conway construction in a larger context which hopefully makes the construction more natural and hence easier to understand. Further it answers questions which will probably occur to the reader as the construction unfolds.

12. Symplectic 2-loops

We begin by recalling some generalities about loops which can be found in [B]. A *loop* is a set L together with a binary operation $(x, y) \mapsto xy$ on L such that

(L1) L has an identity 1.

(L2)   For each $x \in L$ the maps $a \mapsto xa$ and $a \mapsto ax$ defined via left and right multiplication by x are permutations of L.

In the remainder of this section assume L is a loop. The loop L is said to be *Moufang* if L satisfies the Moufang condition (MF):

(MF)   For all $x, y, z \in L$, $(xy)(zx) = (x(yz))x$.

Further L is *diassociative* if the subloop generated by each pair of elements of L is a group.

Lemma 12.1: *Assume L is Moufang and let $x, y \in L$. Then*

(1) $x(yx) = (xy)x$.

(2) $y \cdot x^2 = (yx)x$.

(3) $x^2 \cdot y = x(xy)$.

Proof: Specialize y to 1 in (MF) to get (1). Then specializing z to x in (MF) and using (1) we get $(xy)x^2 = (x(yx))x = ((xy)x)x$. But by (L2), each $u \in L$ can be written in the form $u = xy$ for some $y \in L$, so $ux^2 = (ux)x$, giving (2). Similarly (3) follows by specializing y to x in (MF).

Theorem 12.2 (Moufang's Theorem): *Moufang loops are diassociative.*

Proof: See Chapter VII, Section 4 of [B]. We require Moufang's Theorem only for symplectic 2-loops, where the result is easy to prove using 12.1 and 12.3; see, for example, Exercise 4.5.

Define L to be a *symplectic 2-loop* if L is the extension of the additive group $\mathbf{Z}_2$ of the field $F = GF(2)$ by a finite-dimensional vector space V over F. That is, here exists a surjective morphism $\phi : L \to V$ of loops such that $ker(\phi) \cong \mathbf{Z}_2$.

In the remainder of this section assume L is a symplectic 2-loop with defining morphism $\phi : L \to V$ and $\{1, \pi\} = ker(\phi)$. Let $n = dim(V)$. Observe

Lemma 12.3: *(1) For $x, y \in L$, $\phi(x) = \phi(y)$ if and only if $y = x$ or πx.*

(2) π is in the center of L. That is, for all $x, y \in L$, $x\pi = \pi x$ and $(xy)\pi = x(y\pi) = (x\pi)y$.

(3) For each $x \in L - \langle \pi \rangle$, the subloop $\langle x, \pi \rangle$ generated by x and π is a group of order 4, and the inverse x^{-1} for x in $\langle x, \pi \rangle$ is the unique left, right inverse for x in L.

Proof: As $\phi(L) = V$ is a group of exponent 2, $\phi(x) = \phi(y)$ if and only if $1 = \phi(x)\phi(y) = \phi(xy)$ if and only if $xy \in ker(\phi) = \{1, \pi\}$. In particular, by (L2) there are exactly two elements $y \in L$ with $\phi(x) = \phi(y)$, and these must be x and $x\pi = \pi x$. So (1) holds. Then as $\phi(xy) = \phi(x(y\pi)) = \phi((x\pi)y)$, (1) implies (2). Finally (2) implies (3).

The usual construction from homological algebra for studying extensions of abelian groups can be used to analyze our loops. Define a *cocycle* on V to be a map $\theta : V \times V \to F$ such that $\theta(v, 0) = \theta(0, v) = 0$ for all $v \in V$. Write $\Theta = \Theta(n)$ for the space of all cocycles on a space V of dimension n. Given a cocycle θ define $L(\theta) = F \times V$ and define a product on $L(\theta)$ via

$$(a, u)(b, v) = (a + b + \theta(u, v), u + v).$$

Lemma 12.4: *(1) For each $\theta \in \Theta$, $L(\theta)$ is a loop and the projection $(a, u) \mapsto u$ is a surjective morphism of loops whose kernel $\{(0, 0), (1, 0)\}$ is in the center of $L(\theta)$ and isomorphic to $\mathbf{Z}_2$.*

(2) There exist $\theta \in \Theta$ and an isomorphism $\alpha : L \to L(\theta)$ such that
$\pi\alpha = (1, 0)$.

Proof: Exercise 4.1.

Remark 12.5. From 12.4 we have a bijection $\theta \mapsto L(\theta)$ between the
F-space $\Theta(n)$ and the set of all symplectic 2-loops defined on $F \times V$.

Because of 12.4 and 12.5, we may take $L = L(\theta)$ for some cocycle θ
and $\pi = (1, 0)$. For $x, y, z \in L$, define

$$P(x) = x^2,$$

$$C(x, y) = (xy)(yx)^{-1},$$

$$A(x, y, z) = (x(yz))((xy)z)^{-1}.$$

Thus P is the *power map*, $C(x, y)$ is the *commutator* of x and y, and
$A(x, y, z)$ is the *associator* of x, y, z. The *parameters* of L are the triple
$par(L) = (P, C, A)$. Write $par(\theta)$ for $par(L(\theta))$.

Notice that

$$P(x\pi) = P(x),$$

$$C(x\pi^i, y\pi^j) = C(x, y),$$

and

$$A(x\pi^i, y\pi^j, z\pi^k) = A(x, y, z)$$

for all $x, y, z \in L$ by 12.3. Further as $L\phi = V$ is an elementary abelian
2-group, $P(x), C(x, y), A(x, y, z) \in \langle \pi \rangle$. Thus we can and will regard the
parameters P, C, A as maps from V^n into F. Further we write $\theta(x, y)$ for
$\theta(\phi(x), \phi(y))$, $\theta(x, y + z)$ for $\theta(\phi(x), \phi(y) + \phi(z))$, etc. Subject to these
conventions we calculate:

Lemma 12.6: *For all $x, y, z \in V$:*

(1) $P(x) = \theta(x, x)$.
(2) $C(x, y) = \theta(x, y) + \theta(y, x)$, so C is symmetric.
(3) $A(x, y, z) = \theta(x, y) + \theta(y, z) + \theta(x, y + z) + \theta(x + y, z)$.

Write F^V for the set of all maps $\epsilon : V \to F$ such that $\epsilon(0) = 0$. For
$\epsilon \in F^V$ denote by θ_ϵ the cocycle defined by

$$\theta_\epsilon(x, y) = \epsilon(x) + \epsilon(y) + \epsilon(x + y).$$

Such cocycles will be termed *coboundaries*. Evidently the set of coboundaries forms a subspace of the space of cocycles. Define an equivalence
relation $\sim$ on Θ via $\theta \sim \theta'$ if $\theta' + \theta$ is a coboundary.

Define $Map(L)$ to be the set of functions $\alpha : L \to L$ permuting the cosets of $\langle \pi \rangle$, acting as the identity on $\langle \pi \rangle$, and such that the induced map on V is linear. We also write α for this induced map; thus $\alpha \in GL(V)$. Observe that there exists $\epsilon_\alpha \in F^V$ such that

$$\alpha(a, u) = (a + \epsilon_\alpha(u), \alpha(u)).$$

Conversely given $\epsilon \in F^V$ and $\alpha \in GL(V)$ we get $\alpha \in Map(L)$ defined by $\alpha(a, u) = (a + \epsilon(u), \alpha(u))$.

Given $\theta, \theta' \in \Theta$, a *central isomorphism* $\alpha : L(\theta) \to L(\theta')$ is an isomorphism acting as the identity on $\langle \pi \rangle$ and such that the induced map on V is the identity. Thus if α is a central isomorphism then $\alpha \in Map(L)$ with the induced map α on V the identity.

Lemma 12.7: *Let $\theta, \theta' \in \Theta$ and $\alpha \in Map(L)$. Then*

(1) $\alpha : L(\theta) \to L(\theta')$ *is an isomorphism if and only if $\theta + \theta' \circ \alpha + \theta_{\epsilon_\alpha} = 0$.*

(2) α *is a central automorphism of $L(\theta)$ if and only if the induced map is the identity on V and $\epsilon_\alpha \in Hom_F(V, F)$.*

Proof: Part (1) is an easy calculation, while (2) is a consequence of (1).

Lemma 12.8: *Let U be a hyperplane of V and $x \in V - U$. Assume $\theta, \theta' \in \Theta$ with $par(\theta) = par(\theta')$ and $\theta = \theta'$ on U and on all lines of V through x. Then $\theta = \theta'$.*

Proof: The proof is by induction on n. If $n = 1$ the result is trivial as $\theta(x, x) = P(x) = \theta'(x, x)$. Similarly if $n = 2$ then V is a line through x, so the lemma holds. Hence $n \geq 3$. If $n > 3$ and $y, z \in V$ then $W = \langle x, y, z \rangle \neq V$, so by induction on n, $\theta = \theta'$ on W. In particular $\theta(y, z) = \theta'(y, z)$ and as this holds for all $y, z \in V$, $\theta = \theta'$.

So let $n = 3$. Let $U = \langle y, z \rangle$. Now as $\theta = \theta'$ on U, $\theta(y, z) = \theta'(y, z)$ and as $\theta = \theta'$ on all lines through x, $\theta(x, y) = \theta'(x, y)$ and $\theta(x, y + z) = \theta'(x, y + z)$. Hence by 12.6.3, $\theta(x + y, z) = \theta'(x + y, z)$. That is,

(a) $\theta(x + u, v) = \theta'(x + u, v)$ for all $u, v \in U$.

Next substituting $z + x$ for z in 12.6.3, we conclude:

Lemma 12.9: $A(x, y, z + x) = \theta(x, y) + \theta(y, z + x) + \theta(x, x + y + z) + \theta(x + y, x + z)$.

Now (a) and 12.9 imply $\theta(x + y, x + z) = \theta'(x + y, x + z)$. That is,

(b) $\theta(x + u, x + v) = \theta'(x + u, x + v)$ for all $u, v \in U$.

Notice that (a), (b), and the fact that $\theta = \theta'$ on U complete the proof of 12.8.

Lemma 12.10: *Assume* $V = \langle x, y \rangle$, *par*$(\theta) = $ *par*(θ'), *and* $\theta(x, y) = \theta'(x, y)$. *Then* $\theta = \theta'$.

Proof: First by 12.6.2, $\theta(y, x) = \theta(x, y) + C(x, y) = \theta'(y, x)$. Next specializing y to x and z to y in 12.6.3, respectively, we obtain:

Lemma 12.11: $A(x, x, z) = \theta(x, x) + \theta(x, z) + \theta(x, x+z)$ *and* $A(x, y, y) = \theta(x, y) + \theta(y, y) + \theta(x + y, y)$ *for all* $x, y, z \in V$.

We conclude from 12.11 that $\theta(x, x+y) = A(x, x, y) + P(x) + \theta(x, y) = \theta'(x, x + y)$ and similarly $\theta(y, x + y) = \theta'(y, x + y)$. So indeed $\theta = \theta'$, completing the proof of 12.10.

Let $I = \{1, \dots, n\}$, $X = \{x_i : i \in I\}$ be a basis for V, and for $k \in \mathbf{Z}$ let $\Delta(k) = \{J \subseteq I : |J| \geq k\}$. For $J \in \Delta(1)$, let $m(J) = max\{j : j \in J\}$, $\bar{J} = J - \{m(J)\}$, and $x_J = \sum_{j \in J} x_j$. Define $\epsilon_J \in F^V$ by $\epsilon_J(x_J) = 1$ and $\epsilon_J(v) = 0$ for $v \neq x_J$.

Lemma 12.12: *Let* $\theta' \in \Theta(n)$ *and* $a : \Delta(2) \to F$. *Then there exists a unique* $\theta \in \Theta(n)$ *with* $par(\theta) = par(\theta')$ *and* $\theta(x_{\bar{J}}, x_{m(J)}) = a(J)$ *for all* $J \in \Delta(2)$. *Further* $\theta \sim \theta'$.

Proof: Let $U(m) = \langle x_i : i \leq m \rangle$. Proceeding by induction on m, we produce $\theta_m = \theta \sim \theta'$ with $par(\theta_{|U(m)}) = par(\theta'_{|U(m)})$ and $\theta(x_{\bar{J}}, x_{m(J)}) = a(J)$ for all $J \in \Delta(2)$ with $m(J) \leq m$. Further $\theta_{|U(m)}$ is unique.

If $m = 1$ then $\theta = \theta'$ works. Assume the result for m. Replacing θ' by θ_m, we may assume $\theta'(x_{\bar{J}}, x_{m(J)}) = a(J)$ for all $J \in \Delta(2)$ with $m(J) \leq m$. For $J \in \Delta(2)$ with $m(J) = m + 1$, let $b(J) = \theta'(x_{\bar{J}}, x_{m+1}) + a(J)$, and set $\epsilon = \sum_J b(J) \epsilon_J$, where the sum is over those $J \in \Delta(2)$ with $m(J) = m + 1$. Then $\theta = \theta' + \theta_\epsilon$ satisfies $\theta(x_{\bar{J}}, x_{m+1}) = a(J)$ for all J with $m(J) \leq m + 1$. Further by induction, $\theta_{|U(m)}$ is unique, so the uniqueness of $\theta_{|U(m+1)}$ follows from 12.8 and 12.10.

Theorem 12.13: *Let* $\theta, \theta' \in \Theta$. *Then the following are equivalent:*

(1) $\theta \sim \theta'$.
(2) $L(\theta)$ *is centrally isomorphic to* $L(\theta')$.
(3) $par(\theta) = par(\theta')$.

Proof: By 12.7.1, (1) and (2) are equivalent. Further if $L(\theta)$ is centrally isomorphic to $L(\theta')$, certainly the two loops have the same parameters. Finally if $par(\theta) = par(\theta')$ then by 12.12 there exists θ^* with $\theta \sim \theta^* \sim \theta'$, so (3) implies (1).

Remark 12.14. Let $\Theta_0(n)$ be the space of all coboundaries and

$$\Theta^0(n) = \Theta(n)/\Theta_0(n).$$

Then by 12.13 the map $[L(\theta)] \mapsto \theta + \Theta_0(n)$ is a bijection between the set of central isomorphism classes of symplectic 2-loops of order 2^{n+1} and $\Theta^0(n)$. Thus we may regard $\Theta^0(n)$ as the *space of all symplectic 2-loops of order* 2^{n+1}. Notice that $dim(F^V) = 2^n - 1$ and the kernel of the map $\epsilon \mapsto \theta_\epsilon$ is $Hom(V, F)$ of dimension n, so $dim(\Theta_0(n)) = 2^n - n - 1$. Therefore as $dim(\Theta(n)) = (2^n - 1)^2$, we conclude:

Lemma 12.15: $dim(\Theta^0(n)) = 2(2^n - 1)(2^{n-1} - 1) + n.$

Write $O(par(\theta)) = O(P, C, A)$ for the subgroup of $GL(V)$ fixing P, C, and A.

Theorem 12.16: *Let $G = Aut(L(\theta))_\pi$. Then*

$$E = C_G(L(\theta)/\langle \pi \rangle) \cong V^* = Hom(V, F),$$

$G/E \cong O(par(\theta))$, *and these isomorphisms are equivariant with respect to the representations of G/E by conjugation on E and the dual action of $O(par(\theta))$ on V^*.*

Proof: Identify V with $L/\langle \pi \rangle$ via ϕ and for $g \in G$, let $\bar{g}$ be the map induced by g on V. For $\epsilon \in V^*$ let α_ϵ be the central isomorphism induced by ϵ as in 12.7.2, and let $E = \{\alpha_\epsilon : \epsilon \in V^*\}$. By 12.7, $E = C_G(V)$ and the map $\epsilon \mapsto \alpha_\epsilon$ is an isomorphism of V^* with E.

Thus E is the kernel of the map $g \mapsto \bar{g}$. Further g fixes P, C, A, so $\bar{G} \leq O(par(\theta))$. Conversely given $d \in O(par(\theta))$, we have $par(\theta \circ d) = par(\theta)$, so by 12.13, there exists $\delta \in F^V$ with $\theta \circ d + \theta_\delta = \theta$. Let $g \in Map(L)$ with $\bar{g} = d$ and $\epsilon_g = \delta$. By 12.7.1, $g \in G$, so as $\bar{g} = d$, $\bar{G} = O(par(\theta))$.

Finally $\epsilon_{g^{-1}}(u) = \delta(g^{-1}(u))$, so

$${}^g\alpha_\epsilon(a, u) = g\alpha_\epsilon g^{-1}(a, u) = g\alpha_\epsilon(a + \delta(g^{-1}(u)), g^{-1}(u))$$

$$= g(a + \delta(g^{-1}(u)) + \epsilon(g^{-1}(u)), g^{-1}(u)) = (a + \epsilon(g^{-1}(u)), u),$$

so ${}^g\alpha_\epsilon = \alpha_{\epsilon \circ g^{-1}} = \alpha_{g(\epsilon)}$. Thus the proof is complete.

Theorem 12.17: *(1) $L(\theta)$ is isomorphic to $L(\theta')$ if and only if $par(\theta)$ is conjugate to $par(\theta')$ under $GL(V)$ if and only if $\theta + \Theta_0(n)$ is conjugate to $\theta' + \Theta_0(n)$ under the action of $GL(V)$ on $\Theta^0(n)$.*

(2) $|GL(V) : O(par(\theta))|$ is the number of central isomorphism classes in the π-stable isomorphism class of $L(\theta)$.

Proof: Notice that $\langle \pi \rangle$ is the intersection of the maximal subloops of $L(\theta)$ unless $par(\theta) = 0$ and $L \cong E_{2^{n+1}}$. In this case (1) is trivial, so we may assume $par(\theta) \neq 0$ and hence each isomorphism $\alpha : L(\theta) \to L(\theta')$ fixes π. Thus by 12.7, $\alpha : L(\theta) \to L(\theta')$ is an isomorphism if and only if $\theta' \circ \alpha + \theta + \theta_{\epsilon_\alpha} = 0$. So $L(\theta)$ is isomorphic to $L(\theta')$ if and only if $\theta + \Theta_0(n)$ is conjugate to $\theta' + \Theta_0(n)$ under $GL(V)$, and by 12.13 this is equivalent to $par(\theta')$ conjugate to $par(\theta)$ under $GL(V)$. That is, (1) holds. Also $O(par(\theta))$ is the stabilizer of the central equivalence class of θ, so (2) holds.

We say that $\theta \in \Theta$ is *diassociative* if θ is bilinear on each 2-dimensional subspace of V; that is, for all $x, y \in V$,

$$\theta(x,x) + \theta(x,y) + \theta(x,x+y) = 0 = \theta(x,x) + \theta(y,x) + \theta(x+y,x).$$

Lemma 12.18: *Let $par(\theta) = (P, C, A)$. Then the following are equivalent:*

(1) $L(\theta)$ is diassociative.
(2) $A(x,y,z) = 0$ for all linearly dependent subsets $\{x,y,z\}$ of V.
(3) θ is diassociative.

Moreover if θ is diassociative then $C(x,y) = C(x,x+y)$ and $C(x,y) = P(x) + P(y) + P(x+y)$ for all $x,y \in V$.

Proof: The equivalence of (1) and (2) is just the definition of diassociativity. Notice (2) implies θ is diassociative by 12.11. Finally assume (3). Summing the equations for diassociativity of θ, we get $C(x,y) = C(x,x+y)$. Then summing the three images of 12.11 under the permutation $(x,z,x+z)$, we get $0 = P(x) + P(z) + P(x+z) + C(x,z) + C(x,x+z) + C(z,x+z)$, so as $C(x,x+z) = C(z,x+z)$, also $C(x,z) = P(x) + P(z) + P(x+z)$.

Specializing z to x in 12.6.3, we get:

Lemma 12.19: $A(x,y,x) = C(x,y) + C(x,x+y)$ *for all $x,y \in V$.*

So as $C(x,y) = C(x,x+y)$, $A(x,y,x) = 0$. Also diassociativity of θ and 12.11 shows $A(x,x,y) = A(x,y,y) = 0$. Thus it remains to show $A(x,y,x+y) = 0$. Specializing z to $x+y$ in 12.6.3 yields:

Lemma 12.20: $A(x,y,x+y) = \theta(x,y) + \theta(y,x+y) + P(x) + P(x+y)$ *for all $x,y \in V$.*

Then by 12.20, diassociativity of θ, and 12.6, $A(x,y,x+y) = C(x,y) + P(y) + P(x) + P(x+y)$. Hence $A(x,y,x+y) = 0$ by an earlier remark.

Remark 12.21. Write $\Theta_1(n)$ for the space of diassociative members of $\Theta(n)$ and let $\Theta^1(n) = \Theta_1(n)/\Theta_0(n)$. By 12.18 and Remark 12.14, we have a bijection between the set of central isomorphism classes of diassociative symplectic 2-loops of order 2^{n+1} and $\Theta^1(n)$. Thus we view $\Theta^1(n)$ as the *space of diassociative symplectic 2-loops of order 2^{n+1}*.

Notice that given $x, y \in V^{\#}$, diassociativity of θ and the elements $\theta(x, y)$, $P(x)$, $P(y)$, and $P(x + y)$ determine θ on $\langle x, y \rangle$, so $dim(\Theta_1(n)) = (2^n - 1)(2^{n-1} - 1)/3 + 2^n - 1$. Hence as $dim(\Theta_0(n)) = 2^n - n - 1$, we conclude:

Lemma 12.22: $dim(\Theta^1(n)) = (2^n - 1)(2^{n-1} - 1)/3 + n$.

13. Moufang symplectic 2-loops

In this section $L = L(\theta)$ is a symplectic 2-loop. We continue the notation of Section 12. We also use the terminology of Section 11.

Lemma 13.1: *Let* $par(\theta) = (P, C, A)$. *Then*

    (1) If L is Moufang then $C = P_2$ is the second derived form of P.
    (2) L is Moufang if and only if $A = P_3$ is the third derived form of P.

Proof: Part (1) follows from 12.2 and 12.18.

Let $x, y, z \in V$ and assume L is diassociative. Then

$$(yz)(xy) = (xy)(yz) + C(x + y, y + z)$$
$$= x(y(yz)) + A(x, y, y + z) + C(x + y, y + z)$$
$$= xz + P(y) + A(x, y, y + z) + C(x + y, y + z)$$

by diassociativity. Similarly

$$(y(zx))y = ((zx)y)y + C(y, z + x) = zx + P(y) + C(y, z + x)$$
$$= xz + C(x, z) + P(y) + C(y, z + x)$$

by diassociativity. Therefore $(y(zx))y = (yz)(xy)$ if and only if

$$A(x, y, y + z) = C(x, z) + C(y, z + x) + C(x + y, y + z)$$
$$= P_3(x, y, y + z)$$

by diassociativity and 12.18. That is, if L is diassociative then L is Moufang if and only if $A = P_3$.

But if L is Moufang then by 12.2, L is diassociative, so $A = P_3$. Conversely if $A = P_3$ then $A(x, x, z) = A(x, y, y) = 0$, so by 12.11 and 12.18, L is diassociative and hence L is Moufang.

Lemma 13.2: *Assume X is a basis for V and f is a symmetric trilinear form on V with $f(x, x, y) = 0$ for all $x, y \in X$. Let $\mathcal{F} = (P, Q, f)$ be the unique 3-form with $P(x) = Q(v, x) = 0$ for all $x \in X$, $v \in V$. Then $L(Q)$ is a Moufang symplectic 2-loop with $par(Q) = (P, C, f)$, where*

$C(u,v) = Q(u,v) + Q(v,u)$ *for all* $u,v \in V$. *Further* $f = P_3$ *is the third derived form of* P.

Proof: Let $v \in V$. Then $v = \sum_{x \in X} a_x x$. Hence from 11.9,

$$Q(v,v) = Q\left(v, \sum_{x \in X} a_x x\right) = \sum_{x<y} a_x a_y f(v,x,y)$$

$$= \sum_{x<y<z} a_x a_y a_z f(x,y,z) = P(v)$$

by symmetry of f and as $f(x,x,y) = 0$ for all $x,y \in X$. That is, $P(v) = Q(v,v)$.

By 12.6.2, $C(u,v) = Q(u,v) + Q(v,u)$. By (F3) in the definition of a 3-form,

$$Q(v, u+w) = Q(v,u) + Q(v,w) + f(v,u,w)$$

and as Q is linear in its first variable, $Q(v+u, w) = Q(v,w) + Q(u,w)$. Therefore

$$f(v,u,w) = Q(v,u) + Q(u,w) + Q(v, u+w) + Q(v+u, w).$$

So by 12.6, $par(L(Q)) = (P, C, f)$.

Finally by (F1),

$$P(x+y+z) = P(x+y) + P(z) + C(x+y, z)$$

$$= P(x) + P(y) + P(z) + C(x,y) + C(x+y, z)$$

$$= P(x) + P(y) + P(z) + C(x,y) + C(x,z) + C(y,z) + f(x,y,z)$$

by (F3). But by (F1), $C(x,y) = P(x) + P(y) + P(x+y)$, so $f(x,y,z) = P_3(x,y,z)$. Hence $L(Q)$ is Moufang by 13.1.

Example 13.3 Let $n = 3$, $X = \{x,y,z\}$ a basis for V, and f the symmetric trilinear form on V with one monomial xyz in the basis X. Form the 3-form $\mathcal{F} = (P, Q, f)$ as in 13.2 and let $L = L(Q)$. Then setting $s = x + y + z$, we have $P(s) = 1$ and $P(v) = 0$ for $v \in V - \langle s \rangle$. Further for distinct $a, b \in X$, $Q = 0$ on $\langle a, b \rangle$, while for $W = \langle x+y, x+z \rangle$, $Q(a,b) = 1$ for all distinct $a, b \in W^{\#}$.

For the purposes of Lemma 13.5 it will be more convenient to consider a different cocycle θ with $\theta \sim Q$. Namely define $\epsilon \in F^V$ by $\epsilon = 1$ on $W^{\#}$ and $\epsilon = 0$ on $V - W$, and let $\theta = Q + \theta_\epsilon$. Observe that $\theta = 0$ on W while $\theta(u,v) = 1$ for all distinct $u,v \in \langle a,b \rangle^{\#}$ and all distinct $a, b \in X$. By 12.12, θ is the unique cocycle such that $par(\theta) = (P, C, f)$, $\theta = 0$ on W, and $\theta(s,w) = 1$ for all $w \in W^{\#}$.

For $0 \leq m \in \mathbf{Z}$, define F_m^V to consist of those $P \in F^V$ such that $deg(P) \leq m$, where $deg(P)$ is defined in Section 11. Then

Lemma 13.4: *(1)* $0 = F_0^V \leq F_1^V \leq \cdots \leq F_n^V = F^V$ *is a* $GL(V)$-*invariant filtration of* F^V.

(2) $dim(F_m^V/F_{m-1}^V) = \binom{n}{m}$.

(3) For each $P \in F_3^V$ *there exists a cocycle* θ *such that* $L(\theta)$ *is a Moufang symplectic 2-loop with* $par(L(\theta)) = (P, P_2, P_3)$ *and* (P, θ, P_3) *is a 3-form.*

Proof: For (1) and (2) observe that the map $\psi : P \mapsto P_m$ is a linear map of F_m^V into the space $R^m(V)$ of all symmetric m-linear forms f on V such that

$$f(x_1, \ldots, x_{n-2}, x_{n-1}, x_{n-1}) = 0$$

for all $x_1, \ldots, x_{n-1} \in V$. Further $ker(\psi) = F_{m-1}^V$, so (1) holds. Also $dim(R^m(V)) = \binom{n}{m}$. Hence $dim(F_m^V) = \sum_{i=1}^{m} d(i)$ with $d(i) = dim(F_i^V/F_{i-1}^V) \leq \binom{n}{i}$. So as $\sum_{i=1}^{n} d(i) = 2^n - 1 = dim(F^V)$, (2) follows.

Thus we may take $P \in F_3^V$ and it remains to prove (3). Observe first that if θ is a cocycle with $par(\theta) = (P, P_2, P_3)$ then (P, θ, P_3) is a 3-form if and only if θ is linear in its first variable. This follows from 12.6. We term such θ *1-linear*. Hence it remains to show there exists a 1-linear cocycle θ with $par(L(\theta)) = (P, P_2, P_3)$.

Next by 13.2 there exists $P' \in F_3^V$ and a 1-linear cocycle θ' such that $P_3' = P_3$ and $par(\theta') = (P', P_2', P_3')$. Thus $\bar{P} = P + P' \in F_2^V$, so if $\bar{\theta}$ is a 1-linear cocycle for $\bar{P}$, then $\theta = \theta' + \bar{\theta}$ is a 1-linear cocycle for P.

So we may assume $P \in F_2^V$. Notice this means P is a quadratic form on V and P_2 is the bilinear form defined by P. Then $V = V_0 \oplus V_1 \oplus \cdots \oplus V_r$, with the summands orthogonal with respect to P_2, $V_0 = Rad(P_2)$, and $dim(V_i) = 2$ for $i > 0$. There is a bilinear cocycle θ_0 for P on V_0 and each cocycle θ_i for P on V_i is diassociative for $i > 0$ by 12.18, and hence bilinear. Therefore $\theta = \sum_i \theta_i$ is bilinear with $\theta(x, x) = P(x)$ and $\theta(x, y) + \theta(y, x) = P_2(x, y)$ for all $x, y \in V$. Bilinearity of θ shows 12.6.3 is satisfied, so by 12.6, θ is a cocycle for P, completing the proof.

Lemma 13.5: *Let* $P \in F^V$. *Then there exists a Moufang symplectic 2-loop* L *with* $par(L) = (P, P_2, P_3)$ *if and only if* $deg(P) \leq 3$.

Proof: If $deg(P) \leq 3$ then L exists by 13.4. Conversely assume $L = L(\theta)$ is a symplectic 2-loop with $par(\theta) = (P, C, A)$ and $A = P_3$, but $deg(P) > 3$. Then restricting P to a 4-dimensional subspace of V on which $P_4 \neq 0$, we may assume $n = 4$. Now by 13.4, F_3^V is a hyperplane of F^V such that for all $\bar{P} \in V_3^F$, there exists a cocycle $\bar{\theta}$ with $\bar{P}$ the power map of the Moufang loop $L(\bar{\theta})$. Hence for any $\tilde{P} \in V^F - F_3^V$, $\tilde{P} = P + \bar{P}$ for some $\bar{P} \in F_3^V$ and $\tilde{\theta} = \theta + \bar{\theta}$ is a cocycle such that $L(\tilde{\theta})$ has power map $\tilde{P}$. So without loss of generality $P(x) = 1$ for a unique $x \in V$.

Let U be a complement to $\langle x \rangle$ in V. Pick a basis $Y = \{x_1, x_2, x_3\}$ for U and let $X = Y \cup \{x\}$. By 12.12 we may assume $\theta(x_i, x_j) = 0 = \theta(x_3, x_1 + x_2)$ for all $1 \le j < i \le 3$ and $\theta(x, u) = 1$ for all $u \in U^\#$. Then as $P = 0$ on U, $\theta = 0$ on U by the uniqueness statement in 12.12.

Let u, v be distinct members of $U^\#$ and $\bar{V} = \langle u, v, x \rangle$. Then $\theta = 0$ on $W = U \cap \bar{V}$ while $\theta(x, w) = 1$ for $w \in W^\#$, so θ satisfies the defining relations of the second cocycle of example 13.3 on $\bar{V}$. Hence from the discussion of that example, $\theta(u, v + x) = 1$.

Let $Z = \langle x_1, x_2, x_3 + x \rangle$. Then $P = 0$ on Z so all associators on Z are 0. Hence

$$0 = A(x_1, x_2, x_3 + x)$$

$$= \theta(x_1, x_2) + \theta(x_2, x_3 + x) + \theta(x_1, x_2 + x_3 + x) + \theta(x_1 + x_2, x_3 + x).$$

But by the discussion above, the first term in this sum is 0, while the last three terms are 1, so $1 = A(x_1, x_2, x_3 + x)$, a contradiction. This completes the proof of the lemma.

Remark 13.6. Write $\Theta_*(n)$ for the space of cocycles θ such that $par(\theta) = (P, P_2, P_3)$ for some $P \in F^V$. By 13.1, $L(\theta)$ is Moufang if and only if $\theta \in \Theta_*(n)$. Thus $\Theta^*(n) = \Theta_*(n)/\Theta_0(n)$ is the *space of Moufang symplectic 2-loops of order* 2^{n+1}.

Theorem 13.7: *(1) There is an isomorphism* $\xi : F_3^V \to \Theta^*(n)$ *such that* $(P, P_2, P_3) = par(L(\xi(P))$ *for each* $P \in F_3^V$.
(2) $dim(\Theta^*(n)) = n(n^2 + 5)/6$.
(3) $L(\xi(P)) \cong L(\xi(P'))$ *if and only if* P *is conjugate to* P' *under* $GL(V)$.
(4) $|GL(V) : O(P)|$ *is the number of central isomorphism classes in the* π-stable isomorphism class of $L(\xi(P))$.

Proof: Part (1) is a consequence of 12.13, 13.5, and 13.6. Then part (2) follows from 13.4.2. Parts (3) and (4) follow from 12.17.

14. Constructing a 2-local from a loop

In this section L is a Moufang symplectic 2-loop with parameters (P, C, A). We continue the notation of Section 13. In particular L consists of all pairs (a, u) with $a \in F$, $u \in V$, where multiplication is defined by

$$(a, u)(b, v) = (a + b + \theta(u, v), u + v)$$

with respect to some cocycle θ such that (P, θ, A) is a 3-form. We write π for the generator $(1, 0)$ of $Z(L)$.

We let $E = C_{Aut(L)}(V)$ and recall that by 12.16, $Aut(L)/E = O(V, P)$ and E consists of all maps α_ϵ, $\epsilon \in V^*$, defined by $\alpha_\epsilon : (a, u) \mapsto (a + \epsilon(u), u)$. In addition let $\Gamma_0 \leq O(V, P)$ and assume $v_0 \in V^\#$ such that Γ_0 fixes v_0 and is faithful on $V/\langle v_0 \rangle$, v_0 is in the radical of A, and $P(v_0) = C(v, v_0) = 0$ for all $v \in V$. Let $E \leq \Gamma \leq Aut(L)$ with $\Gamma/E = \Gamma_0$, and pick $s \in L$ such that $\phi(s) = v_0$, where $\phi : (a, u) \mapsto u$ is the projection of L onto V.

As v_0 is in the radical of A and $C(v, v_0) = 0$ for each $v \in V$, s is in the center of L. As $P(v_0) = 0$, $s^2 = 1$.

We term $\alpha \in \Gamma$ to be *even* if α fixes s. Notice the even automorphisms form a subgroup of Γ of index 2. Define $\epsilon \in V^*$ to be *even* if α_ϵ is even; that is, if $v_0 \in ker(\epsilon)$.

Remark 14.1. Each pair of elements $u, w \in V$ determines a unique $\epsilon_{\langle u, w \rangle} \in V^*$ via $\epsilon_{\langle u, w \rangle}(v) = A(u, w, v)$. As A is symmetric and $A(x, x, y) = 0$ for all $x, y \in V$, this definition is independent of the generators u, w for $\langle u, w \rangle$. Write $\alpha_{\langle u, w \rangle}$ for the corresponding element $\alpha_{\epsilon_{\langle u, w \rangle}} \in E$.

Similarly if $d, e \in L$, write $\epsilon_{\langle d, e \rangle}$, $\alpha_{\langle d, e \rangle}$ for the maps $\epsilon_{\langle \phi(d), \phi(e) \rangle}$, $\alpha_{\langle \phi(e), \phi(d) \rangle}$, respectively. Observe that $\alpha_{\langle d, e \rangle}$ is even and if

$$dim(\langle \phi(e), \phi(d) \rangle) < 2$$

then $\alpha_{\langle d, e \rangle} = 1$.

Let $\Omega = L \cup \{0\}$ and decree that $0 \cdot d = d \cdot 0 = 0$ for all $d \in L$. Write Ω^3 for the set of all ordered 3-tuples from Ω. We now associate to each $d \in L$ three permutations $\psi_i(d)$, $i = 1, 2, 3$, of Ω^3 by decreeing that (a, b, c) is mapped by $\psi_i(d)$ for $i = 1, 2, 3$ to

$$(d^{-1}ad^{-1}, db, cd), \quad (ad, d^{-1}bd^{-1}, dc), \quad (da, bd, d^{-1}cd^{-1}),$$

respectively. Next we associate three permutations $\psi_i(\alpha)$ of Ω^3 to each $\alpha \in \Gamma$. If α is even we define $(a, b, c)\psi_i(\alpha) = (a\alpha, b\alpha, c\alpha)$ for each $i = 1, 2, 3$. If α is odd we define the image of (a, b, c) under $\psi_i(\alpha)$ to be

$$(a^{-1}\alpha, c^{-1}\alpha, b^{-1}\alpha), \quad (c^{-1}\alpha, b^{-1}\alpha, a^{-1}\alpha), \quad (b^{-1}\alpha, a^{-1}\alpha, c^{-1}\alpha),$$

for $i = 1, 2, 3$, respectively.

We write $\psi_i(a)$ for $\psi_i(a,0)$, where $(a,0) \in L$. Let S_3 act on Ω^3 via

$$\sigma : (a_1, a_2, a_3) \mapsto (a_{1\sigma^{-1}}, a_{2\sigma^{-1}}, a_{3\sigma^{-1}}),$$

for $\sigma \in S_3$.

Define N to be the subgroup of $Sym(\Omega^3)$ generated by the maps $\psi_i(d), \psi_i(\alpha)$, $d \in L$, $\alpha \in \Gamma$, and $i = 1, 2, 3$. Define N^+ to be the subgroup generated by the maps $\psi_i(d), \psi_i(\alpha)$, $d \in L$, $\alpha \in \Gamma$ even. Set $k_i = \psi_{i-1}(s)\psi_{i+1}(s\pi)$ (where the indices are read modulo 3), $K = \langle k_1, k_2, k_3 \rangle$, $z_i = \psi_i(\pi)$, and $Z = \langle z_1, z_2, z_3 \rangle$.

Lemma 14.2: *(1) For each $\sigma \in A_3$ and $x \in L \cup \Gamma$, $\psi_i^\sigma(x) = \psi_{i\sigma}(x)$.*

(2) $\psi_i : \Gamma \to Sym(\Omega^3)$ is a faithful permutation representation.

(3) $\psi_i(d)^2 = \psi_i(d^2) = \psi_i(P(d))$ and $\psi_i(d)^{-1} = \psi_i(d^{-1})$ for each $d \in L$.

(4) For each $\alpha \in \Gamma$ and $d \in L$, $\psi_i(d)^{\psi_i(\alpha)} = \psi_i(d\alpha)$, $\psi_i(d\alpha)^{-1}$, for α even, odd, respectively.

(5) If $d \in L$ and $\alpha \in \Gamma$ is odd then $\psi_1(d)^{\psi_2(\alpha)} = \psi_3(d\alpha)^{-1}$.

(6) z_i centralizes $\psi_j(d)$ for all $d \in L$ and all i, j.

(7) For $d, e \in L$, $\psi_1(d)\psi_1(e) = \psi_1(de)\psi_3(C(e,d))\psi_1(\alpha_{\langle e,d \rangle})$.

(8) For $d \in L$ and $\alpha_\epsilon \in E$, $[\psi_i(d), \psi_i(\alpha_\epsilon)] = \psi_i(\epsilon(d))$, $\psi_i(P(d) + \epsilon(d))$ for ϵ even, odd, respectively.

(9) $[\psi_i(e), \psi_i(d)] = \psi_i([e,d]) = \psi_i(C(e,d))$.

(10) For $d, e \in L$, $\psi_1(d)^{\psi_2(e)} = \psi_1(d)\psi_3(C(e,d))\psi_1(\alpha_{\langle e,d \rangle})$.

Proof: Parts (1) and (2) are straightforward, as are (3)–(6) once we recall that L is diassociative. Also, in (8)

$$[\psi_i(d), \psi_i(\alpha_\epsilon)] = \psi_i(d)^{-1}\psi_i(d)^{\psi_i(\alpha_\epsilon)} = \psi_i(d)^{-1}\psi_i(d + \epsilon(d))^{sgn(\epsilon)}$$

by (4). Hence (8) follows from (1), (3), and (7).

Similarly in (9),

$$[\psi_1(e), \psi_1(d)] = \psi_1((de)^{-1})\psi_1(de) = \psi_1([e,d]) = \psi_1(C(e,d))$$

by repeated applications of (7) and keeping in mind that $\psi_3(\pi)$ and $\psi_1(\alpha_{\langle d,e \rangle})$ centralize each other and $\psi_1(x)$ for each $x \in \langle e, d \rangle$.

So it remains to prove (7) and (10). We prove (7); the proof of (10) is similar and left as Exercise 4.2. First

$$\psi_1(d)\psi_1(e) : (a,b,c) \mapsto (e^{-1}(d^{-1}ad^{-1})e^{-1}, e(db), (cd)e),$$

while

$$\psi_1(de) : (a,b,c) \mapsto ((de)^{-1}a(de)^{-1}, (de)b, c(de)).$$

Next

$$(de)^{-1}a(de)^{-1} = ((e^{-1}d^{-1})a)(e^{-1}d^{-1})$$
$$= (e^{-1}(d^{-1}a))(d^{-1}e^{-1}) + C(d,e) + A(d,e,a)$$
$$= e^{-1}((d^{-1}a)(d^{-1}e^{-1}))$$
$$+ C(d,e) + A(d,e,a) + A(e,d+a,d+e)$$
$$= e^{-1}(d^{-1}ad^{-1})e^{-1} + C(d,e)$$
$$+ A(d,e,a) + A(e,d+a,d+e) + A(d,e,d+a).$$

Also as A is symmetric trilinear with $A(x,x,y) = 0$ for all $x, y \in V$, we have $A(e, d+a, d+e) = A(d,e,d+a)$. Therefore $\psi_1(de)$ maps (a,b,c) to

$$(e^{-1}(d^{-1}ad^{-1})e^{-1} + C(d,e) + A(d,e,a),$$
$$e(db) + C(d,e) + A(d,e,b), (cd)e + A(d,e,c)).$$

Now (7) follows.

Lemma 14.3: *(1)* $z_1 z_2 z_3 = 1 = \psi_1(d)\psi_2(d)\psi_3(d)$ *for all* $d \in L$.
(2) Z *and* K *are normal 4-subgroups of* N *with centralizer* N^+.
(3) $N/N^+ \cong S_3$.

Proof: Part (1) is an elementary calculation. By 14.2.6, Z is centralized by $\psi_i(d)$ for all $d \in L$ and all $i = 1, 2, 3$. Similarly as s is in the nucleus of L, K is also centralized by these elements. By 14.2.4, KZ is centralized by $\psi_i(\alpha)$ for each even $\alpha \in \Gamma$. So $KZ \le Z(N^+)$.

Next let $\alpha \in \Gamma$ be odd. By 14.2.4, $\psi_1(\alpha)$ centralizes z_1 and $[\psi_1(s), \psi_1(\alpha)] = z_1$, while by 14.2.5, $[z_1, \psi_2(\alpha)] = z_2$ and $[\psi_1(s), \psi_2(\alpha)] = \psi_2(s)z_3$. Therefore $Z \trianglelefteq N$, and $N/C_N(Z) \cong S_3$. Moreover by 14.2, $N^+ \trianglelefteq N$ and $\psi_i(\alpha)^2 \in N^+$, while we calculate that $\psi_2(\alpha)^{\psi_1(\alpha)} = \psi_3(\alpha)$, so also $N/N^+ \cong S_3$. Therefore (3) holds and $N^+ = C_N(Z)$.

Finally from the commutators in the previous paragraph,

$$\psi_1(s)\psi_2(\pi) = \psi_2(s)\psi_3(s)\psi_2(\pi) = \psi_2(s\pi)\psi_3(s) = k_1$$

is centralized by $\psi_1(\alpha)$ and $[k_1, \psi_2(\alpha)] = k_2$, completing the proof of (2).

For $g \in N$, let $\bar{g} = Kg$ and adopt the bar convention. Define

$$E^+ = \{\psi_i(\alpha) : \alpha \in E \text{ even}\}Z,$$

$$Q_i = \langle \psi_i(d), \psi_i(\alpha) : d \in L, \alpha \in E \rangle,$$

$$N^- = \langle Q_1, Q_2 \rangle.$$

$$\Delta_1 = \{(d,0,0) : d \in L\},$$

$$\Delta_2 = \{(0,d,0) : d \in L\},$$

$$\Delta_3 = \{(0,0,d) : d \in L\},$$

and $\Delta = \Delta_2 \cup \Delta_3$.

Lemma 14.4: *(1) $Q_1 \trianglelefteq N_1 = C_N(z_1)$ and $k_1 \in Z(N_1)$.*
(2) $\bar{Q}_1 \cong D_8^n$ is extraspecial and $C_{\bar{N}_1}(\bar{Q}_1) = \langle \bar{z}_1 \rangle$.
(3) $E_{2^{n+1}} \cong E^+ \trianglelefteq N$.
(4) Each $g \in N_1$ can be written uniquely in the form

$$g = \psi_1(d)\psi_2(e)\psi_1(\alpha),$$

for $d, e \in L$, $\alpha \in \Gamma$; and $g \in Q_1$ if and only if $e \in \langle \pi \rangle$ and $\alpha \in E$.
(5) $K \cap Q_1 = \langle k_1 \rangle$.
(6) $Q_2 \cap Q_1 = E^+$ and N_1/Q_1 is the split extension of $(Q_2 \cap N_1)/E^+ \cong$
V by $\psi_1(\Gamma)/\psi_1(E) \cong \Gamma_0$.

Proof: As v_0 is Γ-invariant, E^+ is a $\psi(\Gamma)$-invariant subgroup of N by
14.2.2 and 14.3.2. By 14.2.8, $\psi_i(d)$ acts on E^+ for each $i = 1, 2, 3$ and
$d \in L$. Hence $E^+ \trianglelefteq N$.

Next by 14.2.2, 14.2.8, and 14.2.9, $[Q_1, Q_1] \leq \langle z_1 \rangle$. Then by 14.2.2,
14.2.7, and 14.3.1, each $g \in Q_1$ can be written as $g = \psi_1(d)\psi_2(f)\psi_1(\alpha)$,
with $d \in L$, $f \in \langle \pi \rangle$, and $\alpha \in E$. For uniqueness observe that E^+ is the
kernel of the action of Q_1 on the space $\Delta/\langle z_1 \rangle$ of orbits of $\langle z_1 \rangle$ on Δ and
Q_1/E^+ is regular on $\Delta/\langle z_1 \rangle$. Similarly E^+ is faithful on Δ of order 2^{n+1}.
So the uniqueness statement in (4) is established for elements of Q_1.

By 14.3, $N_1 = N^+ \langle \psi_1(\alpha) \rangle$ for any odd α and $k_1 \in Z(N_1)$. Next Q_1
is the kernel of the action of N_1 on $\Delta_1/\langle k_1 \rangle \cong V$ with $(Q_2 \cap N^+)/E^+$
acting by translation and $\psi_1(\Gamma)$ acting naturally. This shows $Q_1 \trianglelefteq N_1$,
establishes (6), and shows $k_2 \notin Q_1$, so (5) holds.

Now (6) gives the existence of an expression for $g \in N_1$ as in (4),
so to complete the proof of (4) suppose $g = g' = \psi_1(d')\psi_2(e')\psi_1(\alpha')$
is a second expression. Then as Q_1 is the kernel of the action on V,
$\psi_2(e)\psi_1(\alpha) \equiv \psi_2(e')\psi_1(\alpha') \mod Q_1$. Then by (6) and the regular action
of $(Q_2 \cap N^+)/E^+$ on V, we have $\phi(e) = \phi(e')$ and $\alpha \equiv \alpha' \mod E$. Then

multiplying by $\psi_1(\alpha)^{-1}\psi_2(e)^{-1}$ on the right and appealing to 14.2.2 and 14.2.7, we may assume $g, g' \in Q_1$. But we handled that case in the previous paragraph, so the proof of (4) is complete.

We have seen that $[Q_1, Q_1] \leq \langle z_1 \rangle$. Further E^+ is an abelian subgroup of Q_1 and for each $d \in L - \langle \pi \rangle$ there exists $\epsilon \in V^*$ with $\epsilon(d) \neq 0$, so by 14.2.9, $[\psi_1(d), \psi_1(\alpha_\epsilon)] = z_1$. Hence $Z(Q_1) \leq E^+$. Similarly for $0 \neq \epsilon \in V^*$ there is $d \in L$ with $\epsilon(d) \neq 0$ and hence $[\psi_1(d), \psi_1(\alpha_\epsilon)] = z_1$. Thus $\bar{Q}_1$ is extraspecial of order 2^{2n+1} and as $\bar{E}^+ \leq \bar{Q}_1$ is isomorphic to $E_{2^{n+1}}$, $\bar{Q}_1 \cong D_8^n$.

Finally by 14.2.4, $(\bar{Q}_2 \cap \bar{N}^+)/\bar{E}^+$ acts as the full group of transvections of $\bar{E}^+/\langle \bar{z}_1 \rangle$ with center $\bar{Z}/\langle \bar{z}_1 \rangle$. Also by the hypothesis of this section, $\psi(\Gamma)/\psi(E) \cong \Gamma_0$ is faithful on $\bar{E}^+/\bar{Z} \cong V/\langle v_0 \rangle$. So from (6), $\bar{Q}_1$ is the kernel of the action of $\bar{N}_1$ on $\bar{E}^+/\langle \bar{z}_1 \rangle$. This completes the proof of (2) and the lemma.

Lemma 14.5: *Let $R = (Q_1 \cap N_2)(Q_2 \cap N_1)$. Then*

(1) $R \trianglelefteq N$.

(2) $N/R = (N^-/R) \times (N^+/R)$ *with* $N^-/R \cong S_3$ *and* $N^+/R \cong \Gamma_0$.

(3) $\bar{R}$ *is class at most 3 with* $Z(\bar{R}) = \bar{Z} \cong E_4$, $Z_2(\bar{R}) \geq \bar{E}^+ = \bar{Q}_1 \cap \bar{Q}_2 \cong E_{2^{n+1}}$, *and* $\bar{R}/\bar{E}^+ = ((\bar{Q}_1 \cap \bar{N}_2)/\bar{E}^+) \times ((\bar{Q}_2 \cap \bar{N}_1)/\bar{E}^+) \cong E_{2^{2(n-1)}}$.

(4) N^- *induces* S_3 *on* $\bar{Z}$ *and centralizes* $\bar{E}^+/\bar{Z}$, *while* N^+ *centralizes* Z *and* $\bar{E}^+/\bar{Z} \cong Hom(V/\langle v_0 \rangle, F)$ *as a module for* $\bar{N}^+/\bar{R} \cong \Gamma_0$.

(5) $\bar{R}/\bar{E}^+$ *is isomorphic to the tensor product of the 2-dimensional irreducible module for* $\bar{N}^+/\bar{R} \cong S_3$ *with the module* $V/\langle v_0 \rangle$ *for* $\bar{N}^+/\bar{R} \cong \Gamma_0$.

Proof: This follows from 14.4.2, 14.3, and 8.15.

Lemma 14.6: *There exists a monomial representation of N_1 of dimension 2^n over $\mathbf{R}$ with kernel $\langle \psi_2(s\pi), \psi_3(s) \rangle$ in which Q_1 acts irreducibly as D_8^n and z_1 acts as -1.*

Proof: Recall the set Δ defined earlier. Then $K_1 = \langle k_1 \rangle$ has orbits

$$d_2 = \{(0, d, 0), (0, ds, 0)\}$$

and

$$d_3 = \{(0, 0, d), (0, 0, ds\pi)\},$$

$d \in L$, on Δ. Further z_1 is fixed point free on Δ/K_1. So by Exercise 4.3, the monomial representation of N over $\mathbf{R}$ with basis Δ/K_1 subject to the constraint that $d_i z_1 = -d_i$ is of dimension $2^n = |\Delta/K_1|/2$. As z_1 is faithful, the kernel of our representation is contained in $K\langle z_1 \rangle$ by 14.4.2, and then an easy calculation shows the kernel is $J = \langle \psi_2(s\pi), \psi_3(s) \rangle$. By construction z_1 acts as -1, so Q_1 acts as $Q_1/(J \cap Q_1) = Q_1/K_1 \cong \bar{Q}_1 \cong D_8^n$. In particular as the minimal dimension of a faithful representation of an extraspecial group of order 2^{2n+1} is 2^n (cf. 34.9 in [FGT]), Q_1 acts irreducibly.

Remarks. In [Co3], Conway uses the Parker loop to construct the normalizer of a 4-subgroup of the Monster and uses this normalizer to construct the Griess algebra and the Monster.

In the terminology of Section 12, the Parker loop is the Moufang symplectic 2-loop $L(\theta)$ with parameters (P, C, A), where V is the 12-dimensional Golay code module of Section 19 for M_{24} and $(P, C, A) = (P, P_2, P_3)$ is the series of derived forms on V defined in Section 20 using 11.8, and (P, θ, A) is a 3-form. By 13.4.3, such a cocycle exists, and by 13.1, $L(\theta)$ is Moufang. By 12.13, the isomorphism type of $L(\theta)$ depends only on P, not on θ.

The construction of the group $\bar{N}$ in Section 14 is essentially the same as Conway's construction in [Co3]. From 14.4 and 14.5, $\bar{N}$ has a large extraspecial 2-subgroup $\bar{Q}_1$ and a 4-subgroup $\bar{Z} = \langle \bar{z}_1, \bar{z}_2 \rangle \leq \bar{Q}_1$ with $\langle \bar{z}_1 \rangle = Z(\bar{Q}_1)$, $\bar{Z} \trianglelefteq \bar{N}$, and $\bar{N}/C_{\bar{N}}(\bar{Z}) \cong S_3$. For the Parker loop $L(\theta)$, $\bar{Q}_1$ turns out to be a large extraspecial subgroup of the Monster and $\bar{N}$ is the normalizer of $\bar{Z}$ in the Monster.

Only a small part of the discussion in Sections 12 and 13 is necessary to construct $\bar{N}$ and make the calculations needed to construct the Monster. We use 13.2 and the first few lemmas in Section 12 to construct the Parker loop L. We need that L is diassociative, which is easily verified using arguments in 12.18. We need that $Aut(L)$ is large enough, which follows from 12.16, which depends in turn on some of the earlier results in Section 12. This information is sufficient to apply the construction of Section 14 to construct $\bar{N}$ from L.

Exercises

1. Prove Lemma 12.4.
2. Prove Lemma 14.2.10.
3. Let $\sigma : G \to Sym(X)$ be a permutation representation of a group G such that $z\sigma$ is fixed point free on X for some involution z with $[z, G] \leq ker(\sigma)$. Prove there exists a monomial $\mathbf{Z}G$-representation ρ such that
   (1) $z\rho$ acts as -1 on the module V for ρ.
   (2) The permutation representation of G on $\pm Y$ is equivalent to σ for some basis Y of V.

(3) $G\rho$ preserves the bilinear form on V with orthonormal basis Y.

4. Let Q be an extraspecial 2-group and q the quadratic form of 8.3.3 on $\tilde{Q} = Q/Z(Q)$. Let $\mu : Q \to SL(V) = G$ be the unique faithful irreducible $\mathbf{R}Q$-representation and $M = N_G(Q\mu)$. Prove $M/Q\mu \cong O(\tilde{Q}, q)$.
   (Hint: Use 12.16 and 1.1.)

5. Prove that if L is a Moufang symplectic 2-loop then L is diassociative, without using Moufang's Theorem.

6. Assume the hypotheses of Section 14, let $D = V \times V^*$ as a set, and define addition on D by

$$(u, \beta) + (v, \gamma) = (u + v + C(u,v)v_0, \beta + \gamma + \epsilon_{\langle u, v \rangle}).$$

Define $q : D \to GF(2)$ by $q(u, \beta) = (1 + s(\beta))P(u) + \beta(u)$ and $\rho : D \times D \to GF(2)$ by

$$\rho((u, \beta), (v, \gamma)) = \beta(v) + \gamma(u) + s(\beta)P(v) + s(\gamma)P(u) + C(u,v),$$

where $s(\beta) = 0, 1$ for β even, odd, respectively. Let $\tilde{Q}_1 = \bar{Q}_1/\langle \bar{z}_1 \rangle$ and define $\mu : Q_1 \to D$ by $\mu : \psi_1(d)\psi_2(b)\psi(\alpha_\epsilon) \mapsto (\phi(d) + t(b)v_0, \epsilon)$, where $t(b) = 0, 1$ for $b = 1, \pi$, respectively. Prove

(1) D is a $GF(2)$-space under this definition of addition.

(2) q is a quadratic form on D with bilinear form ρ.

(3) μ is a group homomorphism with kernel $\langle z_1, k_1 \rangle$ inducing an isometry $\mu : \tilde{Q}_1 \to D$, where $\tilde{Q}_1$ is an orthogonal space via 8.3.

(4) μ induces an equivalence of the actions of N_1 on $\tilde{Q}_1$ and D, when we define $\psi_1(e) = 1$ on D and $\psi_1(\alpha)$ and $\psi_2(e)$ on D, for $\alpha \in \Gamma$ and $e \in L$, by

$$\psi_1(\alpha) : (u, \beta) \mapsto (u\alpha, \beta^\alpha),$$

$$\psi_2(e) : (u, \beta) \mapsto (u + s(\beta)\phi(e) + (1 + s(\beta))C(e,u)$$
$$+ \beta(e)v_0, \beta + (1 + s(\beta))\epsilon_{\langle e, u \rangle}).$$

(Hint: See the proofs of 27.2 and 23.10.)

7. Assume the hypotheses of Section 14 with L associative. Prove $\bar{\psi}_1 : L \to \bar{Q}_1$ is an injective group homomorphism and $\bar{\psi}_1(L) \trianglelefteq \bar{N}$ with $\bar{Q}_1 = \bar{E}^+\bar{\psi}_1(L)$ and $\bar{E}^+ \cap \bar{\psi}_1(L) = \bar{Z}$.

Chapter 5

The Discovery, Existence, and Uniqueness of the Sporadics

In this chapter we discuss the history of how the sporadic groups were discovered and were first shown to exist uniquely. Then we provide a general outline of how *Sporadic Groups* proves the existence of the twenty sporadics which are realized as sections of the largest sporadic group, the Monster. Finally we outline our approach for proving the uniqueness of the sporadics.

Chapter 5 includes several tables. At the end of Section 15 is a table listing the twenty-six sporadic groups with the notation and name used in *Sporadic Groups* to label the group. This table also lists the group order. The table at the end of Section 16 summarizes the existence results proved in *Sporadic Groups*.

15. History and discovery

Group theory had its beginnings in the early nineteenth century, with Galois making some of the most important early contributions. For example, it was Galois who first defined the notion of a simple group in 1832. During most of the nineteenth century the term "group" meant "permutation group." Thus it is natural that the first sporadic groups were discovered by Mathieu as multiply transitive permutation groups; papers describing this work appeared in 1860 and 1861 [M1] [M2].

More than a century passed before the discovery of the next sporadic group. This is perhaps not too surprising, since with a few notable exceptions in the work of mathematicians such as Burnside, Frobenius, Brauer, and Phillip Hall, there was little progress in the theory of finite groups during the first half of the twentieth century. However, in the mid fifties three crucial events occurred in finite group theory.

First, at the International Congress of Mathematicians in Amsterdam in 1954, Brauer proposed that finite simple groups might be characterized by the centralizers of involutions. Brauer's program was inspired by the Brauer–Fowler Theorem [BF] (cf. Section 8), which says that the order of a finite simple group G is bounded by the order of $C_G(z)$ for z an involution in G. Brauer's program lead to the discovery of many of the sporadic groups and eventually to the classification.

The second major event in finite group theory in the fifties was Chevalley's Tohoku paper of 1955 [Ch] in which he gave a uniform construction of Chevalley groups over all fields, and began modern work on the finite groups of Lie type. The Lie theoretic point of view provided a uniform perspective from which to view most of the simple groups known in the fifties, and thus gave hope that the nonabelian finite simple groups might consist of the alternating groups, the groups of Lie type, plus the five "sporadic groups" of Mathieu.

The third major event was Thompson's thesis of 1959 establishing the conjecture of Frobenius that Frobenius kernels are nilpotent. Thompson's thesis marked the beginnings of serious local group theory: the study of finite groups G from the point of view of normalizers of p-subgroups of G, soon termed *p-local subgroups*.

All three of these events contributed to the discovery of the next sporadic group by Janko in a paper appearing in 1965. First, Brauer's program had taken hold and group theorists were busy characterizing simple groups via the centralizers of involutions. One class of groups for which such a characterization seemed difficult was the Ree groups $^2G_2(3^{2n+1})$, one of the families of groups of Lie type. The problem studied by Janko, Thompson, and Ward was to determine all finite simple groups G possessing an involution z such that $C_G(z) \cong \mathbf{Z}_2 \times L_2(q)$, q odd, and with Sylow 2-subgroups of G abelian. Using a mixture of character theory and local group theory, Janko, Thompson, and Ward [JT],[Wa] showed that under these hypotheses, q is an odd power of 3, or $q = 5$. In [J1], Janko showed that there exists a unique group in the last case: the sporadic group now denoted by J_1.

The next two sporadic groups were also discovered by Janko. Again Janko studied groups with a certain involution centralizer. This time the centralizer satisfied Hypothesis $\mathcal{H}(2, A_5)$, in the notation of the Preface.

In a paper appearing in 1968 [J2], Janko showed that if G satisfies Hypothesis $\mathcal{H}(2, A_5)$, then G has one of two possible structures. In particular in each of the two cases, Janko showed that $|G|$ is determined, as is the local structure of G. On the other hand Janko did not show that a group existed in either case, nor did he show that such a group was unique up to isomorphism. The existence and uniqueness of the Janko groups J_2 and J_3 were established by M. Hall and D. Wales in the first case [HW], and by G. Higman and J. MacKay [HM] in the second.

This was the first instance of a fairly typical pattern. Namely many sporadic groups were discovered via considering groups with a certain centralizer, often satisfying Hypothesis $\mathcal{H}(w, L)$ for suitable w and L. Further the mathematician "discovering" the group often only showed that the order and local structure of a group satisfying his hypotheses was determined, but did not demonstrate the existence or uniqueness of the group.

Hall and Wales constructed the group J_2 as a rank 3 permutation group (cf. Section 3). Very soon after the Hall–Wales construction, three more sporadic groups were discovered as rank 3 groups: HS discovered by D. Higman and C. Sims [HS]; Mc discovered by J. McLauglin [Mc]; and Suz discovered by M. Suzuki [Suz]. Indeed Higman and Sims constructed their group in just a few hours, shortly after hearing Hall lecture on the construction of J_2.

Somewhat later, Rudvalis discovered the group Ru as a rank 3 group and Conway and Wales constructed a 2-fold cover of the group as a subgroup of $GL_{28}(\mathbf{C})$ [CW].

J. Conway discovered the three Conway groups Co_1, Co_2, and Co_3 as automorphism groups of the Leech lattice; this work appears in his 1969 paper [Co2] (cf. Chapter 8). Also visible as stabilizers of sublattices of the Leech lattice or local subgroups of Co_1 were J_2, HS, Mc, and Suz. Thus if Conway had become interested in the Leech lattice a little earlier, he would have discovered a slew of sporadic groups. In any event these embeddings supplied convenient existence proofs for the groups.

Bernd Fischer discovered the three Fischer groups by studying groups generated by 3-transpositions. This work appears in his Warwick preprint, the first few sections of which make up the 1971 paper [F]. As defined by Fischer, a *set of 3-transpositions* of a group G is a G-invariant set D of involutions such that for all $a, b \in D$, $|ab| \leq 3$. In addition to the three Fischer groups, the symmetric groups and several families of groups of Lie type are generated by 3-transpositions. The theory of groups generated by 3-transpositions is beautiful and useful. Indeed it is probably the best way to study and prove the uniqueness of the Fischer groups. The theory is not considered here, as it does not fit well with the main topics of this book. But the author is planning a book on 3-transposition groups to cover the topic.

Fischer was also lead to the Fischer group F_2 (also known as the Baby Monster) in 1973 via the study of groups generated by $\{3,4\}$-transpositions. The existence and uniqueness of the Baby Monster were established later by Sims and Leon using extensive machine calculation discussed in [LS].

The remaining sporadics were discovered via the centralizer of involution approach. For example, Held discovered He. In particular in his 1969 paper [He], Held studied groups satisfying Hypothesis $\mathcal{H}(3, L_3(2))$ and showed each such simple group is isomorphic to $L_5(2)$ or M_{24}, or has a uniquely determined order and local structure. Soon after, G. Higman and MacKay showed there exists a unique group He in this last case. In Chapter 14 we give a simplified treatment of groups satisfying Hypothesis $\mathcal{H}(3, L_3(2))$. This is perhaps the most accessible illustration in *Sporadic Groups* of how to prove the uniqueness and derive the basic structure of a sporadic group.

The group Ly was discovered by Lyons (with the paper [Ly] describing this work appearing in 1972) and shown to exist uniquely by Sims [Si2] using the machine. O'Nan discovered $O'N$; his work appeared in 1976 in [ON].

Fischer and R. Griess independently were led to consider groups satisfying Hypothesis $\mathcal{H}(12, Co_1)$. One quickly sees that if G is a group satisfying this hypothesis then G has a second class of involutions whose centralizer is the covering group of the Baby Monster. During one of his annual meetings in Bielefeld, Fischer discussed the possibilities of such a group with mathematicians then at Cambridge, including Conway, Thompson, K. Harada, and S. Norton. The Cambridge mathematicians began to study the potential group, which they dubbed the Monster. They showed that if the Monster existed then the centralizers of certain elements of order 3 and 5 had to contain sections which were new sporadic groups. In Griess' notation these sporadics are F_3 and F_5.

Harada [Ha] concentrated on the Harada group F_5, pinning down its order and local structure. By some oversight, no uniqueness proof for the group was generated until Y. Segev [Se] took up the problem in 1989.

Thompson studied the Thompson group F_3, generating the group order, local structure, and even the character table. With the character

Table 1 The sporadic groups

Notation	Name	Order
M_{11}	Mathieu	$2^4 \cdot 3^2 \cdot 5 \cdot 11$
M_{12}		$2^6 \cdot 3^3 \cdot 5 \cdot 11$
M_{22}		$2^7 \cdot 3^2 \cdot 5 \cdot 7 \cdot 11$
M_{23}		$2^7 \cdot 3^2 \cdot 5 \cdot 7 \cdot 11 \cdot 23$
M_{24}		$2^{10} \cdot 3^3 \cdot 5 \cdot 7 \cdot 11 \cdot 23$
J_1	Janko	$2^3 \cdot 3 \cdot 5 \cdot 7 \cdot 11 \cdot 19$
J_2		$2^7 \cdot 3^3 \cdot 5^2 \cdot 7$
J_3		$2^7 \cdot 3^5 \cdot 5 \cdot 17 \cdot 19$
J_4		$2^{21} \cdot 3^3 \cdot 5 \cdot 7 \cdot 11^3 \cdot 23 \cdot 29$ $\cdot 31 \cdot 37 \cdot 43$
HS	Higman–Sims	$2^9 \cdot 3^2 \cdot 5^3 \cdot 7 \cdot 11$
Mc	McLaughlin	$2^7 \cdot 3^6 \cdot 5^3 \cdot 7 \cdot 11$
Suz	Suzuki	$2^{13} \cdot 3^7 \cdot 5^2 \cdot 7 \cdot 11 \cdot 13$
Ly	Lyons	$2^8 \cdot 3^7 \cdot 5^6 \cdot 7 \cdot 11 \cdot 31 \cdot 37 \cdot 67$
He	Held	$2^{10} \cdot 3^3 \cdot 5^2 \cdot 7^3 \cdot 17$
Ru	Rudvalis	$2^{14} \cdot 3^3 \cdot 5^3 \cdot 7 \cdot 13 \cdot 29$
$O'N$	O'Nan	$2^9 \cdot 3^4 \cdot 5 \cdot 7^3 \cdot 11 \cdot 19 \cdot 31$
Co_3	Conway	$2^{10} \cdot 3^7 \cdot 5^3 \cdot 7 \cdot 11 \cdot 23$
Co_2		$2^{18} \cdot 3^6 \cdot 5^3 \cdot 7 \cdot 11 \cdot 23$
Co_1		$2^{21} \cdot 3^9 \cdot 5^4 \cdot 7^2 \cdot 11 \cdot 13 \cdot 23$
$M(22)$	Fischer	$2^{17} \cdot 3^9 \cdot 5^2 \cdot 7 \cdot 11 \cdot 13$
$M(23)$		$2^{18} \cdot 3^{13} \cdot 5^2 \cdot 7 \cdot 11 \cdot 13 \cdot 17 \cdot 23$
$M(24)'$		$2^{21} \cdot 3^{16} \cdot 5^2 \cdot 7^3 \cdot 11 \cdot 13 \cdot 17 \cdot 23 \cdot 29$
F_3	Thompson	$2^{15} \cdot 3^{10} \cdot 5^3 \cdot 7^2 \cdot 13 \cdot 19 \cdot 31$
F_5	Harada	$2^{14} \cdot 3^6 \cdot 5^6 \cdot 7 \cdot 11 \cdot 19$
F_2	Baby Monster	$2^{41} \cdot 3^{13} \cdot 5^6 \cdot 7^2 \cdot 11 \cdot 13 \cdot 17 \cdot 19 \cdot 23 \cdot 31 \cdot 47$
F_1	Monster	$2^{26} \cdot 3^{20} \cdot 5^9 \cdot 7^6 \cdot 11^2 \cdot 13^3 \cdot 17 \cdot 19 \cdot 23$ $\cdot 29 \cdot 31 \cdot 41 \cdot 47 \cdot 59 \cdot 71$

table, Thompson had a proof of the existence of an irreducible representation of degree 248. This representation makes possible a one page proof of the uniqueness of the Thompson group. Using machine calculation done by P. Smith, Thompson [Th1] showed F_3 to be a subgroup of the exceptional group $E_8(3)$ of Lie type, hence also establishing the existence of the group.

Thompson [Th2] also showed that *if* the Monster F_1 possesses an irreducible representation of degree 196,883, and several other fairly weak constraints are satisfied, then the Monster is unique. Norton went on to verify the existence of this representation assuming certain other fairly weak properties, but he did not publish this work. Finally in 1989, Griess, Meierfrankenfeld, and Segev [GMS] published the first complete uniqueness proof for the Monster.

In the meantime, working independently from the Cambridge group, Griess had been generating properties of the Monster too. In 1980 he was able to construct the Monster as the group of automorphisms of a real algebra in 196,883 dimensions [Gr1]. Later in [Gr2], Griess worked with the *Griess algebra*, a 196,884-dimensional real algebra with an identity. Griess' existence proof established at the same time the existence of the twenty sporadic groups which are sections of the Monster. Conway [Co3] and Tits [T2] later produced simplifications of portions of Griess' arguments.

It is fitting that Janko, who discovered the first of the modern sporadic groups, also discovered the last of the sporadics, J_4. Janko [J4] discovered J_4 in 1975 while studying groups satisfying Hypothesis $\mathcal{H}(6, Aut(M_{22}/\mathbf{Z}_3))$. (Our notational convention is that $H_1/H_2/\cdots/H_n$ denotes a group with normal series

$$1 = G_n \trianglelefteq G_{n-1} \trianglelefteq \cdots \trianglelefteq G_0 = G$$

with $G_{i-1}/G_i \cong H_i$.) As usual he determined the order and local structure of such a group. Norton [N] established the existence and uniqueness of J_4 using the machine.

Table 1, on page 69, lists the twenty-six sporadic groups and their orders.

16. Existence of the sporadics

In this section we give a very broad outline of how *Sporadic Groups* proves the existence of the twenty sporadics realized as sections of the Monster. We also speculate briefly on how best to establish the existence of the remaining six sporadics.

Our first step is to construct the Mathieu group M_{24}. This is accomplished in Chapter 6 by realizing M_{24} as the group of automorphisms of its Steiner system $S(24, 8, 5)$. A *Steiner system* $S(v, k, t)$ is a rank 2 geometry (cf. Section 4), whose objects are called points and blocks, such that the geometry has v points, each block is incident with exactly k points, and each set of t points is incident with a unique block. Steiner systems provide a geometric point of view for studying multiply transitive groups. Witt was the first to study M_{24} via its Steiner system in his 1938 paper [W1]. His approach is rather different from the one used here, which proceeds by constructing a tower of extensions of Steiner systems beginning with the projective plane of order 4. The five Mathieu groups are then realized as stabilizers of subsystems of the Steiner system for M_{24}. This gives a constructive existence proof for the Mathieu groups, which makes it easy to establish various properties of the groups such as their order and local structure. In later instances, some of our existence proofs are nonconstructive, so we get relatively little information about the group. In particular in many cases we do not calculate the group order. Table 2, on page 74, indicates which sporadics have their order calculated in *Sporadic Groups*; for each of these groups, either our existence proof is constructive, or we calculate the group order as part of the uniqueness proof.

Next the Steiner system is used to construct two 11-dimensional modules for M_{24} over the field of order 2, which in turn provide a realization for the 2-local geometry of M_{24}. One of these modules is associated to an error correcting code (cf. Section 10) for M_{24}, the *binary Golay code*.

The Steiner system and Golay code for M_{24} are then used in Chapter 8 to construct the Leech lattice Λ. The Leech lattice is a certain 24-dimensional $\mathbf{Z}$-submodule of 24-dimensional Euclidean space. Its group of automorphisms is the covering group $\cdot 0$ of the largest Conway group Co_1. The sporadic groups Co_2, Co_3, HS, and Mc are shown to exist as stabilizers of certain sublattices of Λ. Again these existence proofs are constructive and allow us to calculate the group orders.

On the other hand we show the existence of J_2 and Suz as sections of Co_1 by nonconstructive methods. Namely we prove that $G = Co_1$ has a large extraspecial 2-subgroup Q (cf. Section 8). Then we produce subgroups A_p of $N_G(Q)$ of order $p = 3, 5$ and use 8.13 to show that $C_Q(A_p)^*$ is a large extraspecial subgroup of $C_G(A_p)^* = C_G(A_p)/A_p$. Then we appeal to 8.12 or Exercise 2.4 to see that $C_G(A_p)^*$ is a simple group satisfying Hypothesis $\mathcal{H}(w, L)$, where w is the width of $C_Q(A_p)$ and $L = (N_G(Q) \cap N_G(A_p))/C_Q(A_p)A_p$. Thus, for example, we prove there exist simple groups satisfying Hypothesis $\mathcal{H}(3, \Omega_6^-(2))$ and Hypothesis $\mathcal{H}(2, A_5)$, living as sections of Co_1, and define these sections to be Suz and J_2, respectively. We postpone to Chapter 16 the problem of proving the uniqueness and deriving the structure of such groups.

The theory of large extraspecial subgroups is also used to show that Co_1 and Co_2 are simple. This theory is used in an analogous manner in

Chapter 11 to establish the existence and simplicity of sporadic sections of the Monster.

The Monster is constructed as the group of automorphisms of the Griess algebra. Unfortunately this object is not so natural, so it gives us little information about the Monster and its subgroups. Following Conway, we begin by constructing the Parker loop and using it to construct a group $\bar{N}$ which will turn out to be the normalizer of a certain 4-subgroup of the Monster. The necessary discussion of loops and of Conway's construction of $\bar{N}$ from the Parker loop is contained in Chapter 4. The discussion in Chapter 4 gives a general method for constructing Moufang symplectic 2-loops given a form of degree 3 over the field of order 2 (cf. Section 11). The Parker loop is constructed from the form defined by the Golay code.

Our next step is to construct a group C which will turn out to be the centralizer of an involution z in the Monster G. The group C is constructed to have a normal extraspecial 2-subgroup Q of order 2^{1+24} with $C/Q \cong Co_1$ and $Q/\langle z \rangle \cong \Lambda/2\Lambda$. Then the Griess algebra is constructed so as to admit C and $\bar{N}$ with $C \cap \bar{N}$ of index 3 in $\bar{N}$. Finally using arguments of Tits we show $G = \langle C, \bar{N} \rangle$ is finite and $C = C_G(z)$. Thus G is constructed so as to satisfy Hypothesis $\mathcal{H}(12, Co_1)$, so in particular G is simple. We define G to be the Monster; it is constructed to satisfy Hypothesis $\mathcal{H}(12, Co_1,)$.

In [FLM], Frenkel, Lepowski, and Meurman use an alternate approach to construct the Monster as the symmetry group of a vertex operator algebra. Vertex operator algebras are infinite-dimensional graded Lie algebras together with a family of "vertex operators." They play a role in string theory but are also of independent interest to mathematicians.

The Frenkel, Lepowski, and Meurman construction has several advantages. For one thing it can be viewed as fairly natural within the context of the theory of vertex operator algebras. For another it "explains" the connection between the Monster and modular functions of genus 0, discussed in [CN]. The reader is directed to the introduction of [FLM] for a nice discussion of these very interesting properties of the Monster.

On the other hand the Frenkel–Lepowski–Meurman construction is quite lengthy and complicated. Thus we have opted here for the shorter, more group theoretic approach of Griess, Conway, and Tits.

The six sporadic groups F_2, F_3, F_5, He, and $M(24)$ are shown to exists as sections of $C_G(A)$ for suitable subgroups A of $N_G(Q)$ using the same approach we used to prove the existence of J_2 and Suz in Co_1. Namely we show the existence of a group satisfying Hypothesis

$\mathcal{H}(w, L)$ for suitable w, L. In most cases we do not obtain any further information about the structure of such groups. However, in Chapter 14 we do investigate groups satisfying Hypothesis $\mathcal{H}(3, L_3(2))$. In the process we derive the order of He and much of its local structure. This illustrates how the structure of a sporadic can be derived beginning only from Hypothesis $\mathcal{H}(w, L)$. In Chapters 16 and 17 we determine the group order and structure of J_2, Suz, and Co_1, and prove the uniqueness of these groups.

The group $M(24)$ is the largest of Fischer's sporadic 3-transposition groups. We leave till later the problem of showing the subgroup $N_G(A)$ of the Monster of type $M(24)$ is generated by 3-transpositions. Once that is accomplished, the theory of 3-transpositions can be used to prove the existence of the $M(22)$ and $M(23)$ sections of $M(24)$ and to determine the order and local structure of these groups.

This is our outline of the existence proofs for the twenty sporadic groups which live as sections of the Monster. We add a few words of speculation as to the best means for demonstrating the existence of the remaining six sporadics. The groups J_1, J_3, and Ru can be shown to exist as subgroups of the exceptional groups of Lie type $G_2(11)$, $E_6(4)$, and $E_7(5)$. The existence proofs for these groups via these embeddings are fairly elegant and satisfactory. I have no good ideas to advance for establishing the existence of $O'N$, Ly, and J_4. The present existence proofs are highly machine dependent.

We close this section with Table 2 summarizing the existence results established in *Sporadic Groups*. Column 1 lists the sporadic group. Column 2 lists the hypothesis $\mathcal{H}$ under which the group is shown to exist; that is, we show that there exists a simple group G satisfying Hypothesis $\mathcal{H}$. However, often very little information is obtained about the group other than its simplicity; for example, sometimes the group order is not calculated. Column 3 indicates if the order of at least one group satisfying Hypothesis $\mathcal{H}$ is calculated in *Sporadic Groups*. Column 4 indicates the page where existence is established.

The question marks in column 2 and asterisks in column 4 for J_1, J_3, J_4, Ly, Ru, and $O'N$ indicate that the existence question for these groups is not addressed in *Sporadic Groups*. Similarly the asterisk in column 4 for $M(22)$ and $M(23)$ indicates that the proof of existence of these groups is postponed until a later book, where the normalizer in the Monster of the subgroup of order 3 described in 32.4.3 is shown to be a 3-transposition group, and 3-transposition theory is used to demonstrate the existence of the smaller Fischer groups. Sometimes more

Table 2   Existence of the sporadic groups

Group	ExistenceHypothesis	Order?	Page
M_{11}	Multiply transitive	Yes	89
M_{12}	group of	Yes	89
M_{22}	automorphisms of	Yes	82
M_{23}	a Steiner system	Yes	84
M_{24}	$\mathcal{H}(3, L_3(2))$	Yes	85
J_1	?	No	*
J_2	$\mathcal{H}(2, A_5)$	Yes	135
J_3	?	No	*
J_4	?	No	*
HS	rank 3 group and stabilizer	Yes	119
Mc	of a sublattice of the Leech lattice	Yes	118
Suz	$\mathcal{H}(4, Sp_6(2))$	Yes	135
Ly	?	No	*
He	$\mathcal{H}(3, L_3(2))$	Yes	173
Ru	?	No	*
$O'N$	?	No	*
Co_3	Stabilizer of a	Yes	116
Co_2	sublattice of the Leech lattice	Yes	116
Co_1	$\mathcal{H}(4, \Omega_8^+(2))$	Yes	116
$M(22)$	Will exist as a subgroup of $M(24)$ using	No	*
$M(23)$	3-transposition theory once a suitable subgroup	No	*
$M(24)'$	of the Monster is shown to be a 3-transposition group	No	173
F_3	$\mathcal{H}(4, A_9)$	No	173
F_5	$\mathcal{H}(4, A_5 \ wr \ \mathbf{Z}_2)$	No	173
F_2	$\mathcal{H}(11, Co_2)$	No	173
F_1	$\mathcal{H}(12, Co_1)$	No	169

than one existence hypothesis is obtained. For example, we show there exists a simple 5-transitive group of automorphisms of a Steiner system $S(24, 8, 5)$ and we also calculate that such a group satisfies Hypothesis $\mathcal{H}(3, L_3(2))$.

17. Uniqueness of the sporadics

The present treatment of the uniqueness of the sporadics is ad hoc. In many cases the uniqueness proofs are machine aided. One goal of

Sporadic Groups is to correct this situation by providing a uniform approach to uniqueness which will work for almost all the sporadics. Our approach is the one introduced by the author and Yoav Segev in [AS1]. Conceptually the approach is as follows:

We consider groups G satisfying a suitable group theoretic hypothesis $\mathcal{H}$; for example, $\mathcal{H}$ will usually be $\mathcal{H}(w, L)$ for suitable w and L. We show that any such G possesses a suitable family $\mathcal{F}(G) = (G_i : i \in I)$ of subgroups. Associated to $\mathcal{F}$ is a simplicial complex K and the topological space $|K|$ of K. We form the free amalgamated product $\tilde{G}$ of the amalgam defined by the family $\mathcal{F}(G)$ and the topological space $|\tilde{K}|$ of the complex $\tilde{K}$ defined by $\mathcal{F}(\tilde{G})$. The universal property of $\tilde{G}$ translates into a covering $|\tilde{K}| \to |K|$ of topological spaces. On the other hand we prove $|K|$ is simply connected, so $|\tilde{K}| = |K|$ and hence $\tilde{G} = G$. As this holds for any G satisfying $\mathcal{H}$, and as the isomorphism type of the amalgam of $\mathcal{F}(G)$ depends only on $\mathcal{H}$ and not on G, we have our uniqueness proof.

While this is conceptually one way of viewing what goes on, in practice we introduce no topology, but instead carry out our proof at a combinatorial level. Thus we consider the coset geometry $\Gamma = \Gamma(G, \mathcal{F})$ of the family $\mathcal{F}$ as defined in Section 4 and form the collinearity graph Δ on G/G_1 defined by Γ. Our simplicial complex is the clique complex $K = K(\Delta)$ of the graph Δ (cf. Section 4). Similarly we have the coset geometry $\tilde{\Gamma} = \Gamma(\tilde{G}, \mathcal{F}(\tilde{G}))$ and its collinearity graph $\tilde{\Delta}$ and clique complex $\tilde{K} = K(\tilde{\Delta})$. The covering of topological spaces corresponds to a covering of clique complexes which corresponds in turn to a covering $d : \tilde{\Delta} \to \Delta$ of graphs. Here $d : \tilde{\Delta} \to \Delta$ is a *covering of graphs* if d is a surjective local isomorphism of graphs (cf. Section 35). Then we can define Δ to be simply connected if it possesses no proper connected coverings, and this turns out to be equivalent to $|K|$ being simply connected. We work only with this combinatorial definition of graph covering.

Part III of *Sporadic Groups* puts in place the machinery for implementing this approach for proving the uniqueness of the sporadics. Thus in Chapter 12 we develop the basic theory of coverings of graphs, including various means for verifying that a graph is simply connected. Then in Chapter 13 we record the theory of group amalgams we will need, including results which will allow us to prove that our group theoretic hypothesis $\mathcal{H}$ determines the isomorphism type of our amalgam $\mathcal{F}(G)$ independently of the group G satisfying $\mathcal{H}$.

The discussion in Section 4 and Chapters 12 and 13 is more extensive than necessary simply to prove the uniqueness of the sporadic groups.

However, the extra material is useful in studying geometries and simplicial complexes associated to finite groups, such as the Quillen complexes [Q].

Finally the last few chapters of the book illustrate how the machinery can be used by proving the uniqueness of various sporadics. In particular in Chapter 14 we prove $L_5(2)$, M_{24}, and He are the only simple groups satisfying the Hypothesis $\mathcal{H}(3, L_3(2))$. In Chapter 15 we prove the uniqueness of $U_4(3)$. While $U_4(3)$ is not a sporadic group it does exhibit sporadic behavior, and its characterization is crucial to proving the uniqueness of various sporadics. In Chapter 16 we prove J_2, Suz, and Co_1 are unique. In the process we also derive the group order of J_2 and Suz and much of the subgroup structure of these groups. Finally Chapter 17 closes the book with a list of the subgroups of prime order and their normalizers in the five sporadic groups M_{24}, He, J_2, Suz, and Co_1 considered here in detail. Most of this information is generated while proving the uniqueness of the groups; the remainder is established in Chapter 17.

PART II

Chapter 6

The Mathieu Groups, Their Steiner Systems, and the Golay Code

The first five sporadic groups were discovered by Mathieu as multiply transitive permutation groups around 1860. There are by now many constructions of the Mathieu groups. Our approach is to construct the groups as automorphism groups of their Steiner systems.

We recall that a *Steiner system* $S(v, k, t)$ is a rank 2 geometry (whose objects are called *points* and *blocks*) with v points, such that each block contains exactly k points and each set of t points is incident with a unique block. For example, the projective plane of order 4 is a $S(21, 5, 2)$. In Section 18 we construct a tower

$$S(21, 5, 2) \leq S(22, 6, 3) \leq S(23, 7, 4) \leq S(24, 8, 5)$$

of Steiner systems admitting the corresponding Mathieu groups. At the same time we establish many of the properties of these Steiner systems needed to analyze the structure of the Mathieu groups and construct the Leech lattice, the Griess algebra, the Conway groups, and the Monster.

The blocks in the Steiner system $S(24, 8, 5)$ are called *octads*. In Section 19 we construct certain subgroups of the largest Mathieu group M_{24} and determine the action of these subgroups on the octads. Further we study two 11-dimensional $GF(2)$-modules for M_{24}, the Golay code module and the Todd module. These modules are sections of the permutation module for M_{24} on the points of its Steiner system. The Golay code module is an image of a $(24, 12)$-code module.

18. Steiner systems for the Mathieu groups

A *Steiner system* $S(v, k, t)$ or *t-design* is a rank 2 geometry $(X, \mathcal{B})$ (cf. Section 4) whose objects are a set X of *points* and a collection $\mathcal{B}$ of k-subsets of X called *blocks* such that each t-subset of X is contained in a unique block. Of course incidence in this geometry is inclusion.

Example (1) Recall the definition of a *projective plane* from Example 2 in Section 4. Each projective plane of order q is a Steiner system $S(q^2 + q + 1, q + 1, 2)$.

Let $(X, \mathcal{B})$ be a Steiner system $S(v, k, t)$. Given a point $x \in X$ define the *residual design* $D(X, x)$ of X at x to be the geometry $(X(x), \mathcal{B}(x))$, where

$$X(x) = X - \{x\},$$

$$\mathcal{B}(x) = \{B - \{x\} : x \in B \in \mathcal{B}\}.$$

Observe:

Lemma 18.1: *If $t > 1$ then the residual design $D(X, x)$ of X at x is a Steiner system $S(v - 1, k - 1, t - 1)$.*

Next define an *extension* of X to be a Steiner system $(Z, \mathcal{A})$ such that $(X, \mathcal{B}) = D(Z, z)$ is the residual design of Z at some point $z \in Z$. Notice if Z is an extension of X then by 18.1, Z has parameters $(v+1, k+1, t+1)$.

Our object is to construct a tower

$$S(21, 5, 2) \leq S(22, 6, 3) \leq S(23, 7, 4) \leq S(24, 8, 5)$$

of extensions of Steiner systems beginning with the projective plane of order 4, such that the Mathieu group M_v is an automorphism group of the Steiner system $S(v, k, t)$. The remainder of this section is devoted to this construction and to the generation of properties of these Steiner systems that we will need to analyze the Mathieu groups.

Define a subset I of X to be *independent* if no $(t + 1)$-subset of I is contained in a block of X.

Example (2) If X is the projective plane of a vector space V, then the independent subsets are just the sets $\mathcal{O}$ of points such that each triple of points in $\mathcal{O}$ is linearly independent in V in the usual sense.

Write $I_m(X)$ for the set of all independent subsets of X of order m. Define an *extension subset* of X to be a subset C of I_{k+1} such that each member of I_{t+1} is contained in a unique member of C.

Lemma 18.2: *Let $t > 1$, $x \in X$, $C(x)$ the set of blocks of X not containing x, and $Y = D(X,x)$ the residual design of X at x. Then $C(x)$ is an extension subset of Y.*

Lemma 18.3: *Let C be an extension subset of X. Then*

  (1) *There exists an extension Z of X such that C is the extension subset $C(z)$ induced by Z and the point z in $Z - X$.*

  (2) *The restriction map $\alpha \mapsto \alpha_{|X}$ defines an isomorphism of $Aut(Z)_z$ with $N_{Aut(X)}(C)$.*

Proof: The blocks of Z are the members of C together with the blocks $\{z\} \cup B$, $B \in \mathcal{B}$.

Remark 18.4. If Z is an extension of X at some point z we identify $Aut(Z)_z$ with $N_{Aut(X)}(C(z))$ via the isomorphism of 18.3.2. For example, this convention is used in the statement of the following hypothesis:

Extension Hypothesis: *Y is a Steiner system $S(v, k, t)$ with $t > 1$, $y \in Y$, and $A \leq Aut(Y)_y$. Further*

  (Ex1) *$Aut(Y)$ is transitive on the extension subsets of Y invariant under some $Aut(Y)$-conjugate of A, and there exists such an extension subset.*

  (Ex2) *If $A \cong A' \leq Aut(Y)_y$ then $A = A'$.*

  (Ex3) *If Y' is an extension of $D(Y,y)$ with $A \leq Aut(Y')$ then there exists an isomorphism $\pi : Y \to Y'$ acting on $D(Y,y)$.*

  (Ex4) *$N_{Aut(Y)}(C)$ is t-transitive on Y for each A-invariant extension subset C of Y.*

Lemma 18.5: *Let Y be a Steiner system $S(v, k, t)$ and $A \leq Aut(Y)$. Then*

  (1) *If A, Y satisfies (Ex1) then, up to isomorphism, there exists a unique extension X of Y with $A \leq N_{Aut(X)}(Y)$. Moreover $N_{Aut(X)}(Y) = N_{Aut(Y)}(C)$, where C is the extension subset of Y induced by X.*

  (2) *Assume $t > 1$ and A, Y satisfies the Extension Hypothesis with respect to some $y \in Y$. Then $Aut(X)$ is $(t+1)$-transitive on X and transitive on the blocks of X.*

Proof: For $i = 1, 2$, let X_i be $(t+1)$-designs, $x_i \in X_i$, $Y_i = D(X_i, x_i)$, and C_i the extension subset of Y_i induced by X_i, and assume $\alpha_i : Y \to Y_i$

is an isomorphism with $A\alpha_i^* \leq N_{Aut(X_i)}(Y_i)$. Notice $\alpha = \alpha_1^{-1}\alpha_2 : Y_1 \rightarrow Y_2$ is an isomorphism. Then $C_1\alpha$ is an extension subset of Y_2 invariant under $(A\alpha_1^*)\alpha^* = A\alpha_2^*$. So by (Ex1), there exists $\beta \in Aut(Y_2)$ with $C_1\alpha\beta = C_2$. Let $\gamma = \alpha\beta$, and extend γ to X_1 by $x_1\gamma = x_2$. Then $C_1\gamma = C_2$. Hence $\gamma : X_1 \rightarrow X_2$ is an isomorphism. Therefore the uniqueness statement in (1) is established while the other parts of (1) follow from 18.3.

So assume the hypothesis of (2) and let $X = Y \cup \{x\}$ be the unique extension of Y with $A \leq Aut(X)_x$ supplied by (1). Let $x_1 = x$, $x_2 = y$, and define Y_i and C_i as in the previous paragraph with respect to $X = X_1 = X_2$. Then Y_2 is an extension of $Z = D(Y,y)$ with $A \leq Aut(Y_2)$, so by (Ex3) there exists an isomorphism $\alpha_2 : Y \rightarrow Y_2$ with $y\alpha_2 = x$ and $Z\alpha_2 = Z$. Then $A\alpha_2^* \cong A \leq Aut(Y_2)$, so by (Ex2), $A = A\alpha_2^*$. Thus $A\alpha_2^* = A \leq N_{Aut(X)}(Y_2)$.

Let $\alpha_1 : Y \rightarrow Y$ be the identity map. Then we have achieved the hypothesis of paragraph one, so by that paragraph there exists an automorphism γ of X with $x\gamma = y$. Hence as $N_{Aut(Y)}(C) = N_{Aut(X)}(Y)$ is t-transitive on Y, $Aut(X)$ is $(t+1)$-transitive on X. Then as each $(t+1)$-subset of X is in a unique block, $Aut(X)$ is also transitive on blocks.

Lemma 18.6: *Let $F = GF(4)$, X the projective plane over F, p a point of X, and A the stabilizer of p in $L_3(F)$. Then, up to isomorphism, X is the unique extension of $D(X,p)$ admitting A.*

Proof: By Lemma 18.5, it suffices to establish (Ex1) for the pair Y, A, where $Y = D(X,p)$.

Let V be a vector space over F with basis $\{x_1, x_2, x_3\}$ and take $X = PG(V)$ and $p = \langle x_1 \rangle$. Let B be the group of transvections with center p and D the stabilizer of p and the line $l = \langle x_2, x_3 \rangle$. From 5.1, $B \cong E_{16}$, $D \cong L_2(4) \cong A_5$, and A is the split extension of B by D.

Let Δ be the set of lines of X through p and Ω the remaining lines. Let $\mathcal{A} = \{k - \{p\} : k \in \Delta\}$. Then $\mathcal{A}$ is the set of five blocks of the residual design Y and Ω is the extension subset induced by X. Observe that the members of $\mathcal{A}$ are the 5 orbits of B on Y of length 4 and that no member of $B^{\#}$ fixes a point in each orbit. Also B is regular on Ω and D is the stabilizer in A of $l \in \Omega$.

Let C be an A-invariant extension subset of Y and $m \in C$. As $|m| = 5 = |\mathcal{A}|$ and each pair of points in m is independent, it follows that $m \cap k$ is a point for each $k \in \mathcal{A}$.

If $b \in B$ fixes m then b fixes the point $m \cap k$ for each $k \in \mathcal{A}$. But then $b = 1$. So all orbits of B on C are regular. However, by a counting argument, $|C| = 16 = |B|$, so B is regular on C. Thus $N_A(m) = E$ is a complement to B in A and $C = mB$. Further A, and hence also E, is transitive on Δ and therefore also on $m = \{m \cap K : k \in \Delta\}$. So $m = qE$ is the orbit of E on Y containing any point $q \in m$. Pick $q \in l$.

Observe that $A_q = B_q D_q$ with $D_q = N_D(k)$, where $q \in k \in A$. Similarly $A_q = B_q E_q$ with $E_q = N_E(k)$. Now there exists an automorphism α of A centralizing B and A/B with $D\alpha = E$ (cf. 17.3 in [FGT]). As α centralizes B, $(B_q)\alpha = B_q$. Also as α centralizes A/B, $N_A(k)\alpha = N_A(k)$. So $(D_q)\alpha = D\alpha \cap N_A(k)\alpha = E \cap N_A(k) = E_q$. Thus $(A_q)\alpha = (B_q)\alpha(D_q)\alpha = A_q$.

Define α on Y by $(qa)\alpha = q(a\alpha)$ for $a \in A$. As $A_q\alpha = A_q$, this action is well defined. Then $l\alpha = (qD)\alpha = q(D\alpha) = qE = m$. As α commutes with B and $\mathcal{A}$ is the set of orbits of B on Y, α permutes $\mathcal{A}$. So $\alpha \in Aut(Y)$. Similarly $\Omega\alpha = (lB)\alpha = l\alpha B = mB = C$. Thus (Ex1) is satisfied, and the proof is complete.

We now begin to construct the Steiner systems for the Mathieu groups. This requires a detailed analysis of the independent subsets in the plane of order 4.

Lemma 18.7: *Let $F = GF(4)$, V be a vector space over F with basis $\{x_1, x_2, x_3\}$, $Y = PG(V)$, $M = PGL(V)$, $G = PSL(V)$, $H = N_G(\langle x_1 \rangle)$, and $\Gamma = P\Gamma(V)$. Then*

(1) *M is transitive on I_4 as a set of ordered 4-tuples.*

(2) *G is transitive on I_3, and has three orbits on I_4 as a set of unordered 4-tuples.*

(3) *Each $S \in I_4$ is contained in a unique $I(S) \in I_6$.*

(4) *For $S = \{\langle x_1 \rangle, \langle x_2 \rangle, \langle x_3 \rangle, \langle x_1 + x_2 + x_3 \rangle\} \in I_4$,*

$$I(S) = S \cup \{\langle x_1 + ax_2 + a^{-1}x_3 \rangle, \langle x_1 + a^{-1}x_2 + ax_3 \rangle\},$$

   where $F^{\#} = \langle a \rangle$.

(5) *M is transitive on I_6 and G has three orbits I_6^i, $1 \leq i \leq 3$, on I_6.*

(6) *Each $S \in I_3$ is contained in a unique member $I^i(S)$ of I_6^i for each $i = 1, 2, 3$.*

(7) *I_6^i, $1 \leq i \leq 3$, are the H-invariant extension subsets of Y and H has two orbits on I_6^i.*

(8) *For $I \in I_6$, $N_G(I) \cong A_6$, $N_\Gamma(I) \cong S_6$, and $N_\Gamma(I)$ is faithful on I.*

Proof: Let $p_i = \langle x_i \rangle$. Clearly each member of I_3 is conjugate to $T_3 = (p_1, p_2, p_3)$ under G. Let $p_4^1 = \langle x_1 + x_2 + x_3 \rangle$, $p_4^2 = \langle x_1 + x_2 + ax_3 \rangle$, and $p_4^3 = \langle x_1 + x_2 + a^{-1}x_3 \rangle$. Then $N_G(T_3) = \langle g \rangle$, where $g = diag(1, a, a^{-1})$. So p_4^i, $1 \le i \le 3$, are representatives for the orbits of $N_G(T_3)$ on points p of Y with $T_3 \cup \{p\} \in I_4$. Hence T_4^i, $1 \le i \le 3$, are representatives for the action of G on I_4 as a set of ordered 4-tuples, where $T_4^i = T_3 \cup \{p_4^i\}$. Also $N_M(T_3) = \langle g, h \rangle$, where $h = diag(1, 1, a)$. So $N_M(T_3)$ is transitive on members of I_4 through T_3, and hence (1) holds.

To prove (3) and (4) it suffices by (1) to prove $I(T_4^1)$ is the unique member of I_6 containing T_4^1. But if $x = x_1 + rx_2 + sx_3 \in V$ with $T_4^1 \cup \{\langle x \rangle\} \in I_5$ then, as $x_1 + x_2 + tx_3 \in \langle p_3, p_4^1 \rangle$ for all $t \in F^{\#}$, $r \in \{a, a^{-1}\}$. Similarly $s = r^{-1}$, so (3) and (4) are established.

Let $I = I(T_4^1)$. By (1) and (3), $N_M(I)$ is 4-transitive on I. As $\langle g \rangle = N_M(I)_{T_3}$, $N_M(I)$ is faithful on I and isomorphic to A_6 by Exercise 1.2. Then $N_M(I) = N_M(I)^\infty \le G$, so $N_M(I) = N_G(I)$. Also the field automorphism f determined by T_3 induces a transposition on I, so $N_\Gamma(I) \cong S_6$. Thus (8) holds.

By (8), $N_G(I)$ is transitive on $I_3 \cap I$, so $N_G(T_3)$ is transitive on the members of IG through T_3. Also as $N_G(I)$ is 4-transitive on I, T_4^i is not G-conjugate to a subset of I for $i = 2, 3$. Therefore (2) holds. Thus $I(T_4^i)G = I_6^i$, $1 \le i \le 3$, are the orbits of G on I_6. Of course M is transitive on I_6 by (1) and (3). Further (6) holds with $I^i(T_3) = I(T_4^i)$.

As $N_G(I)$ is transitive on the points of I, H is transitive on the set A^i of members of I_6^i through p_1. Also if $J \in I_6$ with $p_1 \notin J$, then J is conjugate under H to a 6-set containing $T = \{p_2, p_3, \langle x_1 + x_2 \rangle\}$. Hence as T is in a unique member of I_6^i, H is transitive on the set B^i of members of I_6^i not through p_1.

Notice by (6) that I_6^i is a G-invariant extension subset of Y for each i. Conversely suppose C is an H-invariant extension set. We may assume $I \in C$, and it remains to show $C = I_6^1$. But $A^1 = IH \subseteq C$ so if $C \ne I_6^1$ there is $J \in C \cap I_6^j$ for $j = 2$ or 3 with $p_1 \notin J$. Then $B^j = JH \subseteq C$. But if $T \in I_3 \cap I$ with $p_1 \notin T$, then $T \subseteq J_i \in I_6^i$. Also $I = I(T \cup \{p_1\})$, so $J_i \in B^i$. Thus T is in two members of C, a contradiction. Hence (7) is established.

Lemma 18.8: *Let $F = GF(4)$, Y the projective plane over F, p a point in Y, and H the stabilizer in $L_3(F)$ of p. Then*

(1)   There exists a unique extension X of Y with $H \le Aut(X)$.

(2)   Let $M = Aut(X)$ and $G = M \cap Alt(X) = M_{22}$. Then $|M : G| = 2$, $N_G(Y) = L_3(F)$, $N_M(Y)$ is $N_G(Y)$ extended by a

field automorphism, G is 3-transitive on X, and G is transitive on $\mathcal{B}(X)$.

(3) M *is transitive on* $I_7(X)$ *and* G *has two orbits* $I_7^i(X)$, $i = 1, 2$, *on* $I_7(X)$.

(4) G *is transitive on* $I_4(X)$, M *is transitive on* $I_5(X)$, *and* G *has two orbits on* $I_5(X)$.

(5) *Each member of* $I_5(X)$ *is contained in a unique member of* $I_7(X)$ *and each member of* $I_4(X)$ *is contained in a unique member of* $I_7^i(X)$, $i = 1, 2$.

(6) *For* $I \in I_7(X)$, $N_G(I) \cong A_7$ *is faithful on* I.

(7) $I_7^i(X)$, $i = 1, 2$, *are the unique* $N_G(Y)$-*invariant extension subsets of* X.

Proof: We claim the pair (Y, H) satisfies the Extension Hypothesis. First (Ex1) and (Ex4) are satisfied by 18.7.5 and 18.7.7. By 5.1 and 5.2, $H = Aut(Y)_p^\infty$, so (Ex2) holds. Finally (Ex3) holds by 18.6. So the claim is established.

Now the claim and 18.5 implies (1), and we may take the extension subset C_X of Y induced by X to be $I_6^3(Y)$. Then by 18.5 and 18.7, $N_M(Y) = N_{Aut(Y)}(C_X)$ is $L_3(F)$ extended by a field automorphism f. Notice f interchanges $I_6^1(Y)$ and $I_6^2(Y)$. Further by 5.2.2, f fixes exactly seven points of Y. So f has seven orbits of length 2 on X, and hence $f \notin G$. Therefore $N_G(Y) = L_3(F)$ and $|M : G| = 2$. By 18.5, M is 3-transitive on X, so G is transitive on X. Then as $L_3(F)$ is 2-transitive on Y, G is also 3-transitive on X. So G is transitive on $\mathcal{B}(X)$. Thus (2) holds.

Let q be the point in $X - Y$ and suppose $I \in I_7(X)$. Conjugating in G, we may take $q \in I$. Then $I_0 = I - \{q\} \in I_6(Y)$ and as $|I_0 \cap B| \leq 3$ for each $B \in C_X, I_0 \in I_6^i(Y)$ for $i = 1$ or 2.

Conversely let $J = \{q\} \cup J_i$, $J_i \in I_6^i(Y)$, $i = 1$ or 2. As $J_i \in I_6(Y)$, no 4-subset of J through q is in a block of X. As $i \leq 2$, $|J_i \cap B| \leq 3$ for each $B \in C_X$ by 18.7.3, so no 4-subset of J_i is in a block of X. Hence $J \in I_7(X)$. That is, the members of $I_7(X)$ through q are the sets $J_i \cup \{q\}$, $J_i \in I_6^i(Y)$, $i = 1, 2$.

If $U \in I_5(X)$, conjugating in G we may take $q \in U$. Then $U_0 = U - \{q\} \in I_4(Y)$, so by 18.7.3, U_0 is in a unique member E_0 of $I_6(Y)$. As $U \in I_5(X)$, $E_0 \in I_6^i(Y)$ for $i \leq 2$. So by the previous paragraph, $E = \{q\} \cup E_0$ is the unique member of $I_7(X)$ containing U. Similarly if $q \in W \in I_4(X)$ then $W_0 = W - \{q\}$ is in a unique member J_i of $I_6^i(Y)$, for $i = 1, 2$.

Further by 18.7.1 and 18.7.2, G is transitive on $I_4(X)$, G has at most two orbits on $I_5(X)$, and M is transitive on $I_5(X)$ as f interchanges the two G-orbits on $I_4(Y)$ whose members are not in a member of $I_6^3(Y)$. As M is transitive on $I_5(X)$ and each member of $I_5(X)$ is in a unique member of $I_7(X)$, M is transitive on $I_7(X)$ and $N_M(I)$ is 5-transitive on $I \in I_7(X)$. As $N_G(Y) \cap N(I) = N_M(Y) \cap N(I) = A_6$ is faithful on I, $N_G(I) = N_M(I) = A_7$ is faithful on I. (I.e., (6) holds.) In particular $N_G(I)$ is transitive on I, so as $\{q\} \cup J_1 = I_1$ is not conjugate to $\{q\} \cup J_2 = I_2$ in G_q, $I_2 \notin I_1 G$. Thus G has two orbits $I_7^i(X)$ on $I_7(X)$ and (3) is established. Indeed the members of $I^i(X)$ containing q are of the form $\{q\} \cup K_i$, $K_i \in I_6^i(Y)$. Similarly as $N_G(U)$ is transitive on $U \in I_5(X)$ but $N_G(q)$ has two orbits on members of $I_5(X)$ through q, G has two orbits on $I_5(X)$. This completes the proof of (4). Also each member of $I_4(X)$ through q is in a unique member of $I_7^i(X)$, so as $N_G(W)$ is transitive on W, the proof of (5) is complete.

Finally by (5), $I_7^i(X)$, $i = 1,2$, are G-invariant extension subsets of X. Conversely suppose C is an $N_G(Y)$-invariant extension subset of X. Let $C_q = \{I \in C : q \in I\}$ and $C_q^* = \{I - \{q\} : I \in C_q\}$. Then C_q^* is an $N_G(Y)$-invariant extension subset of Y, so by 18.7, $C_q^* = I_6^i(Y)$. As $C_X = I_6^3(Y)$, $i \le 2$.

Now $N_G(Y)$ is transitive on C_q. Similarly, for $y \in Y$, $N_G(Y)_y$ is transitive on $\{S \in C : y \in S, q \notin S\}$ from the penultimate paragraph of the proof of 18.7. Thus as $N_G(Y)$ is transitive on Y, $N_G(Y)$ is transitive on $C - C_q$. Now an argument in the last paragraph of the proof of 18.7 completes the proof of (7).

The following two lemmas can be proved along the lines of the proof of 18.8:

Lemma 18.9: *Let Y be the 3-design of M_{22} and H the stabilizer $L_3(4)$ in M_{22} of a point of Y. Then*

(1) *There exists a unique extension X of Y with $H \le \mathrm{Aut}(X)$.*
(2) *Let $G = \mathrm{Aut}(X) = M_{23}$. Then $N_G(Y) = M_{22}$, G is 4-transitive on X, and G is transitive on $\mathcal{B}(X)$.*
(3) *G is transitive on $I_8(X)$.*
(4) *G is transitive on $I_m(X)$ for $m = 5$ and 6, and each member of $I_m(X)$ is contained in a unique member of $I_8(X)$.*
(5) *For $I \in I_8(X)$, $N_G(I) = A_8$ is faithful on I.*
(6) *$I_8(X)$ is the unique extension subset of X.*

Lemma 18.10: *Let Y be the 4-design of M_{23} and H the stabilizer M_{22} in M_{23} of a point of Y. Then*

    (1)   There exists a unique extension X of Y with $H \leq Aut(X)$.

    (2)   Let $G = Aut(X) = M_{24}$. Then $N_G(Y) = M_{23}$, G is 5-transitive on X, and G is transitive on $\mathcal{B}(X)$.

    (3) $I_7(X)$ is empty.

    (4) G is transitive on $I_6(X)$.

Notice that by 18.10.3, the M_{24}-design has no extension subsets, and hence the M_{24}-design cannot be extended.

Lemma 18.11: *(1) The groups M_{22}, M_{23}, and M_{24} are simple.*

    (2) $|M_{22}| = 2^7 \cdot 3^2 \cdot 5 \cdot 7 \cdot 11$.

    (3) $|M_{23}| = 2^7 \cdot 3^2 \cdot 5 \cdot 7 \cdot 11 \cdot 23$.

    (4) $|M_{24}| = 2^{10} \cdot 3^3 \cdot 5 \cdot 7 \cdot 11 \cdot 23$.

Proof: Let G be M_{22}, M_{23}, or M_{24}, and let X be the design of G. Then G is t-transitive on X for $t = 3, 4$, or 5. Suppose $1 \neq K \trianglelefteq G$, let $p \in X$, and let $H = G_p$. By 2.2.2, $G = KH$. By 2.2.3, $H \cap K \neq 1$. But if $G = M_{22}$ then $H = L_3(F)$ is simple, while if G is M_{23} or M_{24} then $H = M_{22}$ or M_{23} is simple by induction. Then as $H \cap K \trianglelefteq H$, we have $H = H \cap K \leq K$. Thus $G = HK = K$.

Of course $|M_{22}| = 22 \cdot |L_3(4)|$, so (2) holds. Similarly (3) and (4) hold.

19. The Golay and Todd modules

In this section $(X, \mathcal{C})$ is the Steiner system $S(24, 8, 5)$ for M_{24} and $G = Aut(X, \mathcal{C})$ is M_{24}, as discussed in Section 18. Let V be the binary permutation module for G on X and proceeding as in Section 10, identify V with the power set of X by identifying $v \in V$ with its support.

The members of $\mathcal{C}$ will be called *octads*. As $(X, \mathcal{C})$ is a 5-design, each 5-subset U of X is contained in a unique octad; denote this octad by $B(U)$. So there are $\binom{24}{5} / \binom{8}{5} = 23 \cdot 11 \cdot 3 = 759$ octads. Recall also from 18.11 that G has order $2^{10} \cdot 3^3 \cdot 5 \cdot 7 \cdot 11 \cdot 23$.

We first prove two lemmas giving fairly complete information about the action of G on the octads.

Lemma 19.1: *Let B be an octad, $M = N_G(B)$, $Q = G_B$, and $x \in X - B$. Then*

    (1) M is the split extension of $Q \cong E_{16}$ by $M_x \cong A_8$.

    (2) Q is regular on $X - B$.

    (3) M_x acts faithfully as $Alt(B)$ on B.

(4) $M_x = GL(Q)$. In particular $A_8 \cong L_4(2)$.

(5) M is 3-transitive on $X-B$, with the representation of M_x on $X-B$ equivalent to the representation of M_x on Q via conjugation.

Proof: Let $Y = D(X,x)$. Then $B \in I_8(Y)$, so by 18.9.5, $M_x = N_{G_x}(B) = Alt(B) = A_8$.

Let U be a 3-subset of B, $H = G_U$, $Y = X - U$, and $l = B - U$. Then Y is a projective plane, l is a line in Y, $H = L_3(4)$, and $Q = H_l$. So $Q \cong E_{16}$, $H \cap M_x = L_2(4)$ is faithful on Q, and Q is regular on $X - B$ by the dual of 5.1.

Next as Q is regular on $X - B$, $M = QM_x$ is the split extension of Q by $M_x \cong A_8$. As M_x is simple and $C_{M_x}(Q) \trianglelefteq M_x$, $C_{M_x}(Q) = 1$ or M_x. The latter is impossible as $H \cap M_x$ is faithful on Q. Thus M_x is faithful on Q, so $M_x \leq GL(Q) \cong L_4(2)$. As $|L_4(2)| = |A_8|$, $M_x = GL(Q)$ and $L_4(2) \cong A_8$.

By 2.1, the representation of M_x on $X - B$ is equivalent to the representation of M_x on Q by conjugation. So as $L_4(2)$ is 2-transitive on $Q^\#$, M is 3-transitive on $X - B$.

For $v \in V$ denote by $C_n(v)$ the set of all octads B such that $|v \cap B| = n$.

Lemma 19.2: *Let A, B be octads, $M = N_G(B)$, $Q = G_B$, and if $A \neq B$ let $x \in A - B$. Then*

(1) $|A \cap B| = 0, 2, 4,$ or 8.

(2) M is transitive on $C_n(B)$ for each n.

(3) $|C_n(B)| = 30, 448, 280, 1,$ for $n = 0, 2, 4, 8,$ respectively.

(4)   If $A \in C_4(B)$ then $N_M(A)$ is the split extension of $N_Q(A) \cong E_4$ by $N_M(A)_x = N_{M_x}(A \cap B) \cong Z_2/(A_4 \times A_4)$.

(5)   If $A \in C_2(B)$ then $N_M(A)$ is a complement to Q in $N_M(A \cap B)$, so $N_M(A) \cong S_6$.

(6)   If $A \in C_0(B)$ then $N_Q(A) \cong E_8$ is a hyperplane of Q and $N_M(A)$ is the split extension of $N_Q(A)$ by $N_{M_x}(N_Q(A)) \cong L_3(2)/E_8$.

(7)   If $A \in C_0(B)$ then $X + A + B = X - (A \cup B)$ is an octad.

(8)   If $A \in C_4(B)$ then $A + B$ is an octad.

Proof: Let A be distinct from B and $x \in A - B$. Notice $N_M(A) \leq N_M(A \cap B) = QN_{M_x}(A \cap B)$, and $N_Q(A)$ is semiregular on $A - (A \cap B)$, so $|N_Q(A)| \leq |A - (A \cap B)|_2 = |A \cap B|_2$.

As each 5-subset is in a unique octad, $|A \cap B| = n \leq 4$. Also if U is a 4-subset of B, then there exists a unique octad $B(U, x)$ containing U and x; further $U = B(U, x) \cap B$. Counting the set Ω of pairs U, x in two ways we get $16\binom{8}{4} = 4|C_4(B)|$, so $|C_4(B)| = 280$. As Q is transitive on

$X - B$ and M_x is transitive on the 4-subsets of B, M is transitive on Ω, and hence on $\mathcal{C}_4(B)$. So $|M : N_M(A)| = 280$ for $A \in \mathcal{C}_4(B)$.

Next $N_{M_x}(A \cap B) \cong Z_2/(A_4 \times A_4)$. Thus $|N_{M_x}(A \cap B)| = |N_M(A \cap B)/Q| \geq |N_M(A)Q/Q| = |N_M(A)/N_Q(A)| \geq |N_M(A)|/4 = |N_{M_x}(A \cap B)|$, as $|N_M(A)| = |M|/280 = 4|N_{M_x}(A \cap B)|$. So all inequalities are equalities and hence $N_Q(A) \cong E_4$ is regular on $A - B$, so $N_M(A) = N_Q(A)N_{M_x}(A)$ with $N_{M_x}(A) = N_{M_x}(A \cap B)$. Thus (4) holds.

Also there exists $t \in M_x$ with $(A \cap B)t = B - A$. Indeed $t \in N_M(N_Q(A))$, so $(A - B)t = (xN_Q(A))t = xN_Q(A) = A - B$. Thus $At = (B - A) + (A - B) = A + B$. So $A + B = At$ is an octad, establishing (8).

If U is a 3-subset of B then $Y = X - U$ is a projective plane and $B - U$ is a line in Y. Then if $U \leq A$, $A - U$ is another line, so $(B - U) \cap (A - U)$ is a point. In particular $n \neq 3$.

Counting the set of pairs (U, C) with U a 2-subset of $C \in \mathcal{C}$ in two ways, we get $759\binom{8}{2} = \binom{24}{2}N_2$, where N_2 is the number of octads through a fixed 2-subset U of B. Hence $N_2 = 77$. Further there are $\binom{6}{2}$ 4-subsets W of B through U, and 4 octads in $\mathcal{C}_4(B)$ through W, so there are 60 members of $\mathcal{C}_4(B)$ through U. This leaves $77 - (60 + 1) = 16$ octads in $\mathcal{C}_2(B)$ through U.

Let $A \in \mathcal{C}_2(B)$ with $A \cap B = U$. If $g \in N_Q(A)^\#$ then $U = Fix_A(g)$, so g induces an odd permutation on A. This is impossible as $N_G(A)^A = Alt(A)$. So $N_Q(A) = 1$. Thus Q is regular on the sixteen members of $\mathcal{C}_2(B)$ through U. So $N_M(A)$ is a complement to Q in $N_M(U)$, and hence $N_M(A) \cong S_6$ and (5) is established. Also as Q is transitive on the members of $\mathcal{C}_2(B)$ through U and M is transitive on 2-subsets of B, M is transitive on $\mathcal{C}_2(B)$. Then $|\mathcal{C}_2(B)| = |M : N_M(A)| = 448$.

We are left with $759 - (1 + 280 + 448) = 30$ octads in $\mathcal{C}_1(B) \cup \mathcal{C}_0(B)$. If $A \in \mathcal{C}_1(B)$ then $N_Q(A) = 1$ as $|A - B|$ is odd. Further $N_M(A) \leq N_M(A \cap B)$, so $|AM| = |M : N_M(A)| \geq |M : N_M(A \cap B)| \cdot 16 \geq 7 \cdot 16 > 30$. So $\mathcal{C}_1(B)$ is empty and $\mathcal{C}_0(B)$ is of order 30. Thus (1) and (3) hold.

Let $A \in \mathcal{C}_0(B)$. As Q is regular on $X - B$, $Q \neq N_Q(A)$. On the other hand as $|\mathcal{C}_0(B)| \equiv 2 \mod 4$, we may choose A so that $|AM|$ is not divisible by 4. Hence a Sylow 2-subgroup of $N_M(A)$ is of index 2 in a Sylow 2-group of M and $N_Q(A)$ is regular on A. Indeed A and $A + B + X$ are the two orbits of $N_Q(A)$ on $X - B$, and $g \in Q - N_Q(A)$ interchanges these orbits, so $A + B + X = Ag$ is also an octad, establishing (7).

Finally as $N_Q(A)$ is regular on A, $N_M(A) = N_Q(A)N_{M_x}(A)$. Also $N_{M_x}(A) \leq N_{M_x}(N_Q(A))$, so

$$|N_{M_x}(A)||N_Q(A)| = |N_M(A)| \leq |M|/30 = |N_{M_x}(N_Q(A))||N_Q(A)|,$$

so all inequalities are equalities. That is, $N_{M_x}(A) = N_{M_x}(N_Q(A))$ and M is transitive on $C_0(B)$. Therefore (2) and (6) hold, and the lemma is established.

Define a *dodecad* of X to be a subset of the form $A + B$, where B is an octad and $A \in C_2(B)$. Thus each dodecad is of order 12.

Lemma 19.3: *Let $D = A + B$ be a dodecad, with B an octad and $A \in C_2(B)$. Let C be an octad. Then*

(1) $|C \cap D|$ *is even.*

(2) D *contains no octads.*

(3) *For each 5-subset U of D, $B(U) \in C_6(D)$ and $B(U) + D$ is an octad.*

Proof: Let $r = |C \cap A \cap B|$, $s = |A \cap C \cap D|$, and $t = |B \cap C \cap D|$. Then $r + s = |A \cap C|$ and $r + t = |B \cap C|$ are even by 19.2.1. Thus $s \equiv r \equiv t$ mod 2, so $s + t = |C \cap D| \equiv 0$ mod 2, and (1) is established.

Suppose $C \subseteq D$. Then $r = 0$ and $s + t = 8$. By 19.2.1, $s, t \le 4$, so $s = t = 4$. By 19.2.8, $C' = C + B$ is an octad. But $|C' \cap A| = 6$, contradicting 19.2.1. Hence (2) holds.

Notice (1) and (2) imply the first part of (3). Let $C = B(U)$, so that $s + t = 6$. Then s or t is at least 3, say t. So by 19.2.1, either $C = B$ or $|B \cap C| = 4$, and we may assume the latter. Then $B + C$ is an octad by 19.2.8. Further by inspection, $|A \cap (B+C)| = 4$, so also $A + (B+C)$ is an octad. But addition in V is associative, so $A + (B+C) = (A+B) + C = D + C$. Hence the proof of (3) is complete.

Given a dodecad D define

$$\mathcal{B}(D) = \{B(U) \cap D : U \text{ is a 5-subset of } D\}.$$

Lemma 19.4: *Let D be a dodecad and $K = N_G(D)$.*

(1) $(D, \mathcal{B}(D))$ *is a Steiner system $S(12, 6, 5)$.*

(2) K *is transitive on $\mathcal{B}(D)$ and 5-transitive and faithful on D.*

(3) G *is transitive on dodecads.*

(4) $|K| = 12 \cdot 11 \cdot 10 \cdot 9 \cdot 8 = 2^6 \cdot 3^3 \cdot 5 \cdot 11$.

(5) *For $S \in \mathcal{B}(D)$, $N_K(S) = S_6$ is faithful on S and $D + S$.*

Proof: By 19.3.3, each 5-subset of D is in a unique member of $\mathcal{B}(D)$, so $(D, \mathcal{B}(D))$ is a Steiner system $S(12, 6, 5)$. By 19.2, G is transitive on pairs A, B of octads with $A + B$ a dodecad, so (3) holds and K is transitive on $\mathcal{B}(D)$. Further if $A + B = D$ then by 19.2.5, $N_K(A)$ acts faithfully as the symmetric group on $A \cap D$ and on $B \cap D = D + (A \cap D)$. Of course

$A \cap D \in \mathcal{B}(D)$, so (5) holds. Now (5) and the transitivity of K on $\mathcal{B}(D)$ show K is 5-transitive on D, completing the proof of (2).

Finally as K is 5-transitive on D, $|K| = 12 \cdot 11 \cdot 10 \cdot 9 \cdot 8 \cdot |K_U|$, for U a 5-subset of $A \cap D$. But $K_U = N_K(A \cap D)_U = 1$, so (4) holds.

Given a dodecad D and subsets D_i of D of order i, $0 \le i \le 4$, denote by M_{12-i} the pointwise stabilizer in $N_G(D)$ of D_i. Thus $M_{12} = N_G(D)$. The groups $M_{24}, M_{23}, M_{22}, M_{12}, M_{11}$ are the *Mathieu groups*. Notice that each Mathieu group is t-transitive on its t-design for $t = 3, 4,$ or 5. It turns out that the Mathieu groups, the symmetric groups, and the alternating groups are the only 4-transitive groups. So far this has only been proved using the classification of the finite simple groups.

We have already seen that the first three Mathieu groups are simple. By Exercise 6.1, M_{12} and M_{11} are also simple, while M_{10} has a unique minimal normal subgroup, and that subgroup is isomorphic to A_6.

Lemma 19.5: *Let D be a dodecad. Then*

(1) $D + X$ is a dodecad.

(2)   There exists $g \in G$ interchanging D and $D + X$.

Proof: By definition $D = A + B$ for suitable octads A, B. Let $C \in \mathcal{C}_0(B)$ and $C' = X + B + C$. Then C' is an octad by 19.2.7. Now $D \cap C = A \cap C$ is of even order, as is $A \cap C'$. Also $(A \cap C) + (A \cap C') = A \cap D$. So either $A \cap C$ or $A \cap C'$ is of order 4, say $A \cap C$. Thus $A + C = C^*$ is an octad. Finally $D + X = (A + B) + (B + C + C') = A + C + C' = C^* + C'$, so $D + X$ is a dodecad.

By 19.4.3 there exists $g \in G$ with $Dg = D + X$. Then $(D + X)g = Dg + X = D + X + X = D$, so g interchanges D and $D + X$.

Lemma 19.6: *Let D be a dodecad, $K = N_G(D)$, and B an octad. Then*

(1) $|D \cap B| = 2, 4,$ or 6.

(2) K is transitive on $\mathcal{C}_n(D)$ for each n.

(3)   If $B \in \mathcal{C}_6(D)$ then $N_K(B) \cong S_6$ is faithful on $B \cap D$ and $D - B$. Also $B + D$ is an octad.

(4)   If $B \in \mathcal{C}_2(D)$ then $N_K(B) \cong S_6$ is transitive on $B \cap D$ and faithful and 2-transitive on $D - B$. Further $B + D + X$ is an octad.

(5)   If $B \in \mathcal{C}_4(D)$ then $N_K(B) = N_K(B \cap D)$ and $B + D$ is a dodecad.

(6) $\mathcal{C}_n(D)$ is of order 132, 495, 132, for $n = 2, 4, 6$, respectively.

Proof: As K is transitive on $\mathcal{B}(D)$, K is transitive on $\mathcal{C}_6(D)$. Further $|\mathcal{C}_6(D)| = \binom{12}{5}/6 = 132$. Also (3) holds by 19.4.5.

Let $D' = D + X$. Then $C_2(D) = C_6(D')$, so $|C_2(D)| = 132$ and $N_K(B) \cong S_6$ for $B \in C_2(D)$. Also $D + B + X = D' + B$ is an octad. By 19.2.5, $N_K(B)$ is transitive on $B - D' = B \cap D$. As K is sharply 5-transitive on D, $K_{B \cap D}$ is 3-transitive on $D - B$ and $|K_{B \cap D}| = 10 \cdot 9 \cdot 8 = |N_K(B)|$. Thus $N_K(B)_{B \cap D}$ is of index 2 in $K_{B \cap D}$ and hence is 2-transitive on $D - B$ by 2.2.4. So (4) is established.

By 19.3, $|B \cap D|$ is even for each octad B, but no octad is contained in D or D'. So (1) holds. Thus as $|C_2(D)| = |C_6(D')| = 132$ and $|C| = 759$, $|C_4(D)| = 495$. But $495 = \binom{12}{4}$, so the map $B \mapsto B \cap D$ is a bijection of $C_4(D)$ with the set of 4-subsets of D. Hence as K is 4-transitive on D, K is transitive on $C_4(D)$. This completes the proof of (2) and (6) and shows $N_K(B) = N_K(B \cap D)$ for $B \in C_4(D)$. Further $B \cap D \leq A \cap D$ for some $A \in C_6(D)$. Now $C = A + D$ is an octad and $A \cap B = B \cap D$ is of order 4, so $A + B = B'$ is an octad. Notice $|B' \cap C| = 2$, so $B' + C$ is a dodecad. Then $B + D = B + (A + C) = (B + A) + C = B' + C$ is a dodecad.

We now use the detailed knowledge of the action of G on its octads and dodecads which we have developed to construct and study several $GF(2)G$-modules. These modules can be studied using the theory of codes and forms in Chapter 3, which in turn endow the modules with a geometric structure in the sense of Section 4. The resulting algebraic and geometric structure is interesting in its own right, but is also an important tool in studying the Mathieu groups.

Denote by V_C the subspace of V generated by C. The error correcting code (cf. Section 10) (V, X, V_C) is the *extended binary Golay code*. As G permutes C, G acts on V_C, so V_C is a $GF(2)G$-module.

Lemma 19.7: *(1) V_C is of dimension 12.*

(2) G has 5 orbits on the vectors of V_C: 0, X, the octads, the dodecads, and the complements of the octads.

Proof: We know G is transitive on octads and on dodecads, so the five subsets of V listed in (2) are orbits of G on V_C. Further there are 759 octads and hence 759 complements. By 19.5, the complement of a dodecad is a dodecad. Also there are $|G : M_{12}| = 2{,}576$ dodecads. Thus the union S of the five orbits of (2) is of order $4{,}096 = 2^{12}$. So to complete the proof, it remains to show each element of V_C is in S.

Let $v \in V_C$. Then $v = v_1 + \cdots + v_n$ with $v_i \in C$. We prove $v \in S$ by induction on n. As $0 \in S$ we may take $n \geq 1$. Then $v = u + v_n$ with $u \in S$ by induction on n. Indeed u or $u + X$ is 0, an octad, or a dodecad, so as $v \in S$ if and only if $v + X \in S$, without loss of generality u is 0, an octad, or a dodecad. If $u = 0$ then $v = v_n$ is an octad. If u is an octad then by 19.2, $u + v_n \in S$. Finally if u is a dodecad then by 19.6, $u + v_n \in S$.

Observe that (V, X, V_C) is a $(24, 12)$-code with minimum weight 8. Thus the code corrects 3 errors. As $G = Aut(X, C)$, G is the group of automorphisms of the code. Pick $x \in X$ and let $X' = X - \{x\}$, $V' = \langle X' \rangle$, and V_C' the projection of V_C on V' with respect to the decomposition $V = \langle x \rangle \oplus V'$. Then (V', X', V_C') is a $(23, 12)$-code with minimum weight 7 which corrects 3 errors. This code is the *binary Golay code*. It turns out the binary Golay code is perfect (in the sense of Section 10) with automorphism group M_{23}. The proof of these remarks is left as an exercise.

Set $\bar{V} = V/\langle X \rangle$. Define $\bar{V}_C$ to be the *11-dimensional Golay code module* for $G = M_{24}$.

Lemma 19.8: *(1) G has exactly two orbits on $\bar{V}_C^{\#}$: the images of the octads, of length 759, and the images of the dodecads, of length 1,288.*
(2) If B is an octad then $N_G(\bar{B}) = N_G(B) \cong A_8/E_{16}$.
(3) If D is a dodecad then $N_G(\bar{D}) = N_G(\{D, D + X\}) \cong \mathbf{Z}_2/M_{12}$.
(4) $\bar{V}_C$ is an irreducible $GF(2)G$-module.

Proof: From 19.7.2, each member of $V_C^{\#}$ is of the form $X + v$, with $v = 0$, an octad, or a dodecad. So each element of $\bar{V}^{\#}$ is the image of an octad or a dodecad. As G is transitive on octads and dodecads, their images are the orbits of G on $\bar{V}_C^{\#}$. If B is an octad then $B + X$ is not an octad, so $N_G(\bar{B}) = N_G(\{B, B + X\}) = N_G(B)$. But if D is a dodecad, 19.5 says $N_G(D)$ is of index 2 in $N_G(\bar{D})$.

If $0 \neq \bar{U}$ is a G-invariant subspace of $\bar{V}_C$ then $\bar{U}^{\#}$ is a union of orbits of G. So if $\bar{U} \neq \bar{V}_C$, then $\bar{U}^{\#}$ is the set of images of octads or dodecads, and hence of order 759 or 1,288, respectively. But then $|\bar{U}|$ is not a power of 2.

Define a *sextet* to be a partition Δ of X into six 4-subsets such that for each pair of distinct S, T from Δ, $S + T$ is an octad.

Lemma 19.9: *Let S be a 4-subset of X. Then*

(1) S is contained in a unique sextet $\Delta(S)$.
(2) For each 3-subset T of S,

$$\Delta(S) = \{S, L - \{s\} : L \in \mathcal{L}\},$$

where $\mathcal{L}$ is the set of lines through $s \in S - T$ in the projective plane $X - T$.

(3)   G is transitive on sextets.

(4)   If Δ is a sextet then $N_G(\Delta)^\Delta = S_6$ and $N_G(\Delta)_\Delta$ is the split extension of E_{64} by a self-centralizing subgroup of order 3.

(5)   There are exactly $\binom{24}{4}/6 = 1{,}771$ sextets.

Proof: Let T be a 3-subset of S, $s \in S - T$, and $\mathcal{L}$ the set of five lines through s in the projective plane π on $X - T$. Then $T + l \in \mathcal{C}$ for each $l \in \mathcal{L}$ by construction of $\mathcal{C}$. Define $\Delta(S)$ as in (2). Then for $S \neq R \in \Delta(S)$, $R + s = l(R) \in \mathcal{L}$, so $S + R = T + l(R)$ is an octad. Also if $R' \in \Delta(S) - \{R, S\}$ then $R + R'$ is the sum of the octads $S + R$ and $S + R'$ with $(S + R) \cap (S + R') = S$ of order 4. Thus $R + R'$ is an octad by 19.2.8. Hence $\Delta = \Delta(S)$ is a sextet.

Suppose Δ' is a sextet through S. For $x \in X - S$ there are unique 4-sets $R \in \Delta$ and $R' \in \Delta'$ containing x. Then $S + R = B(S + x) = S + R'$, so $R = R'$. Hence $\Delta = \Delta'$. This completes the proof of (1) and (2). Moreover (1) and 4-transitivity of G on X imply G is transitive on pairs (S, Δ) such that S is a 4-set and Δ a sextet through S. In particular (3) holds and $N_G(\Delta)$ is transitive on Δ.

By (1) there are $\binom{24}{4}/6 = 1{,}771$ sextets; that is, (5) holds.

Let $T = \{t_1, t_2, t_3\}$ and $H = G_S$. Notice $G_T = L_3(4)$ acts faithfully on π with H the split extension of $E = H_\Delta \cong E_{16}$ by $A_5 = Alt(\Delta - \{S\})$. As $S + R = B$ is an octad, $N_G(B)^B = A_8$, so there is an involution $g \in N_G(B)$ acting as (t_1, t_2) on T and fixing exactly 2 points of R. Then $Tg = T$, so g acts on π and fixes exactly 3 points on the line $\{s, R\}$. So by Exercise 1.3, $\langle g, G_T \rangle^\pi$ is $L_3(4)$ extended by a field automorphism and $\langle H, g \rangle^\Delta = S_5 = Sym(\Delta - \{S\})$. Thus as $N_G(\Delta)$ is transitive on Δ, $N_G(\Delta)^\Delta = Sym(\Delta) = S_6$.

Let $K = N_G(\Delta)_\Delta$ be the kernel of the action of $N_G(\Delta)$ on Δ. Then $K_B = E_B \cong E_4$, so by 2-transitivity of $N_G(\Delta)$ on Δ, $K_A \cong E_4$, where $A = R + R'$, $R' \in \Delta - \{S, R\}$. As E is faithful on A, $K_A \cap E = 1$. Also K_A and E are normal in K, so $M = \langle K_A, E \rangle = K_A \times E \cong E_{64}$ and $M \trianglelefteq K$. But by (5), $|N_G(A)| = 3 \cdot 2^6 \cdot |S_6|$, so as $N_G(\Delta)/K \cong S_6$, $|K| = 3 \cdot 2^6$. Hence $M \in Syl_2(K)$ and K is the split extension of M by P of order 3. By 4-transitivity of G, $N_G(S)^S = S_4$. Also $M^R \cong E_4$, so $M^S \cong E_4$ by symmetry. Thus $(PM)^S = K^S \cong A_4$ since $E_4 \trianglelefteq K^S \leq S_4$. Thus $C_M(P) \leq K_S$. Then by transitivity of $N_G(\Delta)$ on Δ, $C_M(P) = 1$.

Set $\tilde{V} = V/V_\mathcal{C}$. The module $\tilde{V}$ is the *12-dimensional Todd module* for $G = M_{24}$. Notice that if $U = core(V)$ is the subspace of vectors of V of even order then $\tilde{U}$ is an 11-dimensional $GF(2)G$-submodule of $\tilde{V}$ called the *11-dimensional Todd module* for M_{24}.

Lemma 19.10: *(1) G has four orbits $\mathcal{V}_n$, $1 \leq n \leq 4$, on $\tilde{V}^{\#}$, consisting of the images of the n-subsets of V.*

(2) $|\mathcal{V}_n| = \binom{24}{n}$ for $n = 1, 2, 3$, and $|\mathcal{V}_4| = \binom{24}{4}/6 = 1{,}771$.

(3) Each coset in $\mathcal{V}_n$ contains a unique n-set for $n \leq 3$, while the 4-sets in a coset of $\mathcal{V}_4$ form a sextet.

(4) Let $U = core(V)$; then $\tilde{U}^{\#} = \mathcal{V}_2 \cup \mathcal{V}_4$.

Proof: Let $\tilde{v} \in \tilde{V}^{\#}$ with v of weight $n \leq 4$. Then for $u \in V_C^{\#}$, $|u + v| \geq |u| - |v| \geq 4$ with equality if and only if $u \in \mathcal{C}$, $n = 4$, and $v \subseteq u$. This proves that if $n \leq 3$ then v is the unique element of weight at most 4 in the coset $\tilde{v} = v + V_C$. Further for $n = 4$, the set of elements in $\tilde{v}$ of weight at most 4 are v plus those of the form $u + v$, $u \in \mathcal{C}$, $v \subseteq u$. But by 19.9.3, these elements are just the members of $\Delta(v)$. So (3) holds. Further as the number of n-sets is $\binom{24}{n}$, we also have (2). By (2),

$$\sum_{n \leq 4} |\mathcal{V}_n| = 2^{12} - 1 = |\tilde{V}^{\#}|,$$

so (1) holds. As the members of $core(V)$ are the vectors of even weight, (4) holds.

Observe that 19.10 shows each vector v in V is of the form $u + w$ for some w of weight at most 4 and some $u \in V_C$. Thus $d(u, v) = |u + v| = |w| \leq 4$. So each vector in V is in a sphere of diameter 4 from some code word in the extended Golay code.

Lemma 19.11: *Let Δ be a sextet, $L = N_G(\Delta)$, and $K = L_\Delta$. Then*

(1) $\mathcal{C}_i(\Delta)$, $1 \leq i \leq 3$, are the orbits of L on $\mathcal{C}$.

(2) $\mathcal{C}_i(\Delta)$ is of order 15, 360, 384, for $n = 1$, 2, 3, respectively.

(3) $\mathcal{C}_1(\Delta)$ consists of those B with $B = S + T$ for some $S, T \in \Delta$. Thus $N_L(B) = N_L(\{S, T\})$.

(4) $\mathcal{C}_2(\Delta)$ consists of those B with $|B \cap R| = 2$ for all $R \in \Delta - \{S, T\}$ for some $S, T \in \Delta$. $N_L(B)$ is the split extension of $N_K(B) \cong E_8$ by $N_L(B)^\Delta = N_L(\{S, T\})^\Delta$.

(5) $\mathcal{C}_3(\Delta)$ consists of those B such that $|B \cap S| = 3$ for some $S \in \Delta$ and $|B \cap T| = 1$ for $T \in \Delta - \{S\}$. $N_K(B) \cong \mathbf{Z}_3$ and $N_L(B)^\Delta = N_L(S)^\Delta$.

Proof: By 19.9, $\mathcal{C}_1(\Delta)$ is an orbit of L of length 15 and (3) holds.

Let $S, T \in \Delta$ and $A = S + T \in \mathcal{C}_1(\Delta)$. Then $\mathcal{C}_0(A)$ is of order 30 by 19.2 and hence the set $\mathcal{C}_2(\Delta, A)$ of octads in $\mathcal{C}_0(A)$ but not in $\mathcal{C}_1(\Delta)$ is of order 24. As $|B \cap C|$ is even for each $B \in \mathcal{C}_2(\Delta, A)$ and $C \in \mathcal{C}_1(\Delta)$, we conclude $|R \cap B| = 2$ for each $R \in \Delta - \{S, T\}$. Thus

$$C_2(\Delta) = \bigcup_{A \in C_1(\Delta)} C_2(\Delta, A)$$

is of order $15 \cdot 24 = 360$.

Let $S_i \in \Delta$ intersect A trivially, $1 \leq i \leq 3$, and let $P = O_2(K)$. Then $N_K(B)$ acts on $B \cap S_i$ of order 2 so $N_K(B) \leq N_P(B \cap S_i)$ and $|P : N_P(B \cap S_i)| = 2$. Then $N_K(B) \leq \bigcap_i N_P(B \cap S_i)$, which is of index 24 in K. So $|K : N_K(B)| \geq 24 = |C_2(\Delta, A)|$, and hence K is transitive on $C_2(\Delta, A)$. Thus as L is transitive on $C_1(\Delta)$, L is transitive on $C_2(\Delta)$ and (4) holds.

Finally let $x_i \in S_i$ and $B = B(S_1 + x_1 + x_2 + x_3)$. As $|B \cap (S_1 + R)|$ is even for all $R \in \Delta$, we conclude $B \cap R$ is a point for each $R \in \Delta - \{S_1\}$ and $S_1 + x_1 = B \cap S_1$ is of order 3. That is, $B \in C_3(\Delta)$. Also K is transitive on triples (x_1, x_2, x_3) with $x_i \in S_i$ and hence transitive on

$$\Omega = \{B \in C_3(\Delta) : |B \cap S_1| = 3\}$$

of order 64. So $|N_K(B)| = |K|/|\Omega| = 3$, and hence as L is transitive on Δ, L is transitive on $C_3(\Delta)$ and (5) holds.

Finally $\sum_i |C_i(\Delta)| = 759 = |C|$, so (1) holds, completing the proof.

Remarks. The Mathieu groups made their first appearance in two papers of Mathieu in 1860 [M1] and 1861 [M2]. However, M_{24} was mentioned only briefly and Mathieu did not supply details about the group until 1873 [M3]. Even then there remained uncertainty about the group. For example, in 1898 [Mi1], G. A. Miller published what purported to be a proof that M_{24} did not exist, although a year later in [Mi2] he had realized his mistake.

In 1938, Witt published two papers [W1], [W2] studying the Mathieu groups from the point of view of their Steiner systems. This may have been the first fairly rigorous existence proof for M_{24}.

J. Todd's papers in 1959 [To1] and 1966 [To2] seem to represent the first serious attempt to study the Steiner systems of the Mathieu groups. In particular Todd begins the study of the Todd module, and indirectly the Golay code module, as the latter module is the dual of the Todd module. It is in these papers that the terminology of *octads*, *sextets*, etc. is introduced and most of the results of Section 19 first proved.

The existence of the invariant submodule V_C of the 24-dimensional binary permutation module was recognized at least implicitly by Paige [Pa] in 1957, two years before Todd's first paper. The Golay code had been discovered in 1949 by Golay in [G], although he did not realize it was invariant under the Mathieu group M_{23}.

Exercises

1. The Mathieu groups M_{11} and M_{12} are simple while $F^*(M_{10}) \cong A_6$.

2. Assume the hypotheses and notation of Lemma 18.7 and let $T \in I_3$. Prove
   (1) There exist $\bar{T}, T', T^* \in I^3$ with $I^2(T) = T + T'$, $I^3(T) = T + \bar{T}$, and $I^3(T') = T' + T^*$.
   (2) $\bar{T} + T^* \in I_6^2$, so $I^2(\bar{T}) = \bar{T} + T^*$.

3. Let $(X, \mathcal{C})$ be the Steiner system $S(24, 8, 5)$ for M_{24} and let x and y be distinct points of x. Prove
   (1) There are exactly 330 octads missing x and y.
   (2) The number of members of $V_{\mathcal{C}}$ containing exactly one of x and y is 2^{11}.

4. Prove the binary Golay code is a perfect $(23, 12)$-code with minimum weight 7 which corrects three errors. Its automorphism group is M_{23}.

5. Let G be a group t-transitive on a set X of order n, T a t-subset of X, and B a k-subset of X with $T \subseteq B$ such that $G(T) \leq G(B)$ and $G(B)$ is t-transitive on B. Let $\mathcal{B} = \{Bg : g \in G\}$. Prove $(X, \mathcal{B})$ is a Steiner system $S(n, k, t)$.

6. Let $X_9 = (X_9, \mathcal{B}_9)$ be the affine plane over $GF(3)$. Prove
   (1) X_9 is a Steiner system $S(9, 3, 2)$.
   (2) $Aut(\mathcal{B}_9) = HE$ is a split extension of E by H, where $E \cong E_9$ is the translation group and $H \cong GL_2(3)$.
   (3) Let $Q_8 \cong A \leq H$. Prove X_9, A satisfies the Extension Hypothesis.
   (4) Let X_{10} be the projective plane over $GF(9)$ and $\mathcal{B}_{10}$ the collection of translates of the projective line $B = \{\infty, 0, 1, -1\}$ over $GF(3)$ under $M = P\Gamma L_2(9)$. Prove $X_{10} = (X_{10}, \mathcal{B}_{10})$ is a Steiner system $S(10, 4, 3)$, $M = Aut(X_{10})$, and X_{10} is the unique extension X of X_9 with $A \leq N_{Aut(X)}(X_9)$.
   (5) Prove X_{10}, AE satisfy the Extension Hypothesis and there exists a unique extension $X_{11} = (X_{11}, \mathcal{B}_{11})$ of X_{10} admitting AE. Moreover $Aut(X_{11}) = M_{11}$.
   (6) Prove X_{11}, M_{11} satisfy the Extension Hypothesis and there exists a unique extension $X_{12} = (X_{12}, \mathcal{B}_{12})$ of X_{11} admitting M_{11}. Moreover $Aut(X_{12}) = M_{12}$.
   (7) M_{11}, M_{12} are the unique 4,5 transitive groups with point stabilizer M_{10}, M_{11}, respectively.
   (Hint: To prove (7) use (5) and (6) plus Exercise 6.5.)

Chapter 7

The Geometry and
Structure of M_{24}

In Lemma 11.8 we saw that if (V, X, U) is a strictly doubly even binary code then there exists a triple (P, C, f) of forms on U inducing a geometric structure on U. In this chapter we study the geometry induced on the Golay code module $\bar{V}_C$ in this fashion. In particular we find that the octads are the singular points in this module with the trios corresponding to the singular lines and the sextets to maximal hyperbolic subspaces. The *2-local geometry* for M_{24} is the geometry of octads, trios, and sextets with incidence defined by inclusion, subject to the identification of these objects with subspaces of $\bar{V}_C$ just described. In Section 20 we derive various properties of this geometry which will be used in Section 21 to investigate the basic local structure of M_{24}, and later to establish the uniqueness of M_{24} and other sporadics.

20. The geometry of M_{24}

In this section we continue the hypothesis and notation of Sections 18 and 19. In particular G is the Mathieu group M_{24} and $(X, \mathcal{C})$ is the Steiner system for G. Recall V is the power set of X regarded as a $GF(2)G$-module via the operation of symmetric difference. Also V_C is the 12-dimensional Golay code submodule of V generated by the octads and $\bar{V}_C = V_C/\langle X \rangle$ is the 11-dimensional Golay code module.

By 19.7, (V, X, V_C) is a doubly even code so by 11.8 there is a triple of forms P, C, f defined on V_C via

$$P(x) = |x|/4 \bmod 2,$$
$$C(x,y) = |x \cap y|/2 \bmod 2, \quad f(x,y,z)$$
$$= |x \cap y \cap z| \bmod 2.$$

Moreover by 11.8:

Lemma 20.1: *(1)* f *is a symmetric trilinear form with* $f(x,x,y) = 0$ *for all* $x, y \in V_C$.

(2) $\mathrm{Rad}(V_C) = \langle X \rangle$, *so we have induced forms on* $\bar{V}_C$ *which we also write as* P, C, f.

For $x \in \bar{V}_C$, f_x is the bilinear form $f_x(y,z) = f(x,y,z)$, C_x is the quadratic form $C_x(y) = C(x,y)$ (which has f_x as its associated bilinear form) and $R(x) = \mathrm{Rad}(f_x)$. In Section 11 we saw how this algebraic structure on $\bar{V}_C$ can be used to define a geometric structure preserved by the isometry group of P. In particular M_{24} preserves this geometry, which is an excellent tool for studying the group.

For example, recall from Section 11 that a subspace U of $\bar{V}_C$ is *singular* if P is trivial on U and U is singular with respect to f_x for all $x \in V$. Further given $S \subseteq \bar{V}_C$, $S\theta$ consists of those $v \in \bar{V}_C$ such that $f_v = 0$ on S. Recall from the discussion in Section 11 that $S\theta$ is a subspace of $\bar{V}_C$. Further a subspace U of $\bar{V}_C$ is *subhyperbolic* if $U\theta$ is a hyperplane of $\bar{V}_C$ and P is trivial on U.

Define a *trio* to be a triple $\{B_1, B_2, B_3\}$ of octads such that X is the disjoint union of B_1, B_2, and B_3. Thus $B_i \in C_0(B_j)$ for each $i \neq j$ and $B_3 = B_1 + B_2 + X$.

Lemma 20.2: *(1)* G *is transitive on the* $3,795 = 23 \cdot 11 \cdot 5 \cdot 3$ *trios.*

(2) The stabilizer L *of a trio* $\{A, B, A + B + X\}$ *is the split extension of* E_{64} *by* $S_3 \times L_3(2)$.

(3) Let Δ *be a sextet such that* $A, B \in C_1(\Delta)$. *Then* $L = N_L(B)N_L(\Delta)$ *with* $|L : N_L(B)| = 3$ *and* $|L : N_L(\Delta)| = 7$.

Proof: Let $\mathcal{T} = \{A, B, A + B + X\}$ be a trio. Part (1) follows as G is transitive on its 759 octads and $C_0(B)$ is an orbit for $M = N_G(B)$ of length 30. Also this shows $|L : L \cap M| = 3$ and $L^{\mathcal{T}} = S_3$. By 19.2.6, $L_{\mathcal{T}}$ is the split extension of $Q_{\mathcal{T}} \cong E_8$ by $L_3(2)/E_8$, so as $L^{\mathcal{T}} = S_3$, $O_2(L_{\mathcal{T}}) = Q_{\mathcal{T}} \times Q_{\mathcal{T}}^g \cong E_{64}$ for $g \in L - N_L(B)$. Then (2) holds. Finally $N_L(\Delta)$ induces S_3 on $\mathcal{T}$ and $|N_G(\Delta) : N_L(\Delta)| = 15$ by 19.9.4, so (3) holds.

Lemma 20.3: *(1) The octads are the singular points of* $\bar{V}_C$ *and the dodecads are the nonsingular points.*

(2) G has three orbits on lines of $\bar{V}_C$ generated by octads: the singular lines, the hyperbolic lines, and the lines $\langle \bar{A}, \bar{B} \rangle$ with $A \in C_2(B)$.

(3) The singular lines of $\bar{V}_C$ are the lines l with $l^\#$ the octads in a trio.

(4) The hyperbolic lines of $\bar{V}_C$ are the lines $\langle \bar{A}, \bar{B} \rangle$ with $A \in C_4(B)$.

Proof: By 19.8, G has two orbits on $\bar{V}_C^\#$: The octads and the dodecads. So as $|v| \equiv 0, 2 \mod 4$ for v an octad, dodecad, respectively, (1) holds.

Next by 19.2, G has three orbits on lines of $\bar{V}_C$ generated by singular points, and those orbits have representatives $\langle \bar{B}, \bar{A}_i \rangle$, where $A_i \in C_i(B)$, $i = 0, 2, 4$. Now $B + A_2$ is a dodecad so $P(\bar{B} + \bar{A}_2) = 1$, while $\bar{B} + \bar{A}_i$ is an octad for $i = 0, 4$. By 19.11 there exists an octad C with $|B \cap A_4 \cap C|$ odd, so $f(\bar{B}, \bar{A}_4, \bar{C}) = 1$ and hence $\langle \bar{B}, \bar{A}_4 \rangle$ is hyperbolic. Finally as $B \cap A_0 = \varnothing$, $|C \cap B \cap A_0|$ is even for all $C \in \mathcal{C}$, so $\langle \bar{B}, \bar{A}_0 \rangle$ is singular.

Lemma 20.4: *Let Δ be a sextet. Then*

(1) $\{\bar{B} : B \in C_1(\Delta)\} = U^\#$ for some 4-dimensional subspace $U = U(\Delta)$ of $\bar{V}_C$.

(2) U is subhyperbolic.

(3) $N_G(U)$ is irreducible on U, $U\theta/U$, and $\bar{V}_C/U\theta$ of dimension 4, 6, 1, respectively.

(4)   Let $f_U = f_v$ for $v \in V - U\theta$. Then (U, f_U) is a 4-dimensional symplectic space over F and $N_G(U)$ induces the symplectic group $O(U, f_u) \cong Sp_4(2)$ on U.

(5)   For $A \leq U$, $R(A) \cap U$ is the subspace of U orthogonal to A under f_U.

(6)   If $P \in Syl_3(C_G(U))$ then $N_G(P) \leq N_G(U)$ and $C_G(P) \cong A_6/\mathbf{Z}_3$ is quasisimple.

Proof: For distinct $A, B \in C_1(\Delta)$, $A + B$ or $A + B + X$ is in $C_1(\Delta)$, so as $|C_1(\Delta)| = 15$, (1) holds.

Let $L = N_G(U)$ and $K = L_\Delta$ as in 19.9, and take $P \in Syl_3(K)$. By 19.9.4, $C_L(P) \cong A_6/\mathbf{Z}_3$ and $O_2(K) \cong E_{64}$. Thus $O_2(L)$ is a faithful 3-dimensional $GF(4)C_L(P)$-module, so a Sylow 3-subgroup R of $C_L(P)$ is Sylow in $SL_3(4)$, and hence is isomorphic to 3^{1+2}. Therefore $C_L(P)$ is quasisimple.

As K fixes each member of $C_1(\Delta)$, $K = C_G(U)$ and K acts on the hyperplane $\langle \bar{A}, \bar{B} \rangle\theta$ of $\bar{V}_C$ for $\langle \bar{A}, \bar{B} \rangle$ a hyperbolic line in U. Thus $dim(C_{\bar{V}_C}(P)) \geq dim(U) + 1 = 5$, so $dim([\bar{V}_C, P]) \leq 6$. So as 6 is the minimal dimension of a faithful $GF(2)R$-module, $dim([\bar{V}_C, P]) = 6$ and

$C_L(P)$ is irreducible on $([\bar{V}_C, P] + U)/U$ of dimension 6. In particular $[\bar{V}_C, P] + U$ is generated by octads.

Now by 19.11, if $C \in \mathcal{C}_3(\Delta)$ then $|A \cap B \cap C|$ is odd, so $f(\bar{A}, \bar{B}, \bar{C}) = 1$ and $\bar{C} \notin U\theta$, while if $C \notin \mathcal{C}_3(\Delta)$ then $f(\bar{A}, \bar{B}, \bar{C}) = 0$, so $\bar{C} \leq U\theta$. Thus as $[\bar{V}_C, P] + U$ is a hyperplane generated by octads, $U\theta = [\bar{V}_C, P] + U$, so (2) and (3) hold. Then (4) and (5) hold by 11.6 and 19.9.

We also see that $dim(C_{\bar{V}_C}(P)) = 5$ and U is the unique subhyperbolic hyperplane of $C_{\bar{V}_C}(P)$, so $N_G(P) \leq L$, completing the proof of (6).

Using 20.3 and 20.4, we identify the octads, trios, and sextets with subspaces of $\bar{V}_C$ of dimension 1,2,4, respectively.

Proceeding as in Section 4, let Γ be the geometry on $I = \{1, 2, 3\}$ whose objects of type i are the octads, trios, sextets, for $i = 1, 2, 3$, respectively, and with incidence equal to inclusion, subject to our identification of the objects of Γ with subspaces of $\bar{V}_C$. The geometry Γ is the *2-local geometry* of M_{24}, since the stabilizers of objects in a chamber are the maximal 2-local subgroups of M_{24} containing a Sylow 2-subgroup of M_{24} fixing the chamber.

The identification of Γ with subspaces of $\bar{V}_C$ gives an injection of geometries $\Gamma \to PG(\bar{V}_C)$ embedding Γ in the projective geometry of $\bar{V}_C$. The algebraic structure supplied by the forms P, C, f allows us to distinguish subspaces of $\bar{V}_C$ and hence use the embedding of Γ in $PG(\bar{V}_C)$ effectively.

We say a pair of octads are *collinear* if they are incident with a common trio and octads or trios are *coplanar* if they are incident with a common sextet. See Section 4 for definitions of geometric terminology such as "residue," etc. See Chapter 7 in [FGT] for a discussion of symplectic and orthogonal spaces.

Lemma 20.5: *(1) Γ is a residually connected string geometry.*

(2) G is flag transitive on Γ.

(3) The residue of an octad is isomorphic to the points and lines of 4-dimensional projective space over $GF(2)$.

(4) The residue of a sextet is isomorphic to the singular points and lines of 4-dimensional symplectic space over $GF(2)$.

(5) There are three octads and seven sextets incident with each trio.

Proof: Part (4) follows from 20.4.4 and 20.4.5 and part (3) follows from 20.6.1 and 20.6.3 below. Moreover these lemmas show the stabilizer of an octad or sextet is flag transitive on its residue. Notice (3) and (4) say the residues of octads and sextets are connected. By 20.2.3 part (5) holds and the residue of a trio is a generalized digon with the stabilizer of the trio flag transitive on the residue. Thus to complete the proof it suffices to prove Γ is connected. But from 19.2, G is primitive on octads, so Γ is connected.

Lemma 20.6: *If x is an octad then*

(1) *$dim(R(x)) = 5$, G_x is the stabilizer of x in $GL(R(x))$, and $R(x)$ is the union of the trios through x.*

(2) *$(\bar{V}_C/R(x), C_x)$ is a 6-dimensional orthogonal space of sign $+1$, G_x acts as $\Omega_6^+(2)$ on $\bar{V}_C/R(x)$, and for $U \in \Gamma_3(x)$, $U+R(x)/R(x)$ is a singular point of $\bar{V}_C/R(x)$ and $U\theta/R(x)$ is the subspace orthogonal to that point.*

(3) *The map $U \to R(x) \cap U$ is a bijection between $\Gamma_3(x)$ and the set of 3-subspaces of $R(x)$ containing x.*

(4) *Each pair of distinct coplaner trios is contained in a unique sextet.*

(5) *If $U \in \Gamma_3(x)$ then $G_{x,U}$ has three orbits on $\Gamma_3(x)$: $\{U\}$, the eighteen sextets W in $U\theta$ with $U \cap W \in \Gamma_2$, and the sixteen sextets Z not contained in $U\theta$ with $Z \cap U = \langle x \rangle$.*

(6) *Each hyperbolic line h is contained in a unique sextet U, and $U = h \oplus R(h)$.*

Proof: Adopt the notation of 19.2 with $B = x$ and (x, l, U) a chamber. Thus, for example, $M = G_x$, and by 19.2, $M/Q \cong L_4(2)$, while by 19.2.6, M_l/Q is the stabilizer of a point in the natural module for M/Q. In particular M is therefore 2-transitive on $\Gamma_2(x)$. But by 20.5.4 and 20.5.5, $\overline{A+C}$ is an octad if $l = \langle \bar{B}, \bar{A} \rangle$ and $k = \langle \bar{B}, \bar{C} \rangle$ are trios in U, so $\overline{A+C}$ is an octad for all distinct $A, C \in C_0(B)$. Hence as $|C_0(B)| = 30$, (1) holds.

Next by 20.4, $R(x) \cap U$ is the hyperplane of U orthogonal to x under f_U, so $\tilde{U}$ is a point in $\tilde{V}_C = \bar{V}_C/R(u)$. Further $(\tilde{V}_C, C_x)$ is a nondegenerate orthogonal space of dimension 6 and G_x induces $\Omega_6^+(2)$ on $\tilde{V}_C$, so the space is of sign $+1$. By definition of C_x, $\tilde{U}$ is a singular point in $\tilde{V}_C$ and $\tilde{U}\theta = \tilde{U}^\perp$. Then as M_U is the parabolic subgroup of M stabilizing $R(x) \cap U$ and $\tilde{U}$, (3) and (5) hold.

If l and m are trios in U then either $l + m = U$ or $l \cap m$ is an octad which we take to be x. In the first case clearly U is the unique sextet containing l and m. In the second $l + m = R(x) \cap U$, so U is unique by (3). Thus (4) is established.

It remains to prove (6). As each sextet contains a hyperbolic line h and G is transitive on hyperbolic lines, we may take $h \leq U$. Then $U\theta \leq h\theta$, so as $U\theta$ is a hyperplane, $U\theta = h\theta$. Then as G is transitive on sextets and $N_G(U) = N_G(U\theta)$, U is the unique sextet containing h. By 20.5, $U \cap R(h) = k$ is the line orthogonal to h in U, so $U = h \oplus k$. Further if $x \in R(h)$ then $x + h \leq W \in \Gamma_3$ by (3), so by uniqueness of U, $W = U$ and hence $x \in k$.

Given a trio l let $\zeta(l) = \langle R(x) : x \in l \rangle$.

Lemma 20.7: *Let* l *be a trio and* $x, y \in l$ *distinct octads. Then*

   (1) $dim(\zeta(l)) = 8$ *and* $\zeta(l) = R(x) + R(y)$ *with* $R(x) \cap R(y) = l$.
   (2) $\zeta(l)/l$ *is the tensor product of natural modules for the factors of* $G_l/O_2(G_l) \cong L_3(2) \times L_2(2)$.
   (3) G_l *acts as* $L_3(2)$ *on* $\bar{V}_C/\zeta(l)$.
   (4)   *Each octad in* $\zeta(l) - l$ *is in a unique* $R(u)$, $u \in l^{\#}$.

Proof: From 20.6, $G_{x,y}$ acts irreducibly on $R(x)/l$ as $L_3(2)$, so $R(x) \cap R(y) = l$. Thus $dim(R(x) + R(y)) = 8$. Let $Q(z) = O_2(G_z)$ for $z = x, y$. By 20.6, $[Q(z), \bar{V}_C] \leq R(z)$, so $S = \langle Q(x), Q(y) \rangle$ acts on $R(x) + R(y)$. Thus as $G_l = G_{x,y}S$, it follows that $\zeta(l) = R(x) + R(y)$ is of dimension 8, and (2) holds. As $R(x) \cap R(y) = l$, each octad in $\zeta(l) - l$ is in at most one $R(u)$, $u \in l^{\#}$. By (2), G_l has two orbits on vectors in $\zeta(l) - l$, and as one of these contains nonsingular points, each such octad is in at least one $R(u)$. Thus (4) is established. Finally $dim(\bar{V}_C/\zeta(l)) = 3$, so from the action of $G_{x,y}$ on $\bar{V}_C$ obtained from 20.6, (3) holds.

Lemma 20.8: *The stabilizer* G_l *of a trio* l *has five orbits* $\delta_i(l)$, $0 \leq i \leq 4$, *on* Γ_2 *as follows:*

   (1) $\delta_0(l) = \{l\}$.
   (2) $\delta_1(l) = \{k : 0 \neq k \cap l \neq l\}$ *is of order 42.*
   (3) $\delta_2(l) = \{k : k + l \in \Gamma_3\}$ *is of order 56.*
   (4) $\delta_3(l) = \{k : dim(k \cap \zeta(l)) = 1\}$ *is of order 1,008.*
   (5) $\delta_4(l) = \{k : k \cap \zeta(l) = 0\}$ *is of order 2,688.*

Proof: First G_l is transitive on the three octads on l and taking $x \in l$, by 20.6, $R(x)$ contains fifteen trios through x and G_x acts 2-transitively as $L_4(2)$ on these trios, so there are $3 \cdot 14 = 42$ trios in $\delta_1(l)$ and they form an orbit under G_l.

Second G_l is transitive on the seven sextets incident with l and if U is such a sextet, then from 20.5, $G_{U,l}$ is transitive on the eight trios of U intersecting l trivially, so there are $7 \cdot 8 = 56$ trios in $\delta_2(l)$ and they form an orbit under G_l.

Third by 20.6, $G_{l,x}$ is transitive on the 28 octads z in $R(x) - l$. By 20.6.3, $z + l$ is contained in a unique sextet U and by 20.6.3, 20.6.6, and 20.7.4, $\zeta(l) \cap R(z) = U \cap R(z)$ is a 3-subspace of $R(z)$ with $G_{z,U} = G_{z,U\cap R(z)}$. Now $C_G(U) \leq G_{z,l,U}$ is transitive on the 12 trios in $R(z) - \zeta(l)$. Thus the $3 \cdot 28 \cdot 12 = 1,008$ trios in $\delta_3(l)$ form an orbit. This leaves $2,688$ trios k with $k \cap \zeta(l) = 0$. Checking the permutation character of G on G/G_l, we find G is rank 5 on lines, and hence conclude that $\delta_4(l)$ forms the fifth orbit.

Lemma 20.9: *The stabilizer G_U of a sextet U has orbits $\Xi_i(U)$, $0 \leq i \leq 3$, on sextets, where:*

(1) $\Xi_0(U) = \{U\}$.

(2) $\Xi_1(U) = \{W : W \cap U \in \Gamma_2\}$ *is of order 90.*

(3) $\Xi_2(U) = \{W : W \cap U \in \Gamma_1\}$ *is of order 240.*

(4) $\Xi_3(U) = \{W : U \cap W = 0\}$ *is of order 1,440.*

Moreover members of $\Xi_i(U)$ are not contained in $U\theta$ for $i = 2, 3$.

Proof: As G_U is transitive on the fifteen trios $l \in \Gamma_2(U)$ and G_l is 2-transitive on the seven sextets in $\Gamma_3(l)$, $\Xi_1(U)$ is an orbit of length $15 \cdot 6 = 90$. As G_U is transitive on the fifteen octads $x \in \Gamma_1(U)$ and by 20.6.5, $G_{U,x}$ is transitive on the sixteen sextets $W \in \Gamma_3(x)$ with $\langle x \rangle = U \cap W$, $\Xi_2(U)$ is an orbit of length $15 \cdot 16 = 240$. Further by 20.6.5, $W \nleq U\theta$. This leaves 1,440 sextets W with $W \cap U = 0$. From the character table of G, G is rank 4 on G/G_U, so $\Xi_3(U)$ is the fourth orbit of G_U on planes.

Lemma 20.10: *Let $U \in \Gamma_3$, $W \in \Xi_1(U)$, and $l \in \Gamma_2(W)$ with $l \cap U = 0$. Then $\zeta(l) \cap U = W \cap U \in \Gamma_2$.*

Proof: If $x \in U \cap \zeta(l)^\#$ then by 20.7.4, $x \in R(y)$ for some $y \in l^\#$. Further if $\langle x, x' \rangle$ is a trio in U then as $U = (R(x) \cap U) + (R(x') \cap U)$, $(R(x) + U) \cap (R(x') \cap U) = U + (R(x) \cap R(x')) = U$, so $R(y) \cap U = \langle x \rangle$. Thus by 20.7.7,

$$\zeta(l) \cap U = (R(y) \cap U) \cup (R(y_1) \cap U) \cup (R(y_2) \cap U) = W \cap U,$$

where $l^\# = \{y, y_1, y_2\}$.

Lemma 20.11: *For each sextet U, G_U is transitive on the 384 octads and 640 nonsingular points in $\bar{V}_C - U\theta$, and if x is such an octad then (U, C_x) is an orthogonal space of sign -1 and $G_{U,x} \cong \mathbf{Z}_3 \times S_5$ induces the isometry group of C_x on U with kernel of order 3.*

Proof: Let $x \in \bar{V}_C - U\theta$ be an octad. Now $O_2(G_U) = E$ is the group of transvections with axis $U\theta/U$ on V/U, so E is regular on $V/U - U\theta/U$. Thus $H = G_{U,U+x}$ is a complement to E in G_U, so $H \cong S_6/\mathbf{Z}_3$. Now H

has two orbits $x_i H$ on $x + U$ of order 6,10, respectively. So as 27 does not divide the order of G_x, we conclude $x = x_1$, G_U is transitive on the $64 \cdot 6$ octads in $V - U\theta$, and $G_{U,x}$ is as claimed.

Lemma 20.12: *The stabilizer of a sextet U has orbits $\kappa_i(U)$, $0 \leq i \leq 3$, on Γ_2, where:*

    (1) $\kappa_0(U) = \Gamma_2(U)$ is of order 15.

    (2) $\kappa_1(U) = \{l : l \cap U \in \Gamma_1\}$ is of order 180.

    (3) $\kappa_2(U) = \{l : 0 = l \cap U \text{ and } l \leq W \in \Xi_2(U)\}$ is of order 720.

    (4) $\kappa_3(U) = \{l : l \not\leq U\theta\}$ is of order 2,880.

Proof: Of course $\kappa_1(U)$ is an orbit of length 15. Next G_U is transitive on the 15 octads $x \in \Gamma_1(U)$ and by 20.6, $G_{x,U}$ is transitive on the twelve trios l with $U \cap l = \langle x \rangle$, so $\kappa_1(U)$ is an orbit of length $15 \cdot 12 = 180$.

Now $W \in \Xi_1(U)$ contains eight trios l with $l \cap U = 0$ and by 20.10, $W \cap U = \zeta(l) \cap U$. Then $W = (\zeta(l) \cap U) + l$ is uniquely determined by l, so $\kappa_2(U)$ is of order $90 \cdot 8 = 720$. This leaves 2,880 trios. From 20.11, G_U is transitive on the 384 octads $x \in V - U\theta$, and $G_{x,U}$ is transitive on the fifteen trios through x, so G_U is transitive on the $384 \cdot 15/2 = 2,880$ trios in $\kappa_3(U)$.

Lemma 20.13: *Trios l, m are coplanar if and only if $m \in \delta_i(l)$ for $i \leq 2$.*

Proof: This follows as G is flag transitive on Γ and if l is a trio in a sextet U then the orbits of $G_{l,U}$ on $\Gamma_2(U)$ are $\delta_i(l) \cap U$ for $i \leq 2$.

21. The local structure of M_{24}

In this section $(X, \mathcal{C})$ is the Steiner system for $G = M_{24}$, V is the binary permutation module for G on X, $V_{\mathcal{C}}$ is the Golay code submodule, and $\bar{V}_{\mathcal{C}} = V_{\mathcal{C}}/\langle X \rangle$ is the 11-dimensional Golay code module. We also use the rest of the notation and terminology of Chapter 6 and Section 20.

Our object in this section is to determine the conjugacy classes of subgroups of G of prime order and the normalizers of such subgroups.

Lemma 21.1: *(1) G has two classes of involutions with representatives z and t.*

    (2) $Fix_X(z) \in \mathcal{C}$ and t has no fixed points on X.

    (3) $F^*(C_G(z)) = Q \cong D_8^3$ is a large extraspecial subgroup of G and $C_G(z)$ is the split extension of Q by $L_3(2)$.

    (4) $C_G(t) \leq N_G(U)$ for some sextet U, and $C_G(t)$ is the extension of E_{64} by S_5.

Proof: Let i be an involution in G. As G is simple, $|Fix_X(i)| \equiv |X| \equiv 0$ mod 4. But the stabilizer L of three points of X is $L_3(4)$, which has one class of involutions with representative z such that $Fix_X(z) \in C$. Further (3) holds by 8.10.

Thus we may assume i has no fixed points on X. Now 40.6 and 40.4.2 complete the proof.

Lemma 21.2: *Let v be a dodecad. Then*

(1) G_v is irreducible on $\bar{V}_C/\langle \bar{v} \rangle$.

(2) $R(\bar{v}) = \langle \bar{v} \rangle$.

Proof: Recall from 19.4 that $G_v \cong M_{12}$ has order divisible by 11. But 10 is the minimal dimension of a faithful $GF(2)\mathbf{Z}_{11}$-module, so as $dim(\bar{V}_C) = 11$, (1) holds. Then as $f_v \neq 0$ is preserved by G_v, we conclude from (1) that $(\bar{V}_C, f_v)$ is nondegenerate and hence (2) holds.

Lemma 21.3: *If l is a trio (regarded as a singular line of $\bar{V}_C$) then l is the radical of f restricted to $\zeta(l)$.*

Proof: From 20.6 and 20.7, $\zeta(l)/R(x)$ is a totally singular 3-space in $\bar{V}_C/R(x)$ with respect to C_x for each octad $x \in l$. Thus $l \leq R = Rad(f_{|\zeta(l)})$. On the other hand by 20.7, G_l is irreducible on $\zeta(l)/l$, so if $R \neq l$ then f is trivial on $\zeta(l)$. But there is a dodecad v with $\bar{v} \in \zeta(l)$, so if f is trivial on $\zeta(l)$ then $\zeta(l)/\langle \bar{v} \rangle$ is a totally singular subspace of $(\bar{V}_C/\langle \bar{v} \rangle, f)$ of dimension 6, whereas by 21.2.2, that space is nondegenerate of dimension 10 and hence has no such subspace.

Lemma 21.4: *(1) G has two classes of elements of order 3 with representatives y and d.*

(2) $|Fix_X(d)| = 6$ and y has no fixed points on X.

(3) $dim(C_{\bar{V}_C}(d)) = 5$ and $dim(C_{\bar{V}_C}(y)) = 3$. Moreover y centralizes no octads in $\bar{V}_C$.

(4) $C_G(d) \cong A_6/\mathbf{Z}_3$ is quasisimple, $N_G(\langle d \rangle)/\langle d \rangle \cong S_6$, and $N_G(\langle d \rangle)$ is a complement to $O_2(N_G(U))$ in $N_G(U)$ for some sextet $U \leq C_{\bar{V}_C}(d)$.

(5) $N_G(\langle y \rangle) \cong S_3 \times L_3(2)$ and $N_G(Y)$ is a complement to $O_2(N_G(l))$ in $N_G(l)$ for some trio l.

Proof: First the stabilizer $L_3(4)$ of three points of X has one class of elements of order 3, so G is transitive on elements of order 3 fixing a point of X. Now by 20.4 if U is a sextet then there is an element d of order 3 fixing each block in the sextet, fixing six points of X, and satisfying (4). Notice as V/V_C is dual to V_C, $dim(C_V(d)) = 2dim(C_{V_C}(d)) = 2(dim(C_{\bar{V}_C}(d)) + 1)$ and as V is the permutation module for G on X,

$dim(C_V(d))$ is the number of orbits of d on X. This gives the dimensions in (3). We use this same argument below without further comment. Notice also that if g of order 3 fixes an octad B it fixes a point of B, so the final remark in (3) holds too.

Next we saw during the proof of 20.4 that a Sylow 3-subgroup P of G is 3^{1+2} and we may take $\langle d \rangle = Z(P)$. Then from (4), $N_G(P)$ has two orbits on elements on $P - \langle d \rangle$. One of these orbits is fused to d in $L_3(4)$, leaving the other fixed point free on X and completing the proof of (2).

Let l be a trio. By 20.7 there is a subgroup Y of order 3 with $N_{G_l}(Y) \cong L_3(2) \times S_3$ a complement to $O_2(G_l)$ in G_l and $[\bar{V}_C, Y] = \zeta(l)$. Then by 21.3, $N_G(Y) \leq G_l$, completing the proof of the lemma.

Lemma 21.5: *(1) G is transitive on subgroups of order p for $p = 5$, 7, 11, and 23.*

(2) $N_G(A)$ is Frobenius of order $11 \cdot 10$, $23 \cdot 11$, for A of order 11, 23, respectively.

(3) If B is of order 5 then $|Fix_X(B)| = 4$, $dim(C_{\bar{V}_C}(B)) = 3$, $C_G(B) \cong \mathbf{Z}_5 \times A_4$, $|N_G(B) : C_G(B)| = 4$, and $N_G(B)$ fixes a sextet.

(4) If E is of order 7 then $|Fix_X(E)| = 3$, $dim(C_{\bar{V}_C}(E)) = 2$, $N_G(E) \cong \mathbf{Z}_3/\mathbf{Z}_7 \times S_3$, and $N_G(E)$ fixes a trio.

Proof: For $p > 3$ a prime divisor of $|G|$, p divides $|G|$ to the first power, so (1) follows from Sylow's Theorem. Similarly if $p = 11$ or 23 then Sylow gives us $|N_G(A) : A|$ modulo p and from 21.1 and 21.4, $|C_G(A)|$ is prime to 6. We conclude (2) holds.

Next the stabilizer $L_3(4)$ of three points contains subgroups B, E of order 5,7, respectively, allowing us to calculate the number of fixed points of our element on X and $\bar{V}_C$. In particular from 20.7, $C_{\bar{V}_C}(E) = l$ is a trio and then (4) holds. Also $Fix_X(B)$ is of order 4 and hence contained in a unique sextet, and this observation together with 20.4 gives (3).

Remarks. I believe Ronan and Smith were the first to discuss the 2-local geometry Γ for M_{24} in [RS]. Ronan and Smith also point out the representation of Γ as points, lines, and 4-subspaces of the Golay code module $\bar{V}_C$. The use of the derived forms (P, C, f) on $\bar{V}_C$ to induce and study the structure of Γ and G comes from [AS2].

Some of the facts about Γ established in Section 20 first appeared in Todd's paper [To2]. Others appear in R. Curtis's thesis [Cu] and still others in [AS2]. The facts about the local structure of M_{24} fall in the realm of the "well known."

The reader may wonder why the 2-local geometry Γ of M_{24} is discussed here in such excruciating detail. It is because Γ appears as a residue of the 2-local geometries of Co_1, J_4, $M(24)$, and F_1. Thus the uniqueness proofs for these groups require detailed knowledge of Γ.

Exercises

1. Let Δ be the collinearity graph of the 2-local geometry Γ of M_{24} and let $x \in \Delta$. Prove
   (1) $\Delta(x) = C_0(x)$, $\Delta^2(x) = C_4(x)$, and $\Delta^3(x) = C_2(x)$, so Δ is of diameter 3.
   (2) $x^\perp = R(x) \cap \Delta$ and $\Delta^{\leq 2}(x) = \{y \in \Delta : C_x(y) = 0\}$.
   (3) The collinearity graph of the residue of a plane is the graph of singular points in a 4-dimensional symplectic space over $GF(2)$.
   (4) Each pair of singular lines through x is incident with a plane.
   (5) If $z \in \Delta^2(x)$ then x and z are incident with a unique plane.
   (6) For each line k of Γ there is a unique point y of k with $d(x,y) = d(x,k)$.
   (7) If U is a plane determined by a sextet S then $d(x,U) = 1$ if and only if $x \in C_2(S)$.
   (8) If $d(x,y) = 3$ then $G_{x,y} \cong Sp_4(2)$ with $R(x)/x$ the 4-dimensional symplectic module for $G_{x,y}$ and $U \in \Gamma_3(x)$ is at distance 1 from y if and only if $(U \cap R(x))/x$ is a singular line in that space. Hence the graph on $\Gamma_2(x)$ with $l * k$ if and only if $d(\pi(l,k),y) = 1$ is connected, where $\pi(l,k)$ is the unique plane through l and k.

2. Let $G = M_{24}$.
   (1) Let P be the binary permutation module for G on the set $\mathcal{C}$ of octads and Q the submodule of P generated by all elements $A + B + C$ such that $\{\bar{A}, \bar{B}, \bar{C}\}$ is a singular or hyperbolic line in the Golay code module $\bar{V}_\mathcal{C}$. Prove P/Q is $GF(2)G$-isomorphic to $\bar{V}_\mathcal{C}$.
   (2) Let M be the stabilizer in G of an octad and V a faithful $GF(2)G$-module such that
       (i) M stabilizes some $z \in V$ with $V = \langle zG \rangle$.
       (ii) M is faithful on some $U \leq V$ with $Z \in U$, $dim(U) = 5$, and $U^\# \subseteq zG$.
   Prove V is $GF(2)G$-isomorphic to $\bar{V}_\mathcal{C}$.

3. Let $\bar{V}_\mathcal{C}$ be the Golay code module and $W \leq \bar{V}_\mathcal{C}$ with $dim(W) > 3$. Prove W contains a singular point.
   (Hint: Use Exercise 3.3.)

4. Let $\bar{V}_C$ be the 11-dimensional Golay code module for $G = M_{24}$. Prove that for each elementary abelian 2-subgroup A of G, $|\bar{V}_C : C_{\bar{V}_C}(A)| > |A|$.

5. Let $G = M_v$, $v = 22, 23$, or 24, and $(X, \mathcal{B})$ be the Steiner system for G. Prove $Aut(G) = Aut(X, \mathcal{B})$, so $Aut(M_{24}) = M_{24}$, $Aut(M_{23}) = M_{23}$, and $|Aut(M_{22}) : M_{22}| = 2$.

Chapter 8

The Conway Groups and the Leech Lattice

The *Leech lattice* is a certain 24-dimensional **Z**-submodule of 24-dimensional Euclidean space $\mathbf{R}^{24}$ discovered by John Leech. John Conway showed that the group $\cdot 0$ of automorphisms of the Leech lattice is a quasisimple group. The central factor group of $\cdot 0$ is the Conway group Co_1. Further other sporadic groups are the stabilizers of sublattices of the Leech lattice; for example, Conway discovered the Conway groups Co_2 and Co_3 in this way.

In Section 22 we construct the Leech lattice Λ and the Conway groups, and establish various properties of Λ and the Conway groups. In Section 23, we consider $\tilde{\Lambda} = \Lambda/2\Lambda$, the *Leech lattice mod 2*. The Leech lattice mod 2 supplies another tool for studying the Conway groups. For example, it allows us to construct McLaughlin's group Mc and the Higman–Sims group HS.

22. The Leech lattice and $\cdot 0$

In this section $M = M_{24}$ and $(X, \mathcal{C})$ is the Steiner system $S(24, 8, 5)$ for M. Let V be the permutation module over $GF(2)$ for M with basis X and $V_{\mathcal{C}}$ the Golay code submodule. Let $\mathbf{R}^{24}$ be the permutation module over the reals for M with basis X and let $(\ ,\ )$ be the symmetric bilinear form on $\mathbf{R}^{24}$ for which X is an orthonormal basis. Thus $\mathbf{R}^{24}$ together with $(\ ,\ )$ is just 24-dimensional Euclidean space admitting the action of M, so for $\sum_x a_x$ and $\sum_x b_x$ in $\mathbf{R}^{24}$,

$$\left(\sum_x a_x, \sum_x b_x\right) = \sum_x a_x b_x.$$

For $v \in \mathbf{R}^{24}$ define $q(v) = (v,v)/16$. Thus q is a positive definite quadratic form on $\mathbf{R}^{24}$. Given $Y \subseteq X$, define $e_Y = \sum_{y \in Y} y \in \mathbf{R}^{24}$. For $x \in X$ let $\lambda_x = e_X - 4x$.

The *Leech lattice* is the set Λ of vectors $v = \sum_x a_x x \in \mathbf{R}^{24}$ such that:

(Λ1) $a_x \in \mathbf{Z}$ for all $x \in X$.
(Λ2) $m(v) = (\sum_x a_x)/4 \in \mathbf{Z}$.
(Λ3) $a_x \equiv m(v) \mod 2$ for all $x \in X$.
(Λ4) $C(v) = \{x \in X : a_x \not\equiv m(v) \mod 4\} \in V_{\mathcal{C}}$.

Lemma 22.1: *Let* $u, v \in \Lambda$. *Then*

(1) $m(u + v) = m(u) + m(v)$ *and* $m(-u) = -m(u)$.
(2) $C(u + v) = C(u) + C(v)$ *and* $C(-u) = C(u)$.
(3) Λ *is a* $\mathbf{Z}$-*submodule of* $\mathbf{R}^{24}$.

Proof: Parts (1) and (2) are easy. Further (1) and (2) together with the fact that $V_{\mathcal{C}}$ is a subspace of V, imply (3).

Let Λ_0 denote the set of vectors $v \in \Lambda$ such that $m(v) \equiv 0 \mod 4$.

Lemma 22.2: *(1)* $4e_J \in \Lambda$ *for each subset* J *of* X *of even order.*
(2) $2e_v \in \Lambda$ *for each* $v \in V_{\mathcal{C}}$.
(3) $8x \in \Lambda$ *for all* $x \in X$.
(4) Λ_0 *is the* $\mathbf{Z}$-*submodule spanned by* $\{2e_B : B \in \mathcal{C}\}$.

Proof: Parts (1)–(3) are straightforward calculations, keeping in mind that $|v| \equiv 0 \mod 4$ for all $v \in V_{\mathcal{C}}$. By 22.1.1, Λ_0 is a $\mathbf{Z}$-submodule of Λ. By (2), $2e_B \in \Lambda_0$ for $B \in \mathcal{C}$, so the $\mathbf{Z}$-submodule L generated by $\{2e_B : B \in \mathcal{C}\}$ is contained in Λ_0. Hence to complete the proof of (4), it remains to show $\Lambda_0 \leq L$.

We claim $4(x - y)$ and $16x \in L$ for all $x, y \in X$. Namely if T is a 4-subset of X then from 19.9.1, T is contained in a unique sextet $\Delta(T)$. Then for $R, S \in \Delta(T) - \{T\}$,

$$4e_T = 2e_{T+S} + 2e_{T+R} - 2e_{R+S} \in L.$$

Then choosing U to be a 5-subset of X containing x, y,

$$4(x - y) = 4e_{U+x} - 4e_{U+y} \in L$$

as $U + x$ and $U + y$ are 4-subsets. Finally

$$16x = \sum_{t \in T} 4(x - t) + 4e_T \in L,$$

where T is a 4-subset with $x \notin T$. Thus the claim is established.

Now let $v = \sum_x a_x x \in \Lambda_0$; we must show $v \in L$. As $m(v) \equiv 0 \mod 4$, a_x is even for all $x \in X$ by (Λ3), and $\mathcal{C}(v) = \{x \in X : a_x \equiv 2 \mod 4\} \in V_{\mathcal{C}}$ by (Λ4). But $\mathcal{C}(v) = \sum_{B \in S} B$ for some $S \subseteq \mathcal{C}$, so replacing v by $v + \sum_{B \in S} 2e_B$, we may assume $\mathcal{C}(v) = 0$. Now taking $X = \{x_1, \ldots, x_{24}\}$ and $b_k = \sum_{i \le k} a_i$, we have

$$v = \sum_{i < 24} b_i(x_i - x_{i+1}) + b_{24}x_{24}.$$

But $a_i \equiv 0 \mod 4$ for all i, so $b_i \equiv 0 \mod 4$, and hence $b_i(x_i - x_{i+1}) \in L$ by the claim. Similarly $b_{24} = 4m(v) \equiv 0 \mod 16$, so $b_{24}x_{24} \in L$ by the claim.

Lemma 22.3: *Let $x_0 \in X$. Then Λ is the **Z**-submodule generated by Λ_0 and λ_{x_0}.*

Proof: Let $\lambda = \lambda_{x_0}$. Evidently $\lambda \in \Lambda$ with $m(\lambda) = 5$. Conversely let $v = \sum_x a_x x \in \Lambda$; we must show v is in the **Z**-span of Λ_0 and λ. If $m(v)$ is odd then as $m(v + \lambda) = m(v) + m(\lambda)$, replacing v by $v + \lambda$ we may take $m(v)$ even. Similarly replacing v by $v + 2\lambda$ if necessary, we may assume $m(v) \equiv 0 \mod 4$. But then $v \in \Lambda_0$, as desired.

Lemma 22.4: *(1) $(x, y) \equiv 0 \mod 8$ for all $x, y \in \Lambda$.*
(2) $q(x) \in \mathbf{Z}$ for all $x \in \Lambda$.

Proof: By 22.2 and 22.3 it suffices to prove the remarks for $x = 2e_B$ or λ_z and $y = 2e_A$, where $A, B \in \mathcal{C}$, $z \in X$. But $(2e_A, 2e_B) = 4|A \cap B| \equiv 0 \mod 8$ by 19.2.1. Also $(2e_B, \lambda_z) = 8$ or 16 for $z \in B$, $z \notin B$, respectively. Finally $q(\lambda_z) = 2$.

Write $O(\mathbf{R}^{24})$ for the subgroup of $GL(\mathbf{R}^{24})$ preserving the bilinear form $(\ , \ )$, or equivalently preserving the quadratic form q. Let G be the subgroup of $O(\mathbf{R}^{24})$ acting on Λ. The group G is the automorphism group of the Leech lattice; it is often denoted by $\cdot 0$.

For $Y \subseteq X$, write ϵ_Y for the element of $GL(\mathbf{R}^{24})$ such that

$$x\epsilon_Y = \begin{cases} x & \text{if } x \in Y, \\ x & \text{if } x \notin Y. \end{cases}$$

Lemma 22.5: *Let* $P = \{\epsilon_Y : Y \in V\}$, $Q = \{\epsilon_Y : Y \in V_C\}$. *Then*

(1)   For $g \in Sym(X)$, $(\epsilon_Y)g = \epsilon_{Yg}$ *and* $(\epsilon_Y)^g = \epsilon_{Yg}$.

(2)   The map $Y \mapsto \epsilon_Y$ *is an isomorphism of* V *with* P *as abelian groups which commutes with the action of* M, *where* M *acts on* P *by conjugation.*

(3) $P \cdot Sym(X) \leq O(\mathbf{R}^{24})$ and $N = M \cdot Q \leq G$.

Proof: Part (1) is straightforward and, together with the observation that $\epsilon_Y \epsilon_Z = \epsilon_{Y+Z}$, implies (2). As $Sym(X)$ permutes X and X is an orthonormal basis for $(\ ,\ )$, $Sym(X) \leq O(\mathbf{R}^{24})$. Similarly $P \leq O(\mathbf{R}^{24})$.

As M preserves V_C, M acts on Λ. So $M \leq G$. Let $B \in C$ and $\epsilon = \epsilon_B$. It remains to show $Q \leq G$, and for this it suffices by 22.2 and 22.3 to show the images $2e_A \epsilon$ and $\lambda_x \epsilon$ of $2e_A$ and λ_x under ϵ are in Λ for each $A \in C$ and $x \in X$. But $2e_A \epsilon = 2e_{A+B} - 2e_B$ and $2e_{A+B}$ and $2e_B$ are in Λ by 22.2. Similarly $\lambda_x \epsilon = \lambda_x - 2e_B + 8x$ if $x \in B$, and $\lambda_x \epsilon = \lambda_x - 2e_B$ if $x \notin B$. But by 22.2, $2e_B$ and $8x$ are in Λ, so the lemma holds.

In the remainder of this section N will be the subgroup of lemma 22.5.3.

Given a positive integer n, write Λ_n for the set of vectors v in Λ with $q(v) = n$. By 22.4.2,

$$\Lambda = \bigcup_n \Lambda_n.$$

For $v = \sum_x a_x x \in \Lambda$ and i a nonnegative integer, let

$$S_i(v) = \{x \in X : |a_x| = i\},$$

and define the *shape* of v to be $(0^{n_0}, 1^{n_1}, \dots)$, where $n_i = |S_i(v)|$.

Let Λ_2^2 be the set of vectors in Λ of shape $(2^8, 0^{16})$, Λ_2^3 the vectors in Λ of shape $(3, 1^{23})$, and Λ_2^4 the vectors in Λ of shape $(4^2, 0^{22})$.

Lemma 22.6: *(1)* Λ_2^i, $1 \leq i \leq 3$, *are the orbits of* N *on* Λ_2.

(2) $|\Lambda_2^2| = 2^7 \cdot 759$, $|\Lambda_2^3| = 2^{12} \cdot 24$, *and* $|\Lambda_2^4| = 2^2 \cdot \binom{24}{2}$.

(3) $|\Lambda_2| = 196{,}560 = 2^4 \cdot 3^3 \cdot 5 \cdot 7 \cdot 13$.

Proof: First observe that by Exercise 8.1, a vector $v = \sum_x a_x x \in \mathbf{R}^{24}$ of shape $(2^8, 0^{16})$ is in Λ if and only if $S_2(v) \in C$ and $T(v) = \{x \in X : a_x = 2\}$ is of even order. Thus there are 759 choices for $S_2(v)$, and given $S_2(v)$ there are 2^7 choices of signs for the coefficients a_x, $x \in S_2(v)$, so that $|T(v)|$ is even. So (2) holds in this case.

Further if $u \in \Lambda_2^2$ then there is $g \in M$ with $S_2(v) = S_2(u)g = S_2(ug)$, so in proving transitivity of N on Λ_2^2, we may take $S_2(u) = S_2(v)$. Next for each subset R of $S_2(v)$ of order 2, there exists $B \in C$ with

$S_2(v) \cap B = R$ by 19.2. Now $v\epsilon_B$ differs from v only by a sign change on R, so adjusting by such elements ϵ_B, we can map v to u, since $T(u)$ and $T(v)$ are each of even order. Therefore we have shown that N is transitive on Λ_2^2.

Similarly by Exercise 8.2, N is transitive on Λ_2^3, while by Exercise 8.1, if $v \in \mathbf{R}^{24}$ is of shape $(3, 1^{23})$, then $v \in \Lambda$ if and only if $S = \{x \in X : a_x \equiv 1 \bmod 4\} \in V_{\mathcal{C}}$. So there are 2^{12} choices for S, and given S there are twenty-four choices for the place $S_3(v)$. Hence (2) holds in this case.

Notice Exercise 8.3 handles the case of Λ_2^4.

Finally suppose $v = \sum_x a_x x \in \Lambda_2$. We must show v has one of the three shapes above. As $32 = 16q(v) = \sum_x a_x^2$, we have $|a_x| \leq 5$ for all x. Indeed by $(\Lambda 3)$, if some a_x is odd, then all are odd, and hence $a_x^2 \leq 32 - 23 = 9$. So $|a_x| \leq 3$, and if $|a_x| = 3$ then the inequality is an equality, so v has shape $(3, 1^{23})$, as desired. On the other hand if v has shape 1^{24} then $q(v) \neq 2$.

So we may take $m(v) = m$ to be even and $|a_x| = 0, 2$, or 4 for all x. By $(\Lambda 4)$, $S_2(v) \in V_{\mathcal{C}}$. But $4|S_2(v)| \leq 32$, so $|S_2(v)| \leq 8$. Now if $S_2(v)$ is empty then as $q(v) = 2$, v has shape $(4^2, 0^{22})$. Thus we may assume $S_2(v) \neq \varnothing$, so as $|S_2(v)| \leq 8$, $S_2(v) \in \mathcal{C}$ and v has shape $(2^8, 0^{16})$.

Lemma 22.7: *If p is a prime divisor of $|G|$ then $p \leq 23$.*

Proof: Let $g \in G$ be of order p, let ω be a primitive pth root of 1, and let m_i be the multiplicity of ω^i as an eigenvalue of g for $0 \leq i < p$. Then the trace of g is $Tr(g) = \sum_i m_i \omega^i$. But as g acts on Λ, the matrix of g is integral, so $Tr(g)$ is also integral. Thus $Tr(g)$ is invariant under each automorphism $\omega \mapsto \omega^i$ of $\mathbf{Q}(\omega)$, so as $\{\omega^i : 0 \leq i < p\}$ is a basis for $\mathbf{Q}(\omega)$ over $\mathbf{Q}$, $m_i = m_j$ for all $i, j > 0$. In particular as $m_i > 0$ for some $i > 0$, $24 = dim(\mathbf{R}^{24}) \geq p - 1$.

Lemma 22.8: *If $g \in G$ and $z \in X$ with $\langle zg \rangle = \langle w \rangle$ for some $w \in X$, then $g \in N$.*

Proof: Let $\hat{X} = X \cup -X$ and assume $g \in G$, $z \in X$, with $\langle zg \rangle = \langle w \rangle$ for some $w \in X$. Then $zg \in \hat{X}$ since $q(z) = q(zg)$. For $x \in \hat{X}$ let

$$S(x) = \{v \in \mathbf{R}^{24} : (v, x) = 0 \text{ and } 4(v + \epsilon x) \in \Lambda_2 \text{ for } \epsilon = \pm 1\}.$$

Observe that using 22.6, $S(x) = \hat{X} - \{\pm x\}$. As G acts on Λ_2, $S(zg) = S(z)g$. Thus as $zg \in \hat{X}$, $\hat{X}g = \{\pm zg\} \cup S(z)g = \{\pm zg\} \cup S(zg) = \hat{X}$.

So it remains to show $N_G(\hat{X}) = N$. But $N_{O(\mathbf{R}^{24})}(\hat{X}) = P \cdot Sym(X)$, where P is defined in 22.5. Thus $g = h\epsilon_Y$ for some $h \in Sym(X)$ and $Y \in V$. Let $B \in \mathcal{C}$ and $v = 2e_B$. Then $v \in \Lambda$ and $supp(vg) = Bh$. Hence by (Λ4), $Bh \in C$. Therefore $h \in N_{Sym(X)}(C) = M$, so we may assume $g = \epsilon_Y$. Similarly for $u \in \Lambda_2^3$, $R(u) = \{x \in X : a_x \equiv 1 \mod 4\} \in V_C$ and $R(ug) = R(u) + Y$, so $R(u) + Y \in V_C$. As $R(u) \in V_C$ and V_C is a subspace of V, we conclude $Y \in V_C$, so $g \in N$.

Lemma 22.9: $N = N_G(\Lambda_2^4)$.

Proof: By 22.6, $N \leq N_G(\Lambda_2^4) = H$. Let $y, z \in X$ and $v = 4(x+y)$. Thus $v \in \Lambda_2^4$. Notice $N_{\langle v \rangle}$ has two orbits I_1 and I_2 on the set $I = \Lambda_2^4 \cap v^\perp$: namely $I_1 = \{\pm 4(x - y)\}$ and $I_2 = \Lambda_2^4 \cap x^\perp \cap y^\perp$. So either $H_{\langle v \rangle}$ acts on I_1 or $H_{\langle v \rangle}$ is transitive on I. However $|I| = 926 = 2 \cdot 463$ and 463 is a prime, so the latter case is out by 22.7.

Hence $H_{\langle v \rangle}$ acts on I_1 and hence also on $\{\langle x \rangle, \langle y \rangle\} = \{\langle r+v \rangle : r \in I_1\}$. Therefore $H_{\langle v \rangle} \leq N$ by 22.8. Then as N is transitive on Λ_2^4, $H = NH_{\langle v \rangle} = N$.

Lemma 22.10: *Let* Δ *be a sextet of* X *and for* $x \in X$ *let* $T(x)$ *be the 4-set in* Δ *containing* x. *Define* η *to be the element of* $GL(\mathbf{R}^{24})$ *with* $x\eta = x - e_{T(x)}/2$ *for each* $x \in X$. *Then* $\xi_T = \eta\epsilon_T \in G - N$ *for each* $T \in \Delta$.

Proof: For $x, y \in X$, $(x\eta, y\eta) = (x - e_{T(x)}/2, y - e_{T(y)}/2) = (x, y)$. So $\eta \in O(\mathbf{R}^{24})$, and hence as $\epsilon_T \in O(\mathbf{R}^{24})$, also $\xi = \xi_T \in O(\mathbf{R}^{24})$. Thus by 22.2 and 22.3, it remains to show $2e_B\xi$ and $\lambda_x\xi$ are in Λ for each $B \in C$ and some $x \in X$.

Notice $\epsilon_T = \epsilon_R\epsilon_{R+T}$ for each $R \in \Delta - \{T\}$, so as $\epsilon_{R+T} \in G$, we have complete freedom in our choice of T, independently of B and x.

By 19.11, $B \in C_i(\Delta)$, $i = 1, 2, 3$. If $B \in C_1(\Delta)$ then $B = R+S$ for some $R, S \in \Delta$, and by the previous paragraph we may assume $R \neq T \neq S$. Then $2e_B\xi = 2(e_B - 2e_S - 2e_R)\epsilon_T = -2e_B \in \Lambda$.

If $B \in C_2(\Delta)$ then $B \cap S_i$ is of order 2 for four $S_i \in \Delta$. Let A be the sum of the remaining two 4-sets in Δ. Then $A \in C$ with $A \cap B = 0$, so $A + B + X = C \in C$. We may assume $T \subseteq A$. Then $2e_B\xi = 2(e_B - \sum_i e_{S_i})\epsilon_T = -2e_C \in \Lambda$.

Finally if $B \in C_3(\Delta)$ then $B \cap S_i$ is a point for five $S_i \in \Delta$ and $B \cap S$ is of order 3 for the sixth S in Δ. We may assume $T = S$. Then

$$2e_B\xi = 2\left(e_B - \left(3e_S + \sum_i e_{S_i}\right)/2\right)\epsilon_S = \left(2e_B - 3e_S - \sum_i e_{S_i}\right)\epsilon_S$$

is the element of Λ_2^3 such that $a_x = 3$ for $x \in S - B$ and

$$\{y \in X : a_y \equiv 1 \mod 4\} = B.$$

Lemma 22.11: *If $N < H \leq G$ then*

    (1) H is transitive on Λ_2.

    (2) H_u is transitive on $\Lambda_2 \cap u^\perp$ for each $u \in \Lambda_2$.

    (3) $|\Lambda_2 \cap u^\perp| = 93{,}150 = 2 \cdot 3^4 \cdot 5^2 \cdot 23$ for $u \in \Lambda_2$.

Proof: By 22.9, H does not act on Λ_2^4. But $\Lambda_2^4 H$ is the union of orbits of N on Λ_2, so by 22.6, either H is transitive on Λ_2 or $\Lambda_2^4 H = \Lambda_2^4 \cup \Lambda_2^i$ for $i = 2$ or 3. In the latter case by 22.6, $\Lambda_2^4 H$ has order $2^4 \cdot 3 \cdot 23 \cdot 89$ or $2^4 \cdot 3 \cdot 19 \cdot 109$. This contradicts 22.7, as 89 and 109 are primes. So (1) is established.

Next let $x_0 \in X$ and $\lambda = \lambda_{x_0}$. Then there is $g \in M$ of order 23 fixing x_0 and transitive on $X - \{x_0\}$. So if $v = \sum_x a_x x \in C_\Lambda(g)$ then $a_x = a$ is independent of $x \neq x_0$. Thus if $v \in L(\lambda) = \Lambda_2 \cap \lambda^\perp$, then $v = \pm\lambda$, contradicting $(\lambda, \lambda) \neq 0$. So all orbits of H_λ on $L(\lambda)$ have length divisible by 23. As H is transitive on Λ_2 and $\lambda \in \Lambda_2$, the same holds for each $u \in \Lambda_2$.

Let $u = 4(x_0 + x_1)$ for some $x_0, x_1 \in X$. Then $u \in \Lambda_2$ and we find that the orbits of $N_{\langle u \rangle}$ on $L(u)$ are:

    (a) $\{4(x_0 - x_1), 4(x_1 - x_0)\}$ of order 2.

    (b) $\{w \in \Lambda_2^4 : x_0, x_1 \notin supp(w)\}$ of order $4\binom{22}{2}$.

    (c) $\{w \in \Lambda_2^2 : x_0, x_1 \notin supp(w)\}$ of order $330 \cdot 2^7$.

    (d) $\{w \in \Lambda_2^2 : x_0, x_1 \in supp(w)\}$ of order $77 \cdot 2^6$.

    (e) $\{w = \sum_x a_x x \in \Lambda_2^3 : x_0 \in S_1(w)$ and $a_{x_1} = -a_{x_0}\}$ of order $22 \cdot 2^{11}$.

For example, we calculated during the proof of 19.2 that there are exactly 77 octads through $\{x_0, x_1\}$; this gives the order of orbit (d). Similarly by Exercise 6.3, there are 330 octads missing both x_0 and x_1, which gives the order of orbit (c). Finally by Exercise 6.3, the number of members of V_C containing exactly one of x_0 and x_1 is 2^{11}. But w in orbit (e) is determined by such a set plus the element x in $S_3(w)$. So as $x_0 \in S_1(w)$ and $a_{x_1} = -a_{x_0}$, there are $22 \cdot 2^{11}$ elements in orbit (e).

Now the orbit lengths modulo 23 are: 2,4,12,6,22. But no proper sum of these is congruent to 0 modulo 23, so $H_{\langle u \rangle}$ is transitive on $L(u)$. Then H_u is transitive on $L(u)$ as $H_{\langle u \rangle} = \langle \epsilon_B \rangle H_u$ with $B \in C$ fixing w in orbit (b).

Lemma 22.12: *(1) G is transitive on Λ_2.*

    (2) N is maximal in G.

    (3) $|G| = 2^{22} \cdot 3^9 \cdot 5^4 \cdot 7^2 \cdot 11 \cdot 13 \cdot 23$.

    (4) If $u \in \Lambda_2$ and $v \in \Lambda_2 \cap u^\perp$, then $G_{u,v} = N_{u,v} \cong M_{22}/E_{2^{10}}$.

    (5) If $u = 4(x + y)$ and $v = 4(x - y)$ for $x, y \in X$ then $G_{u,v} = N_{u,v}$.

Table 3

Orbit	Λ_3^2	Λ_3^3	Λ_3^4	Λ_3^5
Shape	$(2^{12}, 0^{12})$	$(3^3, 1^{21})$	$(4, 2^8, 0^{15})$	$(5, 1^{23})$
Length	$2^{11} \cdot 2{,}576$	$2^{12} \cdot \binom{24}{3}$	$2^8 \cdot 759 \cdot 16$	$2^{12} \cdot 24$

Orbit	Λ_4^{2+}	Λ_4^{2-}	Λ_4^3	Λ_4^4	Λ_4^{4+}
Shape	$(2^{16}, 0^8)$	$(2^{16}, 0^8)$	$(3^5, 1^{19})$	$(4^4, 0^{20})$	$(4^2, 2^8, 0^{14})$
Length	$2^{11} \cdot 759$	$2^{11} \cdot 759 \cdot 15$	$2^{12} \cdot \binom{24}{5}$	$2^4 \cdot \binom{24}{4}$	$2^9 \cdot 759 \cdot \binom{16}{2}$

Orbit	Λ_4^{4-}	Λ_4^5	Λ_4^6	Λ_4^8
Shape	$(4, 2^{12}, 0^{11})$	$(5, 3^2, 1^{21})$	$(6, 2^7, 0^{16})$	$(8, 0^{23})$
Length	$2^{12} \cdot 2{,}576 \cdot 12$	$2^{12} \cdot \binom{24}{3} \cdot 3$	$2^7 \cdot 759 \cdot 8$	48

Proof: By 22.10, N is proper in G, so G is transitive on Λ_2 and on $S = \{(u, v) : u \in \Lambda_2, v \in \Lambda_2 \cap u^{\perp}\}$ by 22.11. Similarly if $N < H \leq G$ then H is transitive on S, so $G = HG_{u,v}$, where $u = 4(x + y)$ and $v = 4(x - y)$. Now $G_{u,v}$ fixes $(u + v)/8 = x$, so by 22.8, $G_{u,v} \leq N$. Thus $G = HN = H$. So (2) is established.

Further $G_{u,v} = N_{x,y} = M_{22}/E_{2^{10}}$ as the subspace of elements of V_C not projecting on x or y is of codimension 2. So

$$|G| = |S||G_{u,v}| = |\Lambda_2||\Lambda_2 \cap u^{\perp}||M_{22}/E_{2^{10}}|$$
$$= (2^4 \cdot 3^3 \cdot 5 \cdot 7 \cdot 13)(2 \cdot 3^4 \cdot 5^2 \cdot 23)(2^{17} \cdot 3^2 \cdot 5 \cdot 7 \cdot 11)$$
$$= 2^{22} \cdot 3^9 \cdot 5^4 \cdot 7^2 \cdot 11 \cdot 13 \cdot 23.$$

Lemma 22.13: *The orbits of N on Λ_3 and Λ_4 are listed in Table 3, along with the length and shape of each orbit. The members of $\Lambda_4^{2\epsilon}$ are the N-conjugates of v_ϵ, where v_+ has sixteen coefficients equal to 2 and v_- has two coefficients equal to 2.*

Proof: This is Exercise 8.4.

Lemma 22.14: *(1) G is transitive on Λ_3 and Λ_4.*
(2) $|\Lambda_3| = 2^{12}(2^{12} - 1) = 2^{12} \cdot 3^2 \cdot 5 \cdot 7 \cdot 13$.
(3) $|\Lambda_4| = 398{,}034{,}000 = 2^4 \cdot 3^7 \cdot 5^3 \cdot 7 \cdot 13$.

Proof: This is Exercise 8.5.

Notice ϵ_X is the scalar map on $\mathbf{R}^{24}$ determined by -1, and hence is in the center of G. Denote by Co_1 or $\cdot 1$ the factor group $G/\langle\epsilon_X\rangle$. Denote by Co_2 or $\cdot 2$ the stabilizer of a vector in Λ_2 and denote by Co_3 or $\cdot 3$ the stabilizer of a vector in Λ_3. The groups Co_1, Co_2, and Co_3 are the *Conway groups*. We will see later that each of the Conway groups is simple.

Lemma 22.15: *(1)* $|Co_2| = 2^{18} \cdot 3^6 \cdot 5^3 \cdot 7 \cdot 11 \cdot 23.$
(2) $|Co_3| = 2^{10} \cdot 3^7 \cdot 5^3 \cdot 7 \cdot 11 \cdot 23.$

Proof: This follows from 22.14 as $|Co_m| = |G|/|\Lambda_m|$.

23. The Leech lattice mod 2

In this section we continue the hypothesis and notation of the previous section. In addition, for $v \in \Lambda$ let $\Lambda_n(v, i)$ denote the set of $u \in \Lambda_n$ such that $(v, u) = 8i$.

Let $2\Lambda = \{2v : v \in \Lambda\}$. Then 2Λ is a G-invariant $\mathbf{Z}$-submodule of Λ, so G acts on the factor module $\tilde{\Lambda} = \Lambda/2\Lambda$. The module $\tilde{\Lambda}$ is the *Leech lattice mod 2*. For $v \in \Lambda$ let $\tilde{v} = v + 2\Lambda$ and for $S \subseteq \Lambda$ let $\tilde{S} = \{\tilde{s} : s \in S\}$.

By construction $2\tilde{v} = 0$ for all $v \in \Lambda$, so $\tilde{\Lambda}$ is an elementary abelian 2-group which we may view as a $GF(2)G$-module. Also ϵ_X is trivial on $\tilde{\Lambda}$, so $\tilde{\Lambda}$ is also a $GF(2)$-module for $\bar{G} = G/\langle\epsilon_X\rangle \cong Co_1$.

Define the *coordinate frames* of Λ to be the G-conjugates of Λ_4^8.

Lemma 23.1: *(1)* $N = N_G(\Lambda_4^8).$

(2) *The coordinate frames form a system of imprimitivity for G on Λ_4 of order* $8{,}292{,}375 = 3^6 \cdot 5^3 \cdot 7 \cdot 13.$

(3) *If S is a coordinate frame, $u \in S$, and $v \in \Lambda_4(u, 0)$ with $(u+v)/2 \in \Lambda_2$ then $v \in S$.*

Proof: Let $u \in \Lambda_4^8$ and $g \in G$ with $ug \in \Lambda_4^8$. Then $u = 8x$ or $-8x$ and $ug = 8y$ or $-8y$ for some $x, y \in X$, and hence $g \in N$ by 22.8. This implies (1) and shows the coordinate frames are a system of imprimitivity for G on Λ_4. Then as $|\Lambda_4^8| = 48$ by 22.13, we complete the proof of (2) using 22.14.3.

Assume the hypothesis of (3). Conjugating in G, we may take $u = 8x$. As $v \in u^\perp$, the projection of v on x is trivial and hence the projection of $w = (u + v)/2$ on x is 4. Then as $w \in \Lambda_2$, 22.6 says w has shape $(4^2, 0^{22})$. Hence $v \in \Lambda_4^8$, as desired.

Lemma 23.2: *(1)* $\tilde{\Lambda}_i$, $i = 2, 3, 4$, *are the orbits of G on* $\tilde{\Lambda}^{\#}$.

(2) *For $i = 2, 3$, each coset in $\tilde{\Lambda}_i$ contains exactly two members u and $-u$ of Λ_i, and no member of Λ_j for $i \neq j \in \{2, 3, 4\}$.*

(3) The members of Λ_4 in a coset in $\tilde{\Lambda}_4$ form a coordinate frame.
(4) $dim_{GF(2)}(\tilde{\Lambda}) = dim_{\mathbf{Z}}(\Lambda) = 24$.

Proof: Let $L = \Lambda_2 \cup \Lambda_3 \cup \Lambda_4$ and suppose $u, v \in L$ with $\langle u \rangle \neq \langle v \rangle$ but $\tilde{u} = \tilde{v}$. Then $u+v = w \in 2\Lambda$, so $q(w) = 4q(w/2) \geq 8$, with equality if and only if $u/2 \in \Lambda_2$ by Exercise 8.8. Next replacing u by $-u$ if necessary, we may assume $(v, u) \leq 0$. Now $q(w) = q(u+v) = q(v) + q(u) + (v, u)/8$. But $(v, u) \leq 0$ and $q(v), q(u) \leq 4$ as $u, v \in L$, so $q(w) \leq 8$ with equality precisely when u and v are in Λ_4, $(u, v) = 0$, and $(u+v)/2 \in \Lambda_2$. Hence by 23.1.3, u and v are in the same coordinate frame.

Conversely if $u, v \in \Lambda_4^8$ then $u = 8x$, $v = 8y$, and $u+v = 2(4(x+y)) \in 2\Lambda$ by 22.2. So $\Lambda_4^8 \subseteq \tilde{u}$. Thus (2) and (3) are established. Further by (2) and (3):

$$|\tilde{L}| = |\Lambda_2|/2 + |\Lambda_3|/2 + |\Lambda_4|/48 = 2^{24} - 1.$$

Thus $24 \leq dim_{GF(2)}(\tilde{\Lambda}) \leq dim_{\mathbf{Z}}(\Lambda)$. We will show $dim_{\mathbf{Z}}(\Lambda) \leq 24$ to complete the proof.

Indeed the $\mathbf{Z}$-span K of X is a 24-dimensional free $\mathbf{Z}$-module, so as Λ is a Z-submodule of K, $dim_{\mathbf{Z}}(\Lambda) \leq dim_{\mathbf{Z}}(K) = 24$.

Lemma 23.3: $\tilde{\Lambda}$ *is a faithful irreducible* $GF(2)\tilde{G}$*-module.*

Proof: If $0 \neq U$ is a G-submodule of $\tilde{\Lambda}$ then $U^{\#}$ is the union of orbits $\tilde{\Lambda}_i$, $i \in \{2, 3, 4\}$. But also $U^{\#}$ has order $2^n - 1$ for some n, which forces $U = \tilde{\Lambda}$.

We have already observed that ϵ_X is in the kernel K of G on $\tilde{\Lambda}$. Conversely $K \leq N$ by 23.1.1. Also K fixes $\tilde{v}$ for each $v = 2e_B$, $B \in \mathcal{C}$, and hence acts on $\{\pm v\}$ by 23.2.2. So $K = \langle \epsilon_X \rangle$.

Define a bilinear form $(\ ,\ )$ and quadratic form q on $\tilde{\Lambda}$ by

$$(\tilde{u}, \tilde{v}) = (u, v)/8 \mod 2 \text{ and } q(\tilde{u}) = q(u) \mod 2.$$

These maps are well defined since by 22.4, $(u, v) \equiv 0 \mod 8$ for all $u, v \in \Lambda$. Notice G preserves the bilinear form $(\ ,\ )$ and quadratic form q on $\tilde{\Lambda}$. Recall a subspace U of $\tilde{\Lambda}$ is *singular* if q is trivial on U.

Lemma 23.4: $\tilde{\Lambda}_2 \cup \tilde{\Lambda}_4$ *is the set of singular points of* $\tilde{\Lambda}$ *and* $\tilde{\Lambda}_3$ *is the set of nonsingular points, with respect to the quadratic form* q. *The bilinear form* $(\ ,\ )$ *is nondegenerate.*

Proof: The first two remarks are evident. For the third, G acts on the radical R of the form, so by 23.3, $R = \tilde{\Lambda}$ or 0. As $q \neq 0$, $R \neq \tilde{\Lambda}$.

Lemma 23.5: *Let $v \in \Lambda_5$. Then*

(1) $v = u + 2w$ *with* $u \in \Lambda_3$, $w, u + w \in \Lambda_2$, $(u, w) = -24$, *and* $(w, u + w) = 8$.

(2) *There is a unique way to write v as the sum of two elements from Λ_2: namely* $v = w + (u + w)$.

(3) $G_v = N_G(\{w, u + w\}) \leq N_G(\langle u \rangle)$.

(4) *G is transitive on Λ_5.*

(5) $|\Lambda_5| = 2^{14} \cdot 3^3 \cdot 5 \cdot 7 \cdot 13 \cdot 23$.

(6) $|G_{u,w}| = 2^7 \cdot 3^6 \cdot 5^3 \cdot 7 \cdot 11$ *and* $|G_v| = 2|G_{u,w}|$.

Proof: By 23.2, $v = u + 2w$ for some $u \in \Lambda_i$, $i = 2, 3, 4$, and some $w \in \Lambda$. Then

$$5 = q(v) = q(u) + 4q(w) + (u, w)/4 \equiv q(u) \mod 2.$$

Thus $u \in \Lambda_3$. Now $(u, w) = 8 - 16n$, where $w \in \Lambda_n$. But by the Schwarz inequality, $(u, w)^2 \leq (u, u)(w, w) = 16^2 \cdot 3n$, so $4n^2 - 16n + 1 \leq 0$. Hence $n \leq 3$. However, if $n = 3$ then $(u, w) = -40$, so $q(u + w) = 1$, contrary to Exercise 8.8. Thus $n = 2$, so $(u, w) = -24$ and $(w, u + w) = 8$.

Next by 23.2.2, $v = u + 2w = -u + 2(u + w)$ are the unique ways to write v as the sum of an element from Λ_3 and 2Λ. So $G_v \leq N_G(\langle u \rangle) \cap N_G(\{w, u + w\})$. Conversely $N_G(\{w, u + w\})$ acts on $(u + w) + w = v$. Thus (3) holds.

Also if $v = s + t$ with $s, t \in \Lambda_2$ then $(s, t) = 8$, so $q(s - t) = 3$. Then as $v = (s - t) + 2t$, $\{s, t\} = \{w, u + w\}$ by the previous paragraph. So (2) holds.

By Exercise 8.6, G_w is transitive on $\Lambda_2(w, 1)$ and $|\Lambda_2(w, 1)| = 2^{11} \cdot 23$. Hence by (2), G is transitive on Λ_5 and

$$|\Lambda_5| = |\Lambda_2||\Lambda_2(w, 1)|/2 = (2^4 \cdot 3^3 \cdot 5 \cdot 7)(2^{11} \cdot 23)/2,$$

so (5) holds. Now $|G_v| = |G|/|\Lambda_5|$ is as claimed.

The group $G_{u,w}$ of 23.5.6 is the *McLaughlin group Mc*. We will see later that the McLaughlin group is simple.

Lemma 23.6: *Let $v \in \Lambda_7$. Then*

(1) *There is a unique way to write v as a sum $v = u + w$ with $u \in \Lambda_2$ and $w \in \Lambda_3$.*

(2) $u - w \in \Lambda_3$, $v = (u - w) + 2w$, *and* $(u, w) = 16$.

(3) $G_v = G_{u,w}$.

(4) *G is transitive on Λ_7.*

(5) $|\Lambda_7| = 2^{13} \cdot 3^7 \cdot 5 \cdot 7 \cdot 13 \cdot 23$.

(6) $|G_v| = 2^9 \cdot 3^2 \cdot 5^3 \cdot 7 \cdot 11$.

Proof: This is Exercise 8.9. The proof is much like that of 23.5 and uses Exercise 8.7.

The group G_v of 23.6.6 is the *Higman–Sims group HS*. We will see later that the Higman–Sims group is simple.

Lemma 23.7: *G has 3 orbits on 2-dimensional subspaces $\langle \tilde{u}, \tilde{v} \rangle$ of $\tilde{\Lambda}$ with $\tilde{u}, \tilde{v} \in \tilde{\Lambda}_2$.*
They consist of:

(1) *Totally singular lines with $\tilde{u} + \tilde{v} \in \tilde{\Lambda}_4$. $C_{\tilde{G}}(\langle \tilde{u}, \tilde{v} \rangle) \cong M_{22}/E_{2^{10}}$ for such a line.*
(2) *Totally singular lines with $\tilde{u} + \tilde{v} \in \tilde{\Lambda}_2$.*
(3) *Nondegenerate lines. $C_{\tilde{G}}(\langle \tilde{u}, \tilde{v} \rangle) \cong Mc$ for such a line.*

Proof: Let $\tilde{v} \in \tilde{\Lambda}_2$. By Exercise 8.6, $G_{\tilde{v}}$ has 3 orbits on $\tilde{\Lambda}_2 - \{\tilde{v}\}$: namely $\tilde{\Lambda}_2(v, i)$, $i = 0, -1, -2$. Then choosing $u \in \Lambda_2(v, i)$, $q(u + v) = 4, 3, 2$, respectively. That is, $\tilde{u} + \tilde{v} \in \tilde{\Lambda}_4, \tilde{\Lambda}_3, \tilde{\Lambda}_2$, respectively. Then by 23.4, $\langle \tilde{u}, \tilde{v} \rangle$ is totally singular in the first and third cases and nondegenerate in the second. Further if $\tilde{u} + \tilde{v} \in \tilde{\Lambda}_4$ then (1) holds by 22.12.4, while if $\tilde{u} + \tilde{v} \in \tilde{\Lambda}_3$, then (3) holds by 23.5.

In the last few lemmas in this section we work toward a description of $\tilde{\Lambda}$ as a $GF(2)N$-module.
For $J \in V$ define

$$f_J = 2e_J,$$

$$f(J) = 2f_J - |J| f_X / 2.$$

Lemma 23.8: *For all $J, K \in V$,*

(1) $|J + K| = |J| + |K| - 2|J \cap K|.$
(2) $f_{J+K} = f_J + f_K - 2f_{J \cap K}.$
(3) $f(J + K) = f(J) + f(K) - 2f(J \cap K).$

Proof: Parts (1) and (2) are straightforward calculations and imply (3).

Lemma 23.9: *The map $f : V \to \tilde{\Lambda}$ defined by $f : J \mapsto f(J) + 2\Lambda$ is a homomorphism with kernel V_C.*

Proof: Observe first that $f(J) \in \Lambda$. If $|J|$ is even this follows from 22.2.1 and 22.2.2, while if $|J|$ is odd then $f(J) = f(K) - \lambda_x$ for $x \in J$ and $K = J + x$ of even order. This also shows $f(x) = -\lambda_x$.
Notice next that $V_C \leq \ker(f)$ since if $J \in V_C$ then $|J|/2$ is even and $f_J \in \Lambda$ by 22.2.2. Thus it remains to show f is linear. But this follows from 23.8 as $f(J \cap K) \in \Lambda$.

Lemma 23.10: *Define P, C, f as in Section 20 and make $V_C \times V/V_C$ into an F-space via*

$$(v, V_C + J) + (u, V_C + K) = (v + u + C(u, v)X, V_C + J + K + (v \cap u)),$$

and define $\phi : V_C \times V/V_C \to \tilde{\Lambda}$ by $\phi : (v, V_C + J) \mapsto f_v + f(J) + 2\Lambda$. Then

(1) ϕ is an isomorphism of F-spaces.

(2) $\phi(W)$ is totally singular, where $W = \{(0, V_C + J) : J \in V\}$.

(3) $q(\phi(v, 0)) = P(v)$ and $(\phi(v, 0), \phi(0, V_C + J)) = |J \cap v| + |J||v|/4$.

(4) N acts on $V_C \times V/V_C$ via

$$\beta : (v, J + V_C) \mapsto (v\beta, J\beta + V_C),$$

$$\epsilon_Y : (v, J + V_C) \mapsto (v + s(J)Y + ((1 + s(J))C(v, Y)$$
$$+ \epsilon_J(Y))X, V_C + J + (1 + s(J))(v \cap Y))$$

so as to make ϕ N-equivariant, where $s(J) = 0, 1$ for $|J|$ even, odd, respectively, $\beta \in M$, and $Y \in V_C$.

(5) N preserves a filtration

$$0 = L_0 < L_1 < L_2 < L_3 < L_4 = \tilde{\Lambda}$$

of $\tilde{\Lambda}$, where $L_1 = \langle \tilde{\Lambda}_4^8 \rangle$ is a point and $L_3 = L_1^\perp$ is a hyperplane of $\tilde{\Lambda}$, and $L_2/L_1, L_3/L_2$ are isomorphic to the 11-dimensional Todd module, 11-dimensional Golay code module for M, respectively. Further $O_2(\bar{N}) = \bar{Q}$ acts on L_2 as the group of transvections with center L_1, and L_4/L_2 is dual to L_2 as an N-module.

Proof: Let $D = V_C \times V/V_C$ and notice first that this construction is a special case of the construction of Exercise 4.6; in particular D is an F-space by Exercise 4.6.6. However, we sketch a proof of that fact in this case. Check first that the definition of addition on D is associative. This follows from a straightforward calculation once we observe that

$$C(v, u) + C(v + u, w) = P(v) + P(u) + P(w) + P(v + u + w),$$

that

$$v \cap u + ((v + u) \cap w) = (v \cap u) \cup (v \cap w) \cup (u \cap w),$$

that $C(X, v) = 0$ for all $v \in V_C$, and that

$$(v + X) \cap w = (v \cap w) + w \equiv v \cap w \mod V_C.$$

Next by 23.9, $\phi : W \to \tilde{\Lambda}$ is a well-defined linear map. Write $\phi(v), \phi(J)$ for $\phi(v, 0), \phi(0, V + J)$, respectively, and observe that $q(\phi(J)) = |J|(|J| + 1) \equiv 0 \mod 2$, so (2) holds.

Further for each $v \in V_C$, $f_v \in \Lambda$ by 22.2.2, so ϕ maps D into $\tilde{\Lambda}$. To check that ϕ is linear amounts to a verification that

$$f_{v+u+C(u,v)X} \equiv f_v + f_u + f(v \cap u) \mod 2\Lambda \qquad (*)$$

for all $u, v \in V_C$. But by 22.2.2 and 23.8.2, $f_{v+X} \equiv f_v + f_X$ for $v \in V$, and then $(*)$ follows from 23.8.2.

Therefore $\phi : D \to \tilde{\Lambda}$ is linear. We observed during the proof of 23.9 that $f(x) = -\lambda_x$. Thus $\tilde{\lambda}_x \in \phi(D)$ for each $x \in X$, so as Λ is spanned by λ_x, $x \in X$, ϕ is a surjection. Then as $dim(D) = 24 = dim(\tilde{\Lambda})$, ϕ is an isomorphism. Hence (1) is established.

If $v \in V_C$ and $J \in V$ then $q(\phi(v)) = |v|/4 = P(v)$ and $(\phi(v), \phi(J)) = (\tilde{f}_v, f(J)) = |v \cap J| - |J||v|/4 \mod 2$. That is, (3) holds.

It is straightforward to see β acts linearly on D and commutes with ϕ. A more elaborate calculation shows ϵ_Y is linear. Further $\phi : (0, x) \mapsto -\tilde{\lambda}_x$ and we check $-\tilde{\lambda}_x \epsilon_Y = \phi((0, x)\epsilon_Y)$ for each x, so as $\tilde{\Lambda} = \langle -\tilde{\lambda}_x : x \in X \rangle$, ϕ commutes with ϵ_Y. Thus ϕ is N-equivariant and (4) is established.

Let $D_1 = \langle (X, 0) \rangle$, $D_2 = D_1 + W_+$, $W_+ = \{(0, V_C + J) : s(J) = 0\}$, and $D_3 = \langle D_2, (v, 0) : v \in V \rangle$. Set $L_i = \phi(D_i)$. To prove (5) given (4), it remains to observe that the map

$$D_1 + (0, V_C + J) \mapsto V_C + J$$

is an M-isomorphism of D_2/D_1 with the Todd module, while the map

$$D_2 + (v, 0) \mapsto v + \langle X \rangle$$

is an M-isomorphism of D_3/D_2 with the Golay code module $\bar{V}_C$. Similarly $\tilde{Q}$ acts on D_2 as the group of transvections with center D_1 and then, as L_2 is a maximal totally singular subspace of $\tilde{\Lambda}$, $\tilde{\Lambda}/L_2$ is dual to L_2 as an N-module.

Remarks. The Leech lattice Λ was discovered by John Leech in [Le2] extending his earlier construction of the sublattice Λ_0 in [Le1]. In [Co2], Conway determined the group $\cdot 0$ of automorphisms of Λ, proved that $\cdot 0$ is quasisimple, determined its order, and determined the stabilizers of various sublattices, hence also discovering the Conway groups Co_2 and Co_3 and finding Mc and HS as subgroups of $\cdot 0$.

Our treatment of Λ and $\cdot 0$ follows Conway's treatment in [Co1] and [Co2].

Exercises

1. Assume the hypotheses of Section 22 and let $v = \sum_{x \in X} a_x \in \mathbf{R}^{24}$
   with $a_x \in \mathbf{Z}$. Then
   (1) If a_y is odd for some $y \in X$ then $v \in \Lambda$ if and only if the following
      hold:
      (a) a_x is odd for all $x \in X$.
      (b) $\{x \in X : a_x \equiv 1 \mod 4\} \in V_{\mathcal{C}}$.
      (c) $\{x \in X : a_x \equiv \pm 3 \mod 8\}$ is of odd order.
   (2) If a_y is even for some $y \in X$ then $v \in \Lambda$ if and only if the
      following hold:
      (a) a_x is even for all $x \in X$.
      (b) $\{x \in X : a_x \equiv 2 \mod 4\} \in V_{\mathcal{C}}$.
      (c) $\{x \in X : a_x \equiv 4 \text{ or } 6 \mod 8\}$ is of even order.

2. Let $v \in \Lambda$ with $m(v)$ odd and $|S_j(v)| \geq 20$ for some j. Then N is
   transitive on the set of elements of shape v.

3. For each $k \leq 5$ and $r = 2^s \geq 4$, N is transitive on the set Ω_k of
   vectors of shape $(r^k, 0^{24-k})$. Further $\Omega_k \subseteq \Lambda$ and $|\Omega_k| = \binom{24}{k} 2^k$.

4. Prove Lemma 22.13.

5. Prove Lemma 22.14.

6. Let $u, v \in \Lambda_2$. Prove
   (1) $(v, u) = 8i$ with $|i| = 0, 1, 2,$ or 4.
   (2) $|\Lambda_2(v, i)| = 47{,}104 = 2^{11} \cdot 23$ or $4{,}600 = 2^3 \cdot 5^2 \cdot 23$ for $|i| = 1$ or
      2, respectively, while $\Lambda_2(v, 4) = \{v\}$ and $\Lambda_2(v, -4) = \{-v\}$.
   (3) G_v is transitive on $\Lambda_2(v, i)$ for each i.

7. Prove that for $v \in \Lambda_2$, G_v is transitive on $\Lambda_3(v, 2)$ and $|\Lambda_3(v, 2)| = 2^9 \cdot 3^4 \cdot 23$.

8. $q(v) \geq 2$ for all $v \in \Lambda^{\#}$.

9. Prove Lemma 23.6.

10. Define the F-space $D = V_{\mathcal{C}} \times V/V_{\mathcal{C}}$ as in 23.10. For $v \in V_{\mathcal{C}}, J \in V$,
    define $\epsilon_J(v) = |v \cap J| \mod 2$ as in Exercise 3.2. Define $q : D \to F$ by

$$q(v, V_{\mathcal{C}} + J) = (1 + |J|)P(v) + \epsilon_J(v)$$

and $\gamma : D \times D \to F$ by

$$\gamma((v, V_{\mathcal{C}} + J), (u, V_{\mathcal{C}} + K)) = \epsilon_J(u) + \epsilon_K(v) + |J|P(u) \\ + |K|P(v) + C(v, u).$$

Prove
(1) q is a quadratic form on D with bilinear form γ.

(2) The map $\phi : D \to \tilde{\Lambda}$ of Lemma 23.10 is an isometry; that is, $q(\phi(d)) = q(d)$ for all $d \in D$.

(3) For $X \neq Y \in V$, $(v, J) \in C_D(\epsilon_Y)$ if and only if $s(J) = 0 = \epsilon_J(Y)$ and $V \cap Y \in V_C$.

(4) $dim(C_{\tilde{\Lambda}}(\epsilon_Y)) = 16, 12$ for Y an octad, dodecad, respectively. (Hint: See Exercise 4.6.)

Chapter 9

Subgroups of $\cdot 0$

In this chapter we use the machinery developed in Chapter 8 to establish the existence of various subgroups of $\cdot 0$ and to establish various properties of these subgroups.

For example, in Section 23 we defined the stabilizer of a nonsingular vector $\tilde{v}$ of $\tilde{\Lambda}$ to be the Conway group Co_3. In Section 24 we prove that Co_3 is 2-transitive on the 276 lines of $\tilde{\Lambda}$ through $\tilde{v}$ generated by points in $\tilde{\Lambda}_2$, with the stabilizer in Co_3 of such a line isomorphic to $\mathbf{Z}_2/Mc$. Further we prove Mc is a primitive rank 3 group on the remaining 275 lines. Similarly we show HS is a primitive rank 3 group of degree 100 with point stabilizer M_{22}. These representations allow us to prove that Co_3, Mc, and HS are simple.

In Section 25 we prove that the groups Co_1 and Co_2 have large extraspecial 2-subgroups. Similarly we find subgroups A of $G = Co_1$ of order 3,5, such that $C_G(A)/A$ has a large extraspecial 2-subgroup; $C_G(A)/A$ is Suz or J_2, in the respective case. Our theory of large extraspecial subgroups developed in Section 9 then allows us to prove that Co_1, Co_2, Suz, and J_2 are simple.

Finally in Section 26 we establish various facts about the local structure of Co_1 which will be used later to construct various sporadic subgroups of the Monster. Chapters 16 and 17 contain much more information about Co_1, Suz, and J_2.

24. The groups Co_3, Mc, and HS

In this section we continue the hypothesis and notation of Chapter 8. In addition let $x_1, x_2 \in X$,

$$v_1 = 4(x_1 + x_2),$$
$$v_2 = -\lambda_{x_1},$$
$$v_3 = v_1 - v_2,$$
$$v_4 = 4(x_1 - x_2).$$

Let $\Xi = \Lambda_2(v_1, 2) \cap \Lambda_2(v_2, 2)$, and define

$$G_3 = C_G(v_3),$$
$$G_2 = C_G(\langle v_1, v_2 \rangle),$$
$$G_1 = C_G(\langle v_3, v_4 \rangle).$$

Then $v_1, v_4 \in \Lambda_2^4$, $v_2 \in \Lambda_2^3$, and $v_3 \in \Lambda_3^5$. Thus

$$G_3 \cong Co_3$$

as G_3 is the stabilizer of a vector in Λ_3. Also $v_2 \in \Lambda_2(v_1, 1)$, so

$$G_2 \cong Mc$$

by 23.5. Similarly $v_4 \in \Lambda_2(v_3, 2)$, so

$$G_1 \cong HS$$

by 23.6.

Finally let $G_4 = C_G(\langle v_1, v_2, v_4 \rangle)$. Then G_4 fixes $v_1 + v_4 = 8x_1$ and $v_1 - v_4 = 8x_2$, so $G_4 \leq N$ by 22.8. Therefore $G_4 = C_N(\langle x_1, x_2, v_3 \rangle) = M_{x_1, x_2} \cong M_{22}$. That is,

$$G_4 \cong M_{22}.$$

For $w \in \Lambda$ and $x \in X$ let w_x be the coefficient of x in the expansion $w = \sum_{x \in X} w_x x$.

Notice $\langle G_1, G_2 \rangle \leq G_3$ and $G_4 \leq G_2$.

We will discover in 24.6 that $G_2 \cong Mc$ acts as a rank 3 group on Ξ of degree 275 with stabilizer $G_7 = C_{G_2}(v_5)$. Further by 24.7 and Exercise 9.5, $G_7 \cong U_4(3)$.

Lemma 24.1: G_4 *has three orbits* Ξ^n, $n = 2, 3, 4$, *on* Ξ, *where*

(1) Ξ^2 *consists of those* $w \in \Lambda_2^2$ *such that* $w_{x_1} = w_{x_2} = 2$ *and* $w_x = -2$ *for all* $x \in S_2(w) - \{x_1, x_2\}$. $|\Xi^2| = 77$.

(2) Ξ^3 *consists of those* $w \in \Lambda_2^3$ *such that* $w_{x_1} = 3$, $w_{x_2} = 1$, *and* $C(w) = \{x \in X : w_x = 1\} \in \mathcal{C}$. $|\Xi^3| = 176$.

(3) Ξ^4 *consists of those* $w \in \Lambda_2^4$ *with* $w_{x_1} = 4$, $w_{x_2} = 0$, *and* $w_x = -4$ *for* $x_1 \neq x \in S_4(w)$. $|\Xi^4| = 22$.

Proof: From the proof of Exercise 8.6, $w \in \Lambda_2$ is in $\Lambda_2(v_1, 2)$ if and only if $(w_x, w_y) = (2, 2)$, $(3, 1)$ or $(1, 3)$, $(4, 0)$ or $(0, 4)$ for w in Λ_2^2, Λ_2^3, Λ_2^4, respectively. Then we find $w \in \Lambda_2(v_1, 2)$ is in $\Lambda_2(v_2, 2)$ if and only if w is described in (1)–(3).

Further, in (1) w is determined by $S_2(w)$, and from the proof of 19.2 there are 77 octads through $\{x_1, x_2\}$ with G_4 transitive on these octads. So Ξ^2 is an orbit of length 77 under G_4. We can complete the proofs of (2) and (3) similarly.

Lemma 24.2: *(1)* $\Lambda_2(v_3, 3)$ *is an orbit under* G_3 *of length* $2 \cdot 276$, *with*

$$\{\{v, v_3 - v\} : v \in \Lambda_2(v_3, 3)\}$$

a system of imprimitivity of order 276.

(2) $\{\langle \bar{v}_3, \bar{v} \rangle : v \in \Lambda_2(v_3, 3)\}$ *is the set of lines of* $\bar{\Lambda}$ *through* $\bar{v}_3$ *generated by elements of* $\bar{\Lambda}_2$. *There are 276 such lines and they form an orbit under* G_3.

(3) $\Lambda_2(v_3, 3) = \{v_1, -v_2, v_1 - w, w - v_2 : w \in \Xi\}$.

Proof: From 23.5, G is transitive on the set S of pairs (u, v) with $u \in \Lambda_3$, $v \in \Lambda_2(u, 3)$. Further $|S| = 2|\Lambda_5|$. So $|\Lambda_2(v_3, 3)| = 2|\Lambda_5|/|\Lambda_3| = 2 \cdot 276$. Thus (1) is established. By 23.7, (1) implies (2). Finally for $w \in \Xi$, $(v_3, w) = (v_1 - v_2, w) = 0$. Thus $(v_3, v_1 - w) = (v_3, v_1) = 24 = (v_3, -v_2) = (v_3, w - v_2)$. So the set R on the right in (3) is contained in $\Lambda_2(v_3, 3)$. But by 24.1, $|\Xi| = 275$, so $|R| = 2 \cdot 276$. Thus $|R| = |\Lambda_2(v_3, 3)|$ by (1), so (3) holds.

Lemma 24.3: *Let* $u \in \Lambda_2$, $v \in \Lambda_2(u, 2)$, *and* $w \in \Lambda_2(v, 2) \cap \Lambda_2(u, 2)$. *Then*

(1) $|G_{u,v}|$ *is not divisible by 25.*

(2) $|G_{u,v,w}|$ *is not divisible by 27.*

Proof: By Exercise 8.6, G_u is transitive on $\Lambda_2(u, 2)$ with $G_{u,v}$ of order

$$|Co_2|/(2^3 \cdot 5^2 \cdot 23) = 2^{15} \cdot 3^6 \cdot 5 \cdot 7 \cdot 11.$$

So (1) holds. Similarly $z = v - w \in \Lambda_2(u,0)$ and by 22.12 $G_{u,z} \cong M_{22}/E_{2^{10}}$, so $|G_{u,z}|$ is not divisible by 27. But of course $G_{u,v,w} \leq G_{u,z}$.

Let $x_3 \in X - \{x_1, x_2\}$, $v_5 = 4(x_1 - x_3)$, and $G_5 = C_{G_4}(x_3) = M_{x_1,x_2,x_3} \cong L_3(4)$. Let $G_7 = C_G(v_5)$ and notice $G_5 \leq G_7$. That is,

$$G_5 \cong L_3(4), \text{ and } G_5 \leq G_7.$$

By 24.1, $v_5 \in \Xi^4$. Let $\Omega_i = \Xi \cap \Lambda_2(v_5, i)$ and $\Omega_i^n = \Xi^n \cap \Lambda_2(v_5, i)$ for $i = 1, 2$, $n = 2, 3, 4$. Notice $G_5 \leq G_4 \leq G_2$ and G_5 fixes v_5.

Lemma 24.4: *The orbits of G_5 on $\Xi - \{v_5\}$ are Ξ_i^n, $i = 1, 2$, $n = 2, 3, 4$, $(n,i) \neq (4,1)$. Moreover Ω_1^4 is empty and*

(1) Ω_1^2 consists of those $w \in \Xi^2$ with $x_3 \notin S_2(w)$, so $|\Omega_1^2| = 56$.

(2) Ω_1^3 consists of those $w \in \Xi^3$ with $x_3 \in C(w)$, so $|\Omega_1^3| = 56$.

(3) Ω_2^2 consists of those $w \in \Xi^2$ with $x_3 \in S_2(w)$, so $|\Omega_2^2| = 21$.

(4) Ω_2^3 consists of those $w \in \Xi^3$ with $x_3 \notin C(w)$, so $|\Omega_2^3| = 120$.

(5) $\Omega_2^4 = \Xi^4 - \{v_5\}$ is of order 21.

Proof: These are easy calculations. Notice in (1) that the collection of 6-sets of the form $S(w) - \{x_1, x_2\}$, $x_3 \notin S_2(w)$, is one of the three G_5-invariant collections of independent 6-sets I_6^i of the projective plane on $X - \{x_1, x_2, x_3\}$, defined in 18.7. Thus this collection is of order 56 by 18.7.8, and therefore Ω_1^2 is also of order 56. Similarly in (2), the collection of 6-sets $C(w) - \{x_2, x_3\}$, $w \in \Omega_1^3$, is an I_6^j with $j \neq i$.

Lemma 24.5: *(1) G_2 is transitive on Ξ.*

(2) $G_7 = C_{G_2}(v_5)$ is of order $2^7 \cdot 3^6 \cdot 5 \cdot 7$ and is transitive on Ω_1 of order 112 and on Ω_2 of order 162.

(3) If $u, w \in \Omega_1$ with $u \in \Lambda_2(w, 2)$, then the order of the stabilizer in G_7 of u and w is not divisible by 27.

Proof: Assume (1) is false. Then from 24.1, Ξ^n is an orbit of G_2 on Ξ for some n. As the length of Ξ^n is not divisible by 5, a Sylow 5-group of G_2 fixes some $w \in \Xi^n$. But then as 5^3 divides the order of $G_2 \cong Mc$ (cf. 23.5.6), $G_{v_1,w}$ has order divisible by 5^3, contradicting 24.3.1.

Thus (1) holds, so $|G_7| = |G_2|/|\Xi| = 2^7 \cdot 3^6 \cdot 5 \cdot 7$.

Suppose G_7 is not transitive on Ω_2. Then from 24.4, Ω_2^n is an orbit of G_7 on Ω_2 for some n. So as the length of Ω_2^n is not divisible by 9, the stabilizer H in G_7 of $w \in \Omega_2^n$ has order divisible by 3^5. But $H \leq G_{v_2,v_5,w}$, contradicting 24.3.2. Indeed the same argument establishes (3); namely the stabilizer in G_7 of u and w also fixes v_1 and by 24.3.2, $|G_{v_1,u,w}|$ is not divisible by 27.

Finally if G_7 is not transitive on Ω_1 then Ω_1^n, $n = 2, 3$, are the orbits of G_7 on Ω_1. We will see below in 24.7.9 that for $w \in \Omega_1^n$, $Y = \Omega_1^n \cap \Lambda_2(w, 2)$ is of order 45. But as $|\Omega_1^n| = 56$ is prime to 3, the stabilizer S of w in G_7 has order divisible by 3^6, and hence by (3) each orbit of S on Y has order divisible by 3^4, a contradiction.

Lemma 24.6: *(1) $G_3 \cong Co_3$ is 2-transitive on the set $\mathcal{L}$ of lines of $\tilde{\Lambda}$ through $\tilde{v}_3$ generated by points of $\tilde{\Lambda}_2$. Further $\mathcal{L}$ is of order 276 and the stabilizer in G_3 of a member of $\mathcal{L}$ is isomorphic to $\mathbf{Z}_2/Mc$.*

(2) $G_2 \cong Mc$ is a primitive rank 3 permutation group on Ξ of degree 275 with parameters $k = 112$, $l = 162$, $\lambda = 30$, $\mu = 56$. The stabilizer G_7 of $v_5 \in \Xi$ is of order $2^7 \cdot 3^6 \cdot 5 \cdot 7$.

Proof: By 24.2.2, G_3 is transitive on $\mathcal{L}$. Further by 24.2.3, $l = \langle \tilde{v}_1, \tilde{v}_2 \rangle \in \mathcal{L}$ and $\mathcal{L} - \{l\} = \{\tilde{v}_1 - \tilde{w} : w \in \Xi\}$. Next $G_2 \le G_{3,l}$ is transitive on Ξ by 24.5, so G_3 is 2-transitive on $\mathcal{L}$. By 23.5, $G_{3,l} \cong \mathbf{Z}_2/Mc$.

Similarly by 24.5, G_2 is rank 3 on Ξ of degree 275 with parameters $k = |\Omega_1| = 112$ and $l = |\Omega_2| = 162$. We will calculate shortly in 24.7.6 that $\lambda = 30$; hence $\mu = k(k - \lambda - 1)/l = 56$ by 3.3.2. Also by 3.3.3, G_2 is primitive on Ξ.

In the remainder of this section let $\Omega = \Omega_1$ and for $w \in \Omega$ let $\Sigma(w) = \Omega \cap \Lambda_2(w, 1)$ and $\Gamma(w) = \Omega \cap \Lambda_2(w, 2)$. Let $\Sigma^n(w) = \Omega_1^n \cap \Sigma(w)$ and $\Gamma^n(w) = \Gamma(w) \cap \Omega_1^n$. Define a *line* of Ω to be a subset l of Ω which is maximal subject to $u \in \Sigma(w)$ for all distinct $u, w \in l$. Let Ω_0 be the set of lines of Ω. Then (Ω, Ω_0) is a rank 2 geometry and G_7 is represented as a group of automorphisms of this geometry. Notice $G_5 = M_{x_1, x_2, x_3} \le G_7$ is transitive on Ω_1^n for $n = 2, 3$.

Remark. The group G_7 of Lemma 24.7 is isomorphic to $U_4(3) \cong P\Omega_6^-(3)$ with the points Ω and lines Ω_0 corresponding to the singular points and lines in the 6-dimensional orthogonal space for $P\Omega_6^-(3)$ over $GF(3)$. This follows from Exercise 9.5. The group $U_4(3)$ is discussed in much more detail in Chapter 14.

Lemma 24.7: *(1) Each line of Ω has exactly four points.*

(2) Each pair of points in Ω is incident with at most one line of Ω.

(3) Each point is incident with exactly ten lines.

(4) There are 280 lines.

(5) If l is a line and v a point not on l, then v is collinear with a unique point of l.

(6) G_7 is a primitive rank 3 group on Ω with parameters $k = 30$, $l = 81$, $\lambda = 2$, and $\mu = 10$.

(7) For $w \in \Omega$, $C_{G_7}(w)$ *is the split extension of the kernel* P *of its action on the set of lines through* w *by* $C_{G_5}(w) \cong A_6$. P *is transitive on the 3 points distinct from* w *on each line through* w. $P \cong E_{81}$ *and* $C_{G_5}(w)$ *is irreducible on* P.

(8) G_7 *is simple.*

Proof: For $w \in \Omega_1^2$ let $B(w) = S_2(w)$ and $c(w) = \{x_1, x_2\}$. For $w \in \Omega_1^3$ let $B(w) = C(w)$ and $c(w) = \{x_2, x_3\}$. In either case let $I(w) = B(w) - c(w)$. We saw in the proof of 24.4 that $I^2 = \{I(w) : w \in \Omega_1^2\}$ and $I^3 = \{I(w) : w \in \Omega_1^3\}$ are distinct G_5-orbits of independent 6-subsets of the projective plane π on $X - \{x_1, x_2, x_3\}$.

Let $w \in \Omega_1^n$ and $G_6 = C_{G_5}(w)$. Then $G_6 = N_{G_5}(I(w)) \cong A_6$ by 18.7.8. Further an easy calculation shows:

(9) $\Sigma^n(w) = \{v \in \Omega_1^n : B(w) \cap B(v) = c(w)\}$ *and* $\Gamma^n(w) = \{v \in \Omega_1^n : |B(w) \cap B(v)| = 4\}$ *are orbits of* G_6 *of length 10 and 45, respectively, while*

$$\Sigma^{5-n}(w) = \{v \in \Omega_1^{5-n} : |I(w) \cap I(v)| = 3\},$$

$$\Gamma^{5-n}(w) = \{v \in \Omega_1^{5-n} : |I(w) \cap I(v)| = 2\}$$

are orbits of G_6 *of length 20 and 36, respectively.*

Next let $T \in I_3(\pi)$. By 18.7.6 there exists a unique $I^n(T) \in I^n$ with $T \subseteq I^n(T)$. Let $w^n(T)$ be the member of Ω_1^n with $I^n(T) = I(w^n(T))$. Define

$$l(T) = \{w^2(T), w^3(T), w^3(I(w^2(T)) + T), w^2(I(w^3(T)) + T)\}.$$

Now by Exercise 6.2, there are $\bar{T}, T', T^* \in I^3(\pi)$ with $I^2(T) = T + T'$, $I^3(T) = T + \bar{T}$, and $I^3(T') = T' + T^*$, and $I^2(\bar{T}) = \bar{T} + T^*$. Therefore

$$\{I(w) : w \in l(T)\} = \{I^2(T) = T + T', I^3(T) = T + \bar{T}, I^3(T')$$

$$= T' + T^*, I^2(\bar{T}) = \bar{T} + T^*\}.$$

Thus by (9), $u \in \Sigma(v)$ for all distinct $u, v \in l(T)$.

Observe that if $w \neq u \in \Sigma^n(w)$ then $\Sigma(w) \cap \Sigma(u) \cap \Omega_1^n$ is empty. For $I(w) + I(u) = D$ is a dodecad so if $v \in \Omega_1^n$ then by 19.6.1, $I(v) \cap D$ is nonempty, and hence $v \notin \Sigma(w) \cap \Sigma(u)$ by (9). Thus if $u, w \in l \in \Omega_0$ then $l - \{u, w\} \subseteq \Omega_1^{5-n}$.

Next if $u \in \Sigma^{5-n}(w)$ then by (9), $T = I(w) \cap I(u)$ is of order 3. We claim $l(T)$ is the unique maximal subset S of Ω with $u, w \in S$ and with $s \in \Sigma(r)$ for all distinct $r, s \in S$. For if $v \in \Sigma(w) \cap \Sigma(u)$ then without loss of generality $v \in \Omega_1^{5-n}$, $I(w) = T + T'$, and $I(u) = T + \bar{T}$. As $I(v) \cap I(u) = 0$, $I(w) \cap I(v) = T'$. But then $v \in l(T') = l(T)$.

This shows $\Omega^* = \{l(T) : T \in I_3(\pi)\} \subseteq \Omega_0$ and if $w \in l \in \Omega_0 - \Omega^*$ then $l \subseteq \Omega_1^n$. But then by an earlier remark, $|l| \leq 2$. Further the map $T \mapsto l(T)$ of $I_3(\pi)$ onto Ω^* has fibers of order 4, so $|\Omega^*| = |I_3(\pi)|/4 = 280$. Also the set $\Omega^*(w)$ of members of Ω^* incident with w is of order $\binom{6}{3}/2 = 10$. As each member of $\Omega^*(w)$ has a unique member of $\Sigma^n(w)$, (9) shows each member of $\Sigma^n(w)$ is in a unique member of $\Omega^*(w)$ and G_6 is transitive on $\Omega^*(w)$. This together with previous remarks shows $\Omega^* = \Omega_0$ and establishes (1)–(4).

Notice we have also shown $|\Sigma(w)| = 30$, which is the parameter λ of the McLaughlin graph of 24.6.

Next if $u \in \Gamma(w)$, then by 24.5.3, 27 does not divide the order of the stabilizer in G_7 of w and u. But $|C_{G_7}(w)| = |G_7|/112 = 2^3 \cdot 3^6 \cdot 5 = |A_6| \cdot 3^4$, so as $|\Gamma(w)| = 27$, $H = C_{G_7}(w)$ is transitive on $\Gamma(w)$.

Similarly by (3) and Exercise 9.1, $H = G_6 P$, where P is the kernel of the action of H on $\Omega^*(w)$ and P is of order 81. Now as G_6 is simple, either $[G_6, P] = 1$ or an element $g \in G_6$ of order 5 is faithful on P. In the latter case $P \cong E_{81}$ and g is irreducible on P. So either $[G_6, P] = 1$ or (7) holds, and we may assume the former. Then P acts on the orbits of G_6 on $\Gamma(w)$ of length 10 and 20 and hence fixes each point of $\Sigma(w)$. But then 24.5.3 supplies a contradiction.

So (7) is established. Next P is transitive on $l - \{w\}$ for $l \in \Omega^*(w)$. For if not, P fixes each point of l, and then by transitivity of G_6 on $\Omega^*(w)$, P fixes each point of $\Sigma(w)$. We observed that this is not the case in the preceding paragraph.

Now H is transitive on $\Sigma(w)$ and $\Gamma(w)$, so G_7 is rank 3 on Ω with $k = |\Sigma(w)| = 30$ and $l = |\Gamma(w)| = 81$. Further for $u \in \Sigma(w)$, $\Sigma(u) \cap \Sigma(w) = l - \{u, w\}$, where l is the line through u and w. Therefore $\lambda = 2$. Then $\mu = k(k - \lambda - 1)/l = 10$ by 3.3.2. G is primitive on Ω by 3.3.3. Thus (6) is established.

Finally suppose $1 \neq K \unlhd G_7$. Then as 112 is not a prime power, $1 \neq H \cap K$ by Exercise 1.4.1. But by (7), P is the unique minimal normal subgroup of H, so $P \leq K$. Further for $v \in \Sigma(w)$, $P(v)_w \not\leq P$, so as H/P is simple, $H \leq K$. Thus $G_7 = K$ is simple.

Lemma 24.8: *McLaughlin's group is simple.*

Proof: By 24.6, G_2 is primitive of rank 3 on Ξ while by 24.7 the stabilizer G_7 in G_2 of a point of Ξ is simple. So as $|\Xi| = 275$ is not a prime power, G_2 is simple by Exercise 1.4.2.

Lemma 24.9: *The third Conway group Co_3 is simple.*

Proof: Let $1 \neq K \unlhd G_3$. Let $\mathcal{L}$ be the set of lines described in 24.6.1. By

24.6.1, G_3 is doubly transitive on $\mathcal{L}$ with G_2 of index 2 in the stabilizer H of some $l \in \mathcal{L}$. Indeed G_2 is simple by 24.8 and primitive on $\mathcal{L} - \{l\}$ by 24.6.2, so G_2 is the unique minimal normal subgroup of H. Thus as $|\mathcal{L}| = 276$ is not a prime power, $G_2 \le K$ by 2.2.3. So $|G_3 : K| \le |G_2 : G_2 \cap K| \le 2$. Thus we may assume $|G_3 : K| = 2$. Let $P \in Syl_{23}(K)$. By a Frattini argument, $N_{G_3}(P)$ has even order. This contradicts Exercise 9.2.

Lemma 24.10: *(1) Let $\theta = \Lambda_2(v_3, 3) \cap \Lambda_2(v_4, 0)$. Then $\theta = \theta^1 \cup \theta^2 \cup \theta^4$, where $\theta^1 = \{v_1\}, \theta^2 = \{v_1 - w : w \in \Xi^2\}$, and $\theta^4 = \{w - v_2 : w \in \Xi^4\}$.*
(2) Let $\Gamma = \Lambda_2(v_3, 3) \cap \Lambda_2(v_4, -1)$. Then $\Gamma = \{v_1 - w, w - v_2 : w \in \Xi^3\}$.

Proof: This is left as Exercise 9.4; use 23.6.

Lemma 24.11: $G_1 \cong HS$ *is a primitive rank 3 group on θ of degree 100 with parameters $k = 22$, $l = 77$, $\lambda = 0$, $\mu = 6$. The stabilizer in G_1 of $v_1 \in \theta$ is $G_4 \cong M_{22}$.*

Proof: By 24.1 and 24.10, $G_4 \cong M_{22}$ has orbits θ^n, $n = 1, 2, 4$, on θ of length 1, 77, 22, respectively. So if G_1 is not transitive on θ then G_1 fixes v_1 or has an orbit of length 23 or $78 = 2 \cdot 3 \cdot 13$. But $G_1 \cong HS$ has order $2^9 \cdot 3^2 \cdot 5^3 \cdot 7 \cdot 11$ by 23.6, which is not divisible by 23 or 13. Further the stabilizer in G_1 of v_1 is $C_G(\langle v_3, v_4, v_1 \rangle) = G_4 \cong M_{22}$. So indeed G_1 is a rank 3 group with $k = 22$ and $l = 77$. Now $\theta^4 \subseteq \Lambda_2(v_1, 1)$, whereas $(w, v) = 16$ for distinct $v, w \in \theta^4$. Thus $\lambda = 0$, and then $\mu = k(k - \lambda - 1)/l = 6$.

Lemma 24.12: *The Higman–Sims group is simple.*

Proof: This follows from 24.11 and the simplicity of M_{22} via Exercise 1.4.2.

Lemma 24.13: $G_1 \cong HS$ *is doubly transitive on the set S of 3-dimensional subspaces $\langle \tilde{v}_3, \tilde{v}_4, \tilde{v} \rangle$ of $\tilde{\Lambda}$ through $\langle \tilde{v}_3, \tilde{v}_4 \rangle$ such that*

$$v \in \Lambda_2(v_3, 3) \cap \Lambda_2(v_4, -1).$$

Proof: Adopt the notation of 24.10. Let T be the set of pairs $\{v, v_3 - v\}$ as v ranges over Γ. Notice $|\Gamma| = 2 \cdot 176$ by 24.10.2 and 24.1.2. Then T is a G_1-invariant partition of Γ of order 176 and by 24.10 and 23.2 the map $\{v, v_3 - v\} \mapsto \langle \tilde{v}_3, \tilde{v}_4, \tilde{v} \rangle$ is a bijection of T with S. So it remains to show G_1 is 2-transitive on T. By 24.1, $G_4 \le G_1$ is transitive on T, so it remains to show $H = N_{G_1}(\alpha)$ is transitive on $T' = T - \{\alpha\}$, for $\alpha = \{v, v_3 - v\}$.

Let $v = v_1 - z$, $z \in \Xi^3$, and $L = C_{G_4}(z)$. By Exercise 9.3, L has two orbits on T' of length 70 and 105. Thus if H is not transitive on T', these are also the orbits of H. In particular a subgroup P of index 5 in a Sylow 5-group of H fixes a point $u \in \Gamma$ with $\{u, v_3 - u\}$ in the orbit of length 105. Then by Exercise 9.3, $u \in \Lambda_2(v, 2)$. But $|H| = |G_1|/176 = 2^5 \cdot 3^2 \cdot 5^3 \cdot 7$ by 23.6.6, so $|P| = 25$, contradicting 24.3.1.

25. The groups Co_1, Co_2, Suz, and J_2

In this section we continue the hypotheses and notation of Chapter 8, except we write E for $\{\epsilon_U : U \in V_C\}$. Pick an octad $B \in C$ and let $\bar{B} = X + B$. Let $\Lambda(B)$ be the sublattice of Λ consisting of all elements of Λ whose support lies in B. Let Q_E be the subgroup of G generated by all elements ϵ_A, $A \in C_0(B)$, and Q_M the subgroup of M fixing B pointwise. Let $Q = Q_E Q_M$ and $z = \epsilon_{\bar{B}}$. Finally let $\tilde{G} = G/\langle \epsilon_X \rangle$. Thus $\tilde{G} \cong Co_1$.

We will show that $\tilde{Q} \cong Q \cong 2^{1+8}$ is a large extraspecial 2-subgroup of $\tilde{G}$ and indeed $\tilde{G}$ satisfies $\mathcal{H}(4, \Omega_8^+(2))$. In Chapter 16 we prove that Co_1 is the unique group satisfying this hypothesis and use the existence of the extraspecial 2-subgroup to study the structure of Co_1. Similarly we will show that Co_2 satisfies $\mathcal{H}(4, Sp_6(2))$, Suz satisfies $\mathcal{H}(3, \Omega_6^-(2))$, and J_2 satisfies $\mathcal{H}(2, \Omega_4^-(2))$. In Chapter 16 we characterize Suz and J_2 by these hypotheses. On the other hand the existence of Suz and J_2 follows from work in this section.

Lemma 25.1: (1) $Q_E \cong E_{32}$, $z \in Q_E$, and $Q_E - \langle z \rangle$ consists of all ϵ_A, $A \in C_0(B)$.

(2) $Q_M \cong E_{16}$.

(3) $Q \cong D_8^4$ is extraspecial with $Z(Q) = \langle z \rangle$.

(4) $N_N(Q) = N_M(B)E$ and $N_N(Q)/Q\langle \epsilon_X \rangle \cong A_8/E_{64}$.

Proof: Part (1) follows from 22.5.2 and 20.6.1.

Next (2) holds by 19.1.1. Of course $N_M(B) \leq N_M(Q_E)$, so $Q = Q_M Q_E$ is a subgroup of G of order 2^9. For $g \in Q_M$ and $A \in C_0(B)$, $\epsilon_A^g = \epsilon_{Ag} = \epsilon_A$ or $\epsilon_{A+\bar{B}} = \epsilon_A \epsilon_{\bar{B}}$ by 22.5, so $[Q, Q] \leq \langle z \rangle$. Also $|Q_M : N_{Q_M}(A)| = 2$ by 19.2.6. Hence Q_M induces the group of all transvections of Q_E with center $\langle z \rangle$, so (3) holds.

Next $N_N(Q) \leq C_N(z) = EN_M(B)$. Further $N_M(B)$ acts on Q_E and Q_M and hence also on Q. Finally for $g \in Q$ and $A \in C$, $[\epsilon_A, g] = \epsilon_{A+Ag} \in Q_E$ because g fixes B pointwise and hence the support of $A + Ag$ is contained in $\bar{B}$. So $E \leq N_N(Q)$ and $N_N(Q) = N_M(B)E$. Finally by 19.1, $N_M(B)/Q_M \cong A_8$, while by (1) and 22.5.2, $E/Q_E\langle \epsilon_X \rangle \cong E_{64}$, so (4) holds.

Lemma 25.2: $\Lambda(B) = C_\Lambda(z)$, $Q = C_G(\Lambda(B))$, and $\tilde{Q} = C_{\tilde{G}}(\tilde{\Lambda}(B))$. Thus $Q \trianglelefteq C_G(z)$.

Proof: Evidently $\Lambda(B) = C_\Lambda(z)$. Next as $4(x \pm y) \in \Lambda(B)$ for $x, y \in B$, $C_G(\Lambda(B))$ and $C_G(\tilde{\Lambda}(B))$ are contained in N by 22.8 and 23.2.2. Indeed $C_N(\Lambda(B))$ fixes $4(x + y)$ for each $x, y \in B$, so $Q = C_N(\Lambda(B)) = C_G(\Lambda(B))$. Similarly $C_G(\tilde{\Lambda}(B)) = Q\langle\epsilon_X\rangle$.

Lemma 25.3: $N_G(Q) \not\leq N$.

Proof: Let Δ be a sextet with $B = T + S$ for some $S, T \in \Delta$. Define $\xi = \xi_T$ as in 22.10. Then for $x \in B$, $x\xi = (x - e_{T(x)}/2)\epsilon_T$, where $x \in T(x) \in \Delta$. So $T(x) = S$ or T. Therefore ξ acts on $\Lambda(B)$ and hence on Q by 25.2.

Lemma 25.4: *Suppose $\tilde{U}$ is a totally singular subspace of $\tilde{\Lambda}$. Then q_U is a quadratic form on $\tilde{U}$ with bilinear form $(\, , \,)_U$, where $q_U(\tilde{u}) = q(u)/2$ mod 2 and $(\tilde{u}, \tilde{v})_U = (u, v)/16$ mod 2.*

Proof: As $\tilde{U}$ is totally singular, $q(u)$ is even and (u, v) is divisible by 16 for all $u, v \in U$. So q_U and $(\, , \,)_U$ are well defined and visibly are a quadratic form and associated bilinear form.

Lemma 25.5: *(1) $\Lambda_4(B)$ is of order $16 \cdot 135$ and $N_G(Q)$ is transitive on $\Lambda(B)_4$.*

(2) Each coordinate frame containing a member of $\Lambda(B)$ contains exactly sixteen members of $\Lambda(B)$.

(3) $\Lambda(B)_2$ is of order 240.

(4) $\tilde{\Lambda}(B)^\# = \tilde{\Lambda}_2(B) \cup \tilde{\Lambda}_4(B)$.

(5) $dim_{\mathbf{Z}}(\Lambda(B)) = dim_{GF(2)}(\tilde{\Lambda}(B)) = 8$.

(6) q_B is an $N_G(Q)$-invariant quadratic form on $\Lambda(B)$ of sign $+1$, where $q_B(\tilde{v}) = q(v)/2$ mod 2 for $v \in \Lambda(B)$.

(7) $Q \cong \tilde{Q} = F^*(C_{\tilde{G}}(\tilde{z}))$ with $C_{\tilde{G}}(\tilde{z})/\tilde{Q} \cong \Omega_8^+(2)$.

Proof: Using 22.13, we calculate that $\Lambda_4(B)$ consists of three orbits under $C_N(z)$:

(a) $\Lambda_4^8(B)$ of order 16.

(b) $\Lambda_4^6(B)$ of order $16 \cdot 2^6$.

(c) $\Lambda_4^4(B)$ of order $\binom{8}{4} \cdot 16 = 70 \cdot 16$.

Next the element $\xi \in C_G(z) = H$, produced during the proof of 25.3, moves orbit (a) into orbit (c). As 71 does not divide the order of G, it follows that (1) holds. Further orbit (a) is the intersection of a coordinate frame with $\Lambda_4(B)$, so (1) implies (2).

Similarly we find $\Lambda_2(B)$ falls into two $C_N(z)$ orbits: $\Lambda_2^2(B)$ of order 2^7 and $\Lambda_2^4(B)$ of order 112. So (3) holds.

Now $\Lambda(B)$ is a sublattice of the 8-dimensional lattice spanned by B, so

$$dim_{GF(2)}(\tilde{\Lambda}(B)) \leq dim_{\mathbf{Z}}(\Lambda(B)) \leq 8.$$

On the other hand by (1)–(3) and 23.2, $\tilde{\Lambda}_2(B) \cup \tilde{\Lambda}_4(B)$ is of order $2^8 - 1$, so (4) and (5) hold. Then (6) follows from 25.4.

By 25.2 and (6), $\tilde{H}/\tilde{Q} \leq O(\tilde{\Lambda}(B), q_B) \cong O_8^+(2)$. Indeed if $v = 8x$, $x \in B$, then the stabilizer in $\tilde{H}$ of $\tilde{v}$ is $C_{\tilde{N}}(\tilde{z})$ and $C_{\tilde{N}}(\tilde{z})/\tilde{Q} = N_M(B)E/Q\langle \epsilon_X \rangle \cong A_8/E_{64}$ by 25.1.4. So as A_8/E_{64} is the stabilizer of $\tilde{v}$ in $\Omega_8^+(2)$, (1) and (6) say $\tilde{H}/\tilde{Q} \cong \Omega_8^+(2)$. In particular $\tilde{H}/\tilde{Q}$ is simple, so either $\tilde{Q} = F^*(\tilde{H})$ or $\tilde{H} = \tilde{Q}C_{\tilde{H}}(\tilde{Q})$. The latter is impossible as $C_{H \cap N}(Q) = \langle z, \epsilon_X \rangle$.

Lemma 25.6: *All involutions in $\tilde{Q}$ are fused to $\tilde{z}$ in $\tilde{G}$.*

Proof: All involutions in $\tilde{Q}_E$ are fused to $\tilde{z}$ in N, while by 25.5.7, $C_{\tilde{G}}(\tilde{z})$ is transitive on noncentral involutions of $\tilde{Q}$.

Lemma 25.7: $\tilde{G} \cong Co_1$ *is simple and $\tilde{Q}$ is a large extraspecial subgroup of $\tilde{G}$.*

Proof: This follows from 8.12 with $\tilde{N}$ in the role of "K."

Lemma 25.8: *Let $v \in \Lambda_2(B)$ and $H = C_G(v)$. Then*

(1) $C_H(z)/Q \cong Sp_6(2)$.
(2) Q *is a large extraspecial 2-subgroup of H.*
(3) $H \cong Co_2$ *is simple.*

Proof: $C_G(z)$ acts as $\Omega(q_B, \tilde{\Lambda}(B))$ on $\tilde{\Lambda}(B)$ with $C_H(z)$ the stabilizer of the nonsingular vector $\tilde{v}$ of $\tilde{\Lambda}(B)$. Hence (1) holds by 22.5 in [FGT]. In particular $C_H(v)/Q$ is simple with $Q = F^*(C_H(v))$ by 25.5.7.

Pick $v = 4(x + y)$, $u = 4(x - y)$, $x, y \in B$, and let $K = N_{v,u}$. By 22.12.4, $\langle z^K \rangle$ is abelian with $\langle z^K \rangle \not\leq Q$. By 25.6 and 8.7, for each involution $u \in Q - Z$, $z \in O_2(C_H(u))$, so (2) and (3) hold by 8.12.

Lemma 25.9: *Let A be of order 3 in $C_G(z)$ with $R = C_Q(A) \cong Q_8^3$. Let $M = C_G(A)$ and $M^* = \tilde{M}/\tilde{A}$. Then*

(1) R^* *is a large extraspecial subgroup of M^*.*
(2) $R^* = F^*(C_{M^*}(z^*))$ *with $C_{M^*}(z^*)/R^* \cong \Omega_6^-(2)$.*
(3) M^* *is simple.*
(4)   *All involutions in R^* are fused to z^* in M^*.*

Proof: By Exercise 2.6, A exists and $C_{\tilde{G}}(\tilde{A}\langle\tilde{z}\rangle) = \tilde{A}\times\tilde{L}$, with $\tilde{R} = F^*(\tilde{L})$ and $\tilde{L}/\tilde{R} \cong \Omega_6^-(2)$ simple. Thus (2) holds and by 25.6 and 8.13, (1) and (4) hold. Finally (3) holds by Exercise 2.4.

The simple group M^* of 25.9 is the *sporadic Suzuki group Suz*. The same proof shows:

Lemma 25.10: *Let A be of order 5 in $C_G(z)$ with $R = C_Q(A) \cong D_8 Q_8$. Let $M = C_G(A)$ and $M^* = \tilde{M}/\tilde{A}$. Then*

(1) *R^* is a large extraspecial subgroup of M^*.*
(2) *$R^* = F^*(C_{M^*}(z^*))$ with $C_{M^*}(z^*)/R^* \cong \Omega_4^-(2) \cong A_5$.*
(3) *M^* is simple.*
(4) *All involutions in R^* are fused to z^* in M^*.*

The simple group M^* of 25.10 is the *Hall–Janko group J_2* or HJ.

26. Some local subgroups of Co_1

In this section we continue the hypotheses and notation of Section 25. We determine the normalizers of certain subgroups of G of prime order. We restrict attention to results needed to establish the existence of certain sporadic groups as sections of the Monster in Chapter 11. Complete results appear in Chapter 17.

Lemma 26.1: *Let $B \leq N$ be cyclic of odd order, $v(2) = 2e_C$, $v(3) = \lambda_x$, and $v(4) = 4e_{xy}$, with $C \in \mathcal{C}$ and $x, y \in X$. Then*

(1) *$dim(C_{\tilde{\Lambda}}(B)) = c$, where c is the number of cycles of B on X.*
(2) *$C_{\tilde{\Lambda}}(B)$ is a nondegenerate subspace of the orthogonal space $\tilde{\Lambda}$.*
(3) *$|C_{\tilde{\Lambda}_2}(B) \cup C_{\tilde{\Lambda}_4}(B)| = (2^{c/2} - \epsilon)(2^{c/2} + \epsilon)$, where ϵ is the sign of the orthogonal space $C_{\tilde{\Lambda}}(B)$.*
(4) *If $B^G \cap G_{\tilde{v}} = B^{G_{\tilde{v}}}$ for some $\tilde{v} \in \tilde{\Lambda}_i$, then $N_{\tilde{G}}(B)$ is transitive on $C_{\tilde{\Lambda}_i}(B)$.*
(5) *If $B^N \cap N_{v(i)} = B^{N_{v(i)}}$ for $i = 2, 3, 4$, then $N_N(B)$ is transitive on $C_{\Lambda^i}(B)$ for each i and $|C_{\tilde{\Lambda}_2}(B)| = \sum_{i=2}^4 |N_N(B) : N_N(B)_{\tilde{v}(i)}|$.*
(6) *$M_{\tilde{v}(i)}$, $E_{\tilde{v}(i)}$ are isomorphic to $N_M(C) \cong A_8/E_{16}$, E_{2^5}; $M_x \cong M_{23}$, $\mathbf{Z}_2$; $M(\{x,y\}) \cong 2/M_{22}$, $E_{2^{11}}$ for $i = 2; 3; 4$, respectively.*

Proof: First $c = dim(C_V(B))$ as V is the permutation module on X. Then as V/V_C is the dual of V_C, $c = 2dim(C_{V_C}(B)) = 2dim(C_{V/V_C}(B))$, so 23.10.5 completes the proof of (1).

Next $\tilde{\Lambda} = [\tilde{\Lambda}, B] \oplus C_{\tilde{\Lambda}}(B)$ as $|B|$ is odd, so (2) follows from 22.1 in [FGT].

Part (3) follows as $C_{\tilde{\Lambda}_2}(B) \cup C_{\tilde{\Lambda}_4}(B)$ is the set of singular points in $C_{\tilde{\Lambda}}(B)$. Parts (4) and (5) follow from a standard argument (cf. 5.21 in [FGT]). Part (6) is an easy calculation.

Remark. If B is of prime order p in a group H and $K \leq H$ with $|K|_p \leq p$, then by Sylow's Theorem, $B^H \cap K = B^K$. In particular if $|B| = 7$ then $|G_{\tilde{v}}|_7 = 7$ for $\tilde{v} \in \tilde{\Lambda}_i$, $i = 2, 4$, by 22.15 and 23.1.2. So by 26.1.4, $N_{\tilde{G}}(B)$ is transitive on $C_{\tilde{\Lambda}_i}(B)$ for $i = 2, 4$. Similarly if $|B| = 5$ then $N_{\tilde{G}}(B)$ is transitive on $C_{\tilde{\Lambda}_4}(B)$. Finally for $p = 5, 7$, $|G_{v(i)}|_p = p$ by 26.1.6, so we can calculate $|C_{\tilde{\Lambda}_2}(B)|$ via 26.1.5. To do so we obtain the structure of $N_N(B)$ from Section 21, while $N_N(B)_{v(i)}$ is easily calculated from 26.1.6.

Lemma 26.2: *Let B be a subgroup of N of order 7. Then $C_{\tilde{G}}(B) \cong \mathbf{Z}_7 \times L_3(2)$ and $|N_G(B) : C_G(B)| = 6$ with an involution-inverting B inducing an outer automorphism on $E(C_{\tilde{G}}(B))$. Further $C_{\tilde{\Lambda}}(B)$ is of dimension 6 and sign $+1$ and $|C_{\tilde{\Lambda}_4}(B)| = 14$.*

Proof: By 21.5.4, $C_{N/E}(B) \cong \mathbf{Z}_7 \times S_3$, $C_{\tilde{E}}(B) \cong E_4$ with B the kernel of the action of $C_{N/E}(B)$ on $C_{\tilde{E}}(B)$, and $|N_N(B) : C_N(B)| = 3$. Therefore $C_{\tilde{N}}(B) \cong \mathbf{Z}_7 \times S_4$.

Next by 26.1.1, $W = C_{\tilde{\Lambda}}(B)$ is of dimension 6 and by 26.1.2, W is nondegenerate. From the structure of $C_N(B)$ described in the previous paragraph and the Remark above, $N_G(B)$ is transitive on $C_{\tilde{\Lambda}_k}(B)$ for $k = 2, 4$, and B fixes 3,12,6, points of $\tilde{\Lambda}_2^i$ for $i = 2, 3, 4$, respectively, and hence $|C_{\tilde{\Lambda}_2}(B)| = 21$. In particular an element of order 7 is induced on W in $C_G(B)$, so the orthogonal space W has sign $+1$. Hence by 26.1.3, $|C_{\tilde{\Lambda}_4}(B)| = 14$, so $|N_{\tilde{G}}(B)| = 14 \cdot |N_{\tilde{N}}(B)| = 2^4 \cdot 3^2 \cdot 7^2$. Thus the non-trivial vectors in two irreducibles W_i, $i = 1, 2$, for a Sylow 7-group of W make up the fourteen vectors of $W \cap \tilde{\Lambda}_4$, and these are the irreducibles for $E(C_{\tilde{N}}(B)) \cong L_3(2)$. From 25.5.7, B is inverted in $C_{\tilde{N}}(z)$ for $z \in C_{\tilde{E}\#}(B)$ by an involution t with $dim(C_{\tilde{\Lambda}}(B\langle t \rangle)) = 3$. So $|Aut_{\tilde{G}}(B)| = 6$ and as $dim(C_{\tilde{\Lambda}}(B\langle t \rangle)) = 3$, t induces an outer automorphism on $E(C_{\tilde{G}}(B))$.

Lemma 26.3: *Let $D \in V_C$ be a dodecad and $t = \tilde{e}_D \in \tilde{E}$. Then $C_{\tilde{G}}(t) \leq \tilde{N}$ is the split extension of $\tilde{E} \cong E_{2^{11}}$ by $\mathbf{Z}_2/M_{12}$.*

Proof: First $C_{\tilde{N}}(t)$ is the split extension of $\tilde{E}$ by $N_M(D) \cong \mathbf{Z}_2/M_{12}$ by 22.5 and 19.9.3. Thus it remains to show $C_{\tilde{G}}(t) \leq \tilde{N}$. Now $e = \epsilon_D$ fixes 24 of the 48 members of Λ_4^8 and inverts the rest. Similarly if $v \in \Lambda_4$ is fixed by e then the support $S(v)$ of v has order at most 12 and hence by 22.13, v has shape $(4^4, 0^{20})$, $(4^2, 2^8, 0^{14})$, $(6, 2^7, 0^{16})$, or $(8, 0^{23})$. But

$S_2(v)$, $S_2(v) \cup S_6(v)$ are octads in the second and third cases, and hence intersect D nontrivially by 19.6.1, a contradiction. Finally if u is in case 1 then the coordinate frame Δ of u consists of those v in case 1 such that $S_2(v)$ is in the same sextet as $S_2(u)$ and v has the same parity of signs as u. In particular by 19.6.1, $S_2(u)$ is the unique member of Δ contained in $X + D$, so e fixes only eight members of the coordinate frame of u.

We have shown Λ_4^8 is the unique coordinate frame containing 24 fixed vectors of e, so the lemma holds by 23.2.3.

Lemma 26.4: *Let $Y = \langle y \rangle$ be a non-3-central subgroup of N. Then $N_{\tilde{G}}(Y) \cong S_3 \times A_9$, $dim(C_{\tilde{\Lambda}}(Y)) = 8$, and $N_{\tilde{G}}(Y)$ is transitive on the 135 singular vectors of $C_{\tilde{\Lambda}}(Y)$, each of which is in $\tilde{\Lambda}_4$.*

Proof: By 21.4, Y has no fixed points on X, so the second statement of the lemma follows from 26.1.1. Similarly by 21.4, $N_{\tilde{N}}(Y) \cong S_3 \times (L_3(2)/E_8)$, with $O_2(N_{\tilde{N}}(Y)) = A = C_{\tilde{E}}(Y)$. By 21.4.3, Y stabilizes no octad, so each element in $A^{\#}$ is of the form ϵ_D for some dodecad D. Hence by 26.3, AY/Y is a TI-set in $H = C_{\tilde{G}}(Y)/Y$ with $K = C_{\tilde{N}}(Y) = N_H(A)$, subject to the convention of identifying A with its image AY/Y in H.

As Y has no fixed points on X, Y stabilizes no member of $\tilde{\Lambda}_2$, so all singular points in $C_{\tilde{\Lambda}}(Y)$ are in $\tilde{\Lambda}_4$ and as $Y^G \cap N = Y^N$, H is transitive on these points. Thus by 26.1.3, there are 135 such points, as 17 does not divide $|G|$. So $|H| = 135 \cdot |K| = 2^6 \cdot 3^4 \cdot 5 \cdot 7$.

Represent H on $W = C_{\tilde{\Lambda}}(Y)$. Then W is an 8-dimensional orthogonal space of sign $+1$ and K is the stabilizer of a singular point v of W. From 23.10, N stabilizes a maximal totally singular subspace $L(v)$ of $\tilde{\Lambda}$ and $L(v)/v$ is the Todd module for N/E. By 21.4, $dim(C_{L(v)/v}(Y)) = 3$, so $L(v) \cap W = W(v)$ is a maximal totally singular subspace of W. Further by 23.10, E induces the group of transvections on $L(v)$ with center v, so by the Thompson $A \times B$ Lemma (cf. 24.2 in [FGT]) A induces the group of transvections on $W(v)$ with center v.

Next by Exercise 8.10.4, for $a \in A^{\#}$, $dim(C_{\tilde{\Lambda}}(a)) = 12$, so $C_{\tilde{\Lambda}}(a) = [\tilde{\Lambda}, a]$. Also as a induces a transvection on $L(v)$, $C_{L(v)}(a)$ is a hyperplane of $L(v)$, so $C_{W(v)}(a)$ is a hyperplane of $C_W(a) = [W, a]$. In particular

(∗) $W(v)/v = C_{W/v}(B)$ for each hyperplane B of A.

Let $v \neq w$ be a point of $W(v)$ and $h \in H$ with $vh = w$. Then $B = C_A(w)$ is a hyperplane of A. As A is a TI-set weakly closed in $K = N_H(A)$, $C_{A^h}(B)$ is a hyperplane of A^h, which we may take to be B^h. Let $z \in B^{\#}$. Then $B^h \leq C_K(z)$, and $R = F^*(C_K(z)) \cong D_8^2$

with $C_K(z)/R \cong S_3$. Hence $B^h \cap R \neq 1$. Further $m(A^h \cap R) \leq 1$ or else $R = \langle C_R(a) : a \in A^h \cap R^\# \rangle \leq K^h$, and then $\langle z \rangle = \Phi(R) \leq A^h$, a contradiction. So we may take $\langle z^h \rangle = B^h \cap R$ and similarly $A \cap R^h = \langle z \rangle$.

By 8.15.7, $R \cap R^h = D \cong E_8$. But $C_R(z^h) \cong \mathbf{Z}_2 \times D_8$ has two E_8-subgroups, so as $A \cap R^h = \langle z \rangle$, D is uniquely determined as the E_8-subgroup distinct from $C_A(z^h)$ and $DA = R$. Now a subgroup F of order 3 in $C_K(z)$ acts faithfully on D and $\langle R, R^h, F \rangle = K_0 \leq N_H(D)$ induces $GL(D) \cong L_3(2)$ on D.

We claim z^h induces a transvection on $W(v)$. For if not, $R = AA_0 = DA_0$, where A_0 is the group of transvections of $W(v)$ with axis $C_{W(v)}(z)$. Thus $C_{W(v)}(D) = \langle z \rangle$, so as $C_{W(v)}(z)$ is a hyperplane of $C_W(z)$, $dim(C_W(D)) \leq 2$. Hence as $K_0/D \cong L_3(2)$, $K_0 \leq C_H(C_W(D)) \leq C_H(v) = K$, a contradiction.

So z^h induces a transvection on $W(v)$ with center w. Then as $N_K(B)$ is transitive on $B^\#$, B induces transvections on $W(v)$ with center w, so by (∗), $W(v) = W(w)$. Thus $L = \langle K, K^h \rangle$ induces $GL(W(v)) = L_4(2)$ on $W(v)$ and as $C_H(W(v)) = C_K(W(v)) = 1$, $L \cong L_4(2)$. Then as $|H| = |A_9|$, $|H : L| = 9$, so we have a faithful permutation representation of H on H/L of degree 9. Therefore as $L \cong L_4(2) \cong A_8$, $H \cong A_9$. That is, the lemma holds.

Lemma 26.5: *Let $B \in Syl_5(N)$. Then $dim(C_{\bar{\Lambda}}(B)) = 8$, $C_{\tilde{G}}(B) \cong \mathbf{Z}_5 \times (A_5 wr \mathbf{Z}_2)$, $|C_{\bar{\Lambda}_4}(B)| = 75$, and $4 = |Aut_{\tilde{G}}(B)|$.*

Proof: The first statement of the lemma follows from 26.1.1.

Using the Remark above, we conclude $N_{\tilde{G}}(B)$ is transitive on $C_{\bar{\Lambda}_4}(B)$ and check that B fixes 32, 64, 24 elements of Λ_2^i for $i = 2, 3, 4$, respectively. So by 26.1.5, $|C_{\bar{\Lambda}_2}(B)| = 60$. Also as $B \in Syl_5(N)$ but B is not Sylow in Co_2, $|C_{\bar{\Lambda}_4}(B)| \equiv 0 \mod 5$, so the number of singular points in $C_{\bar{\Lambda}}(B)$ is divisible by 5, and hence by 26.1.3, that space has sign $+1$ and contains 135 singular points. Thus $|C_{\bar{\Lambda}_4}(B)| = 75$. Therefore $|N_{\tilde{G}}(B)| = 75 \cdot |N_{\tilde{N}}(B)| = 2^7 \cdot 3^2 \cdot 5^3$ by 21.5.3.

Let $H = C_{\tilde{G}}(B)$ and $H^* = H/B$. By 21.5.3, $|Aut_{\tilde{G}}(B)| = 4$ and $C_{\tilde{N}}(B)/\tilde{B}C_{\tilde{E}}(B) \cong A_4$ is faithful on $C_{\tilde{E}}(B) \cong E_8$ and stabilizes a hyperplane Z of $C_{\tilde{E}}(B)$. Thus $|H^*| = 2^5 \cdot 3^2 \cdot 5^2$ by the previous paragraph. Also a Sylow 2-group T of $C_{\tilde{N}}(B)$ is isomorphic to $E_4 wr \mathbf{Z}_2$. Moreover elements in $C_{\tilde{E}}(B) - Z$ are conjugate to $\tilde{z}$ while those in $Z^\#$ are of the form $\tilde{e}_D$, D a dodecad, so $C_H(s) \leq C_{\tilde{N}}(B)$ for each $s \in Z^\#$ by 26.3. Therefore $T \in Syl_2(H)$.

Let $J = J(T) \cong E_{16}$. By the previous paragraph we may assume $\tilde{z} \in T - J$. Next $C_Q(B) = \langle z \rangle$, so from 25.1, B is determined up to conjugacy in $C_G(z)$ and $C_{\tilde{G}}(B\langle z\rangle) \cong \mathbf{Z}_{10} \times A_5$. In particular $|C_H(\tilde{z})|_2 = 8$, so $\tilde{z}^H \cap J = \varnothing$, and hence by Thompson transfer H has a subgroup K of index 2 with $\tilde{z} \notin K$. As $J = [J,R]$ for $R \in Syl_3(C_{\tilde{N}}(B))$, $J \in Syl_2(K)$.

Let $I = N_K(J)$ and recall that $C_{K^*}(s) = J^*$ for all $s \in Z^\#$. Further by 7.7, I controls fusion in J, so by 7.3, s fixes a unique point of K/I. Also $I \neq K$ as $C_K(z) \not\leq I$. Hence by Exercise 2.10, there exists $U \leq J$ such that $C_I(U)/U$ is strongly embedded in $C_K(U)/U$, $U \cap Z = 1$, and $C_I(U)$ is transitive on $(J/I)^\#$. As 7 does not divide $|H|$, transitivity of $C_I(U)$ on $(J/I)^\#$ implies $|U| \neq 2$, so as $U \cap Z = 1$, either $|U| = 4$ and $J = Z \times U$, or $U = 1$. But in the latter case I is transitive on $J^\#$, so as $C_{K^*}(s^*) = J^*$, Exercise 16.6 in [FGT] says $K^* \cong L_2(16)$. This is impossible as 17 does not divide $|H|$.

Thus $J = U \times Z$, so by Exercise 16.6 in [FGT], $C_K(U)/BU \cong L_2(4)$. Let $J_1 = U$ and $R_1 \in Syl_3(C_K(J_1) \cap N_K(J))$ be z-invariant. Then $RR_1 \in Syl_2(K)$ by an order argument and $\langle z\rangle RR_1$ has two orbits on $J^\#$ of length 6 and 9, with elements of $J_1^\#$ in the orbit of length 6. In particular $J = J_1 \times J_2$, where $J_2 = J_1^z$ and J_1 is RR_1-invariant.

Now $K_1 = RC_K(J_1)$ is of index 5 in K, so we have a permutation representation $\alpha : K^* \to S_5$. We conclude $K_1^* = C_K(J_1)^*$ is the kernel of this representation, so $K^* = K_1^* \times K_2^*$, where $K_2^* = K_1^{*z}$. That is, $H^* \cong A_5 wr Z_2$.

Lemma 26.6: *Let Y be a 3-central subgroup of N of order 3. Then $dim(C_{\tilde{\Lambda}}(Y)) = 12$, $|Aut_{\tilde{G}}(Y)| = 2$, $|C_{\tilde{G}}(Y) : E(C_{\tilde{G}}(Y))| = 2$, and $E(C_{\tilde{G}}(Y)) \cong U_4(3)/E_9$.*

Proof: As usual 21.4 and 26.1.1 imply the first two remarks.

Next by 21.4, N has two classes of elements of order 3 and from 26.4 the dimensions of the fixed point spaces on $\tilde{\Lambda}$ are different for the two classes, so they are not fused in $\tilde{G}$. Thus we can apply 26.1.4. As in the Remark above, we check that Y fixes 96, 30 vectors in Λ_2^i for $i = 3, 4$, and $N_{\tilde{N}}(B)$ has two orbits of length 96 and 60 on $\tilde{\Lambda}_2^2$. So $|C_{\tilde{\Lambda}_2}(Y)| = 378$.

Next Y centralizes a subgroup Y_1 of order 3 such that $C_{\tilde{G}}(Y_1)/Y_1 \cong Sz$ by 25.9. By 46.6, $K = C_{\tilde{G}}(YY_1)$ is quasisimple with $K/\tilde{Y}\tilde{Y}_1 \cong U_4(3)$. By Exercise 9.6, Y_1 has no fixed points on $\tilde{\Lambda}$, so $C_{\tilde{\Lambda}}(Y)$ has sign $+1$, and then by 26.1.3, $|C_{\tilde{\Lambda}_4}(Y)| = 3^5 \cdot 7$ and $|N_{\tilde{G}}(Y)| = 3^5 \cdot 7 \cdot |N_{\tilde{N}}(Y)| = 2^9 \cdot 3^8 \cdot 5 \cdot 7$. So as $|U_4(3)| = 2^7 \cdot 3^6 \cdot 5 \cdot 7$ and $|Aut_N(Y)| = 3$, we conclude $K = E(N_{\tilde{G}}(Y))$ is of index 2 in $C_{\tilde{G}}(Y)$ and $|N_{\tilde{G}}(Y) : K| = 4$.

Remarks. Conway was the first to find Co_2, Co_3, Mc, and HS as subgroups of Co_1 in [Co2]. Some of our arguments are roughly the same as

his, although Conway felt free to use classification theorems in the literature to identify various subgroups, whereas here we keep our treatment
self-contained. This will cause small problems later too.

The McLaughlin group Mc and the Higman–Sims group HS were discovered as rank 3 permutation groups by McLaughlin [Mc] and D. Higman and C. Sims [HS], respectively. Conway showed that these rank 3
representations could be realized on vectors in the Leech lattice as in
24.6 and 24.11. After Higman and Sims discovered HS, G. Higman hap-
pened upon HS as a 2-transitive group [HiG]. Sims proved the group
discovered by G. Higman was isomorphic to HS [Si1]. Conway showed
this representation could be realized on the Leech lattice as in 24.13.

In [Co2], Conway attributes the calculation of various local subgroups
of Co_1 to Thompson. This work was never published, but a fairly complete description of the local structure of Co_1 was derived using different
methods by Nick Patterson in his thesis [P]. In many cases we follow
Patterson's treatment, although again Patterson felt free to quote the
literature, whereas we do not. Also when possible, we use the theory
of large extraspecial subgroups in Chapter 2 to replace Patterson's
arguments.

Exercises

1. Let X be a set of order 10 and $A_6 \cong A \le H \le Sym(X)$ with A
   transitive on X and $|H| = |A| \cdot 3^i$. Prove $H = A$.
2. Let $G = Co_1$ and $P \in Syl_{23}(G)$. Prove $|N_G(P)| = 23 \cdot 11$.
3. Adopt the notation of Section 24 and let $v = v_1 - z$ for some $z \in \Xi^3$.
   Let $L = C_{G_4}(z)$, $\Gamma_1 = \{v_1 - w : w \in \Xi^3\}$, and $\gamma = \Gamma_1 - \{v\}$. Prove L
   has two orbits γ_1 and γ_2 on γ, where $\gamma_i = \Lambda_2(v, i)$. Further $|\gamma_1| = 70$
   and $|\gamma_2| = 105$.
4. Prove Lemma 24.10. Use 23.6.
5. Let $\Omega = (\Omega_1, \Omega_0)$ be a rank 2 geometry satisfying the properties of
   24.7 and G a group of automorphisms of Ω satisfying the properties
   of G_7 in 24.7. Denote by Δ the collinearity graph of Ω. Let (x, l) be
   a flag of Ω, $z \in \Delta^2(x)$, $\{y\} = \Delta(z) \cap \Omega(l)$, and $P = P(x)$ the kernel
   of the action of G_x on the set $\Omega_0(x)$ of lines through x. Prove
   (1) G_x is faithful on $\Delta(x)$.
   (2) P is regular on $\Delta^2(x)$ and $G_{x,z}$ is a complement to P in G_x.
   (3) $G_{x,z}$ preserves an orthogonal space structure of sign -1 on P and
   $G_{x,z,l}$ is the stabilizer in $G_{x,z}$ of a singular point $W(l)$ of P.

(4) The kernel $Q(l)$ of the action of G_l on $\Omega(l)$ is of order $2 \cdot 3^5$ with $P(l) = O_3(Q(l))$ the unique normal 3^{1+4} subgroup of $G_{x,l}$ and with $W(l) = Z(P(l))$.

(5) $P \cap P(y) = W(l)$.

(6) For $h \in G_x$ and $p \in P$, $G_{x,y} = G_{x,z,l}(P \cap P(l))$; $yh \in \Delta(y)$ if and only if $h \in G_l = G_{x,z,l}P$; and $zp \in \Delta(y)$ if and only if $p \in P \cap P(l)$.

(7) For $p \in P^{\#}$, $zp \in \Delta(z)$ if and only if $p \in W(k)$ for some $k \in \Omega_0(x)$.

(8) $G \cong U_4(3)$.

(Hint: See the proof of 45.11 and use Exercise 15.3.)

6. Assume the hypothesis of Lemma 25.9. Prove $C_{\tilde{\Lambda}}(A) = 0$.

(Hint: Prove $\tilde{\Lambda}_4 \cap C_{[\tilde{\Lambda},z]}(A) = \varnothing$, and use this to show A is fixed point free on $[\tilde{\Lambda}, z]$ and hence also on $\tilde{\Lambda}/C_{\tilde{\Lambda}}(z)$. Conclude $C_{\tilde{\Lambda}}(A) \leq C_{\tilde{\Lambda}}(C_G(A))$. Finally observe $|C_G(A)|_2$ does not divide $|Co_3|$, and hence $C_G(A)$ fixes no nonsingular point of $\tilde{\Lambda}$.)

Chapter 10

The Griess Algebra
and the Monster

In this chapter we construct the Griess algebra and its automorphism group, which is the largest sporadic group: the Monster. We begin in Section 27 by specializing the construction of Section 14 to the Parker loop L. The subgroup $\bar{N} = N/K$ supplied by this construction is the normalizer of the 4-group $Z = \langle z_1, z_2 \rangle$ in the Monster.

We saw in Section 14 that $\bar{N}$ contains a large extraspecial 2-subgroup $\bar{Q} \cong D_8^{12}$. In Section 27 we construct a group C with $F^*(C) = \bar{Q}$ and $C/\bar{Q} \cong Co_1$. The group C is the centralizer of the involution z_1 in the Monster and $\bar{Q}$ is a large extraspecial subgroup of the Monster. In Section 27 we construct a 196,884-dimensional $\mathbf{R}C$-module B for C admitting a C-invariant bilinear form γ.

Next in Section 28, we define an algebra map τ on B preserved by C. The algebra (B, τ) is the *Griess algebra* first constructed by Griess. From Section 9, the algebra map τ is equivalent to a C-invariant trilinear form β. The construction follows Conway in [Co3] and uses the Parker loop L.

Next $\bar{N}_1 = C_{\bar{N}}(z_1)$ is a subgroup of C. In Section 29 we extend the representation of $\bar{N}_1$ to $\bar{N}$ in such a way that $\bar{N}$ preserves γ. Thus $G_0 = \langle C, \bar{N} \rangle \le O(B, \gamma)$, with $C \cap \bar{N} = \bar{N}_1$.

In Section 30 we prove that $\bar{N}$ preserves β. Thus $G_0 \le G = O(B, \gamma, \beta)$, and from Section 9, G is a group of automorphisms of the Griess algebra. Finally in Section 31 we prove that $C = C_G(z_1)$ and that G is a finite simple group. We define G to be the Monster.

The Remarks at the end of this chapter contain a brief discussion of the approaches of Griess, Conway, and Tits to constructing the Monster, and a comparison of those approaches to the treatment given here.

Since this chapter is replete with specialized notation, we close this introduction with a list of notation used in Chapter 10. Column 1 of Table 4 lists the symbol and column 2 the page where the symbol is defined.

27. The subgroups C and N of the Monster

In this section we assume the hypotheses of Section 14 with $V = V_C$ the Golay code module for the Steiner system $(X, \mathcal{C})$ and $P(v) = |v|/4$ mod 2 as discussed in Sections 18, 19, and 20. Write U for the power set of X regarded as a vector space over the field F of order 2 under symmetric difference; thus $V \le U$. Notice this differs slightly from the notational conventions of Sections 19 and 20.

Recall from Section 14 that L is a Moufang symplectic 2-loop with parameters (P, C, A), where from 11.8 the commutator map $C : V \times V \to F$ is given by

$$C(u,v) = |u \cap v|/2 \mod 2$$

and the associator map $A : V \times V \times V \to F$ is given by

$$A(u, v, w) = |u \cap v \cap w| \mod 2.$$

The loop L is the *Parker loop.*

Recall also from Section 14 that the center of L is $\langle \pi \rangle = \{1, \pi\}$ and $\phi :$ $L \to V$ is a surjection of loops with kernel $\langle \pi \rangle$. Further the centralizer E in $Aut(L)$ of $L/\langle \pi \rangle$ consists of the maps α_ϵ, $\epsilon \in V^*$, with $\alpha_\epsilon(d) = d\pi^{\epsilon(d)}$ for $d \in L$. For $d \in L$ and $f = 0$ or 1 in $GF(2)$ we sometimes write $d + f$ for $d\pi^f$. So $\alpha_\epsilon(d) = d + \epsilon(d)$ under this convention.

We choose the distinguished element v_0 of Section 14 to be $v_0 = X$ and take $\Gamma_0 = M_{24}$. Recall $E \le \Gamma \le Aut(L)$ with $\Gamma/E = \Gamma_0$. In particular the natural map $\Gamma \to \Gamma_0$ gives us a permutation representation of Γ on X with kernel E and for $\alpha \in \Gamma$ we write $x\alpha$ for the image of $x \in X$ under α via this representation. Hence, for example, for each $\epsilon \in V^*$, α_ϵ is in the kernel of the action of Γ on X.

Recall from Section 14 that $\Omega = L \cup \{0\}$ with $0 \cdot d = d \cdot 0 = 0$ for all $d \in L$, and to each $d \in L$ and $\alpha \in \Gamma$ there are associated permutations $\psi_i(d)$, $\psi_i(\alpha)$, $i = 1, 2, 3$, of Ω^3. See Section 14 for the definition of these permutations and lemma 14.2 for an extensive list of properties of these permutations and how they multiply.

Recall N is the subgroup of $Sym(\Omega^3)$ generated by these permutations. Further $z_i = \psi_i(\pi)$, $s \in L$ with $\phi(s) = X$, $k_i = \psi_{i-1}(s)\psi_{i+1}(s\pi)$, $Z = \langle z_1, z_2 \rangle$, and $K = \langle k_1, k_2 \rangle$. By 14.3, $1 = \psi_1(d)\psi_2(d)\psi_3(d)$ for each $d \in L$, so in particular $z_1 z_2 z_3 = k_1 k_2 k_3 = 1$. Also by 14.2, Z and K are normal in N. Further $N^+ = C_N(Z) = C_N(K)$ and $N/N^+ \cong S_3$ by 14.3. We write $\bar{N} = N/K$. Recall $N_i = C_N(z_i)$ and $Q_i = \langle \psi_i(d), \psi_i(\alpha) : d \in L, \alpha \in E \rangle$. By 14.4, $Q_i \trianglelefteq N_i$, $z_i, k_i \in Z(N_i)$, and $Q_i/\langle k_i \rangle \cong \bar{Q}_i \cong D_8^{12}$ is extraspecial of width 12.

For $J \in U$, define the elements f_J and $f(J)$ of the Leech lattice Λ as in Section 23; thus

$$f_J = 2e_J, \quad f(J) = 2f_J - |J| f_X/2.$$

By Exercise 3.2.1 we have a surjection $J \mapsto \epsilon_J$ of U onto V^* with kernel V, where $\epsilon_J(v) = |J \cap v| \mod 2$. Write α_J for $\alpha_{\epsilon_J} \in E$, and observe using Exercise 3.2 and Remark 14.1 that:

Lemma 27.1: *(1) The map $J \mapsto \alpha_J$ is an isomorphism of U/V with E.*
(2) For all $d, e \in L$, $\alpha_{\langle d, e \rangle} = \alpha_{\phi(d) \cap \phi(e)}$.

Lemma 27.2: *Let $\bar{Q}_1 = Q_1/\langle z_1, k_1 \rangle$ and regard $\bar{Q}_1$ as an orthogonal space over F. Then there exists a surjective group homomorphism*

$\xi_1 : Q_1 \to \tilde{\Lambda}$ *with kernel* $\langle z_1, k_1 \rangle$ *defined by*

$$\xi_1 : \psi_1(d)\psi_2(b)\psi_1(\alpha_J) \mapsto f_{\phi(d)+t(b)}X + f(J) + 2\lambda,$$

where $t(b) = 0, 1$ *for* $b = 1, \pi$, *respectively, which induces an isometry*
$\tilde{\xi} : \tilde{Q}_1 \to \tilde{\Lambda}$.

Proof: Let $D = V \times U/V$ and regard D as an F-space as in 23.10 with quadratic form q defined in Exercise 8.10. Then 23.10 and Exercise 8.10 give us an isometry $\iota : D \to \tilde{\Lambda}$ defined by

$$\iota : (v, V + J) \mapsto f_v + f(J) + 2\Lambda.$$

Also Exercise 4.6 gives us a group homomorphism $\mu : Q_1 \to D$ inducing an isometry $\tilde{\mu} : \tilde{Q}_1 \to \tilde{\Lambda}$ defined by

$$\mu : \psi_1(d)\psi_2(b)\psi_1(\alpha_J) \mapsto (\phi(d) + t(b)X, V + J).$$

Then $\xi_1 = \mu\iota$ is our homomorphism inducing the isometry $\tilde{\xi} = \tilde{\mu}\iota$. Exercises 4.6 and 8.10 apply in a general setting; we give a proof now that μ is a homomorphism and $\tilde{\xi}$ an isometry in this special case.

First the map μ is well defined by 14.4.4. By 14.2.7,

$$\psi_1(d)\psi_1(e) = \psi_1(de)\psi_3(C(e,d))\psi_1(\alpha_{\langle e,d\rangle}).$$

By 27.1.2, $\psi_1(\alpha_{\langle e,d\rangle}) = \psi_1(\alpha_{\phi(d)\cap\phi(e)})$. Also for $b \in \langle \pi \rangle$,

$$[\psi_1(\alpha_J), \psi_1(d)] \equiv [\psi_1(\alpha_J), \psi_2(b)] \equiv 1 \mod \langle z_1 \rangle$$

by 14.2.4 and 14.2.5, since $\psi_2(b)\psi_3(b) = \psi_1(b)$. Therefore

$$\psi_1(d)\psi_2(a)\psi_1(\alpha_J) \cdot \psi_1(e)\psi_2(b)\psi_1(\alpha_K)$$

$$\equiv \psi_1(de)\psi_2(ab + C(d,e))\psi_1(\alpha_J\alpha_K\alpha_{\phi(d)\cap\phi(e)}).$$

Then as z_1 is in the center of Q_1 and the kernel of μ, the check that μ is a homomorphism is a straightforward calculation.

So $\tilde{\xi} : \tilde{Q}_1 \to \tilde{\Lambda}$ is an isomorphism of F-spaces. To check that $\tilde{\xi}$ is an isometry, we first observe that as $\psi_1(E)$ is elementary abelian, $\xi_1(E)$ is totally singular. Further by 23.10.2, $\tilde{\xi}(\psi_1(E))$ is totally singular. Thus it suffices to show

(a) $g^2 = 1$ if and only if $q(\tilde{\xi}(\tilde{g})) = 0$ for $g \in \psi_1(L)$, and
(b) $[g,h] = 1$ if and only if $(\tilde{\xi}(\tilde{g}), \tilde{\xi}(\tilde{h})) = 0$ for $g \in \psi_1(L), h \in \psi_1(E)$.

But if $d \in L$ then $\psi_1(d)^2 = \psi_1(P(d))$, so (a) holds by 23.10.3. Similarly

$$[\psi_1(d), \psi_1(\alpha_J)] = \psi_1(\epsilon_J(d) + |J|P(d)) = \psi_1(|J \cap d| + |J||d|/4)$$

by 14.2.8, so 23.10.3 also implies (b).

We next construct several **R**-modules for N_1 and N^+.

Recall from Section 14 that Δ_i consists of those members of Ω^3 with ith entry in L and 0 elsewhere. For $d \in L$, write $\delta_i(d)$ for the element of Δ_i determined by d. For $i \neq j$ let $B^j(\Delta_i)$ be the **R**-space with basis the orbit space $\Delta_i/\langle k_j \rangle$ subject to the convention that $\delta(d)\psi_j(\pi) = -\delta(d)$, where $\delta(d) = \delta_i^j(d)$ denotes the orbit under $\langle k_j \rangle$ of $\delta_i(d)$. Observe that by 27.3 and Exercise 4.3, $B^j(\Delta_i)$ is a 2^{11}-dimensional monomial module for N^+ with respect to the basis $\delta(d)$, $d \in L$, via $g : \delta(d) \mapsto \delta(d)g$. Regard $B_j(\Delta_i)$ as an orthogonal space with orthonormal basis $\delta(d)$, d varying over a suitable transversal. By Exercise 4.3, N^+ preserves this inner product. Also an easy calculation shows:

Lemma 27.3: *For all $b, d \in L$ and $\alpha \in \Gamma$*

  (1) $\psi_1(d) : \delta_2(b) \mapsto \delta_2(db)$.
  (2) $\psi_2(d) : \delta_2(b) \mapsto \delta_2(b + P(d) + C(b, d))$.
  (3) $\psi_2(\alpha) : \delta_2(b) \mapsto \delta_2(b\alpha)$ *if α is even.*
  (4) $\psi_2(\alpha) : \delta_2(b) \mapsto \delta_2(b\alpha + P(b))$ *and* $\psi_1(\alpha) : \delta_2(b) \mapsto \delta_3(b\alpha + P(b))$ *if α is odd.*
  (5) $\langle \psi_1(s), \psi_2(\pi), \psi_2(s) \rangle$ *is the kernel of the action of N^+ on $B^1(\Delta_2)$.*
  (6) $\langle \psi_1(s\pi), \psi_2(\pi), \psi_2(s) \rangle$ *is the kernel of the action of N^+ on $B^3(\Delta_2)$.*

Next for $x \in X$, $\epsilon \in \{0, 1\}$, define

$$L_i^\epsilon(x) = \{\delta_i(d) : d \in L \text{ and } \epsilon_x(\phi(d)) = \epsilon\}$$

and using Exercise 4.3, define B_{24}^i to be the 24-dimensional **R**-space with basis $L_i^\epsilon(x)$, $x \in X$, $\epsilon \in \{0, 1\}$, subject to the convention that $L_i^{\epsilon+1}(x) = -L_i^\epsilon(x)$. Make B_{24}^i into an orthogonal space by decreeing that $L_i^1(x)$, $x \in X$, be an orthonormal basis. By Exercise 4.3 and 27.4, B_{24}^i is a monomial module for N_i and N_i preserves the form on B_{24}^i as N_i permutes the vectors $L_i^\epsilon(x)$ via right multiplication; indeed:

Lemma 27.4: *For each $x \in X$, $d \in L$, $\alpha \in \Gamma$, and $\epsilon \in \{0, 1\}$:*

  (1) $\psi_i(d)$ *fixes $L_i^\epsilon(x)$ and* $\psi_j(d) : L_i^\epsilon(x) \mapsto L_i^{\epsilon+\epsilon_x(\phi(d))}(x)$ *for $j \neq i$.*
  (2) $\psi_i(\alpha) : L_i^\epsilon(x) \mapsto L_i^\epsilon(x\alpha)$.
  (3) $\psi_2(\alpha) : L_1^\epsilon(x) \mapsto L_3^\epsilon(x\alpha)$ *if α is odd.*
  (4) Q_i *is the kernel of the action of N_i on B_{24}^i and k_j inverts B_{24}^i for $j \neq i$.*

Proof: These are easy computations. For example, $(a, 0, 0)\psi_2(d) = (ad, 0, 0)$ and $\epsilon_x(\phi(ad)) = \epsilon_x(\phi(a) + \phi(d)) = \epsilon_x(\phi(a)) + \epsilon_x(\phi(d))$, so $L_1^\epsilon(x)\psi_2(d) = L_1^{\epsilon+\epsilon_x(d)}(x)$.

Next by 27.3, $B^1(\Delta_i)$, $i = 2, 3$, are monomial N^+-modules with respect to the bases $\delta_i(d)$, $d \in L$, with $B^1(\Delta_2)\psi_1(\alpha) = B^1(\Delta_3)$ for α odd, so $B^1_- = B^1(\Delta_2) \oplus B^1(\Delta_3)$ is a monomial N_1-module. Indeed B^1_- is the N_1-module constructed in 14.6. Notice that by 14.6, z_1 acts as -1 on B^1_-. Let $B_- = B^1_{24} \otimes B^1_-$, regarded as an N_1-module. Also B_- is an orthogonal space whose form is the tensor product of the forms on the factors. We write $x \otimes \delta_i(d)$ for the basis vector $L^1(x) \otimes \delta^1_i(d)$ of B_-. Observe that a suitable subset of these vectors forms an orthonormal basis for the form on B_-. Notice also that $-\delta^1_i(d) = \delta^1_i(d)z_1 = \delta^1_i(d\pi) = \delta^1_i(d+1)$, so $L^{1+\epsilon}(x) \otimes \delta^1_i(d) = L^\epsilon(x) \otimes \delta^1_i(d+1) = -(L^\epsilon(x) \otimes \delta^1_i(d))$.

Lemma 27.5: *(1)* B_- *is a monomial module for* N_1 *of dimension* $2^{12} \cdot 24$ *with respect to the basis* $x \otimes \delta_i(d)$, $x \in X$, $d \in L$, $i = 2, 3$, *and* N_1 *preserves the form on* B_-.

(2) z_1 *acts as* -1 *on* B_-.

(3) K *is the kernel of the action of* N_1 *on* B_-.

(4) $\psi_2(s\pi)$ *is trivial on* B^1_- *and inverts* B^1_{24}.

Proof: We have observed that B_- is a monomial N_1-module, so (1) holds. As z_1 centralizes B^1_{24} and inverts B^1_-, (2) holds. From 14.6, $\langle \psi_2(s\pi), \psi_3(s) \rangle$ is the kernel of the action of N_1 on B^1_-. In particular together with 27.4.4 this means that k_1 is trivial on B^1_- and B^1_{24} and hence also on B_-. Further $k_2 = \psi_2(s\pi)z_1$ inverts B^1_- and B^1_{24}, so k_2 is trivial on B_- and (4) holds. That is, K is contained in the kernel J of the action of N_1 on B_-. By 27.4, $[Q_1, B^1_{24}] = 0$, so as $\langle \psi_2(s\pi), \psi_3(s) \rangle$ is the kernel of the action of Q_1 on B^1_1, $Q_1 \cap \langle \psi_2(s\pi), \psi_3(s) \rangle = \langle k_1 \rangle$ is the kernel of the action of Q_1 on B_1. Thus $[Q_1, J] \leq Q_1 \cap J = \langle k_1 \rangle$. Then $J \leq C_{N_1}(Q_1K/K) = \langle z_1, K \rangle$ by 14.4.2, so $J = K$.

We next construct the centralizer C of a 2-central involution in the Monster and an amalgam

$$C \leftarrow \bar{N}_1 \rightarrow \bar{N}.$$

Let $\mu_1 : N_1 \rightarrow SL(B^1_-)$ be the representation of N_1 on B^1_-. Then by 14.6, $Q_1\mu_1$ is extraspecial and irreducible on B^1_-, so $N_1\mu_1 \leq M_1$ the normalizer in $SL(B^1_-)$ of $Q_1\mu_1$. In particular by Exercise 4.4, $M_1/Q_1\mu_1$ is the isometry group of the quadratic form on $\tilde{Q}_1 = Q_1K/K\langle z_1 \rangle$. We use the isomorphism of 27.2 to give $\tilde{Q}_1$ the structure of $\tilde{\Lambda}$ and let C_1 be the subgroup of M_1 preserving this structure with $C_1/Q_1\mu_1 \cong Co_1$. We will see that $N_1\mu_1 \leq C_1$. Let $\nu_1 : N_1\mu_1 \rightarrow O(\tilde{Q}_1)$ be the natural map with kernel $Q_1\mu_1$.

Similarly let $\mu_2 : N_1 \rightarrow SL(B^1_{24})$ be the representation of N_1 on B^1_{24}.

The map $L_1^1(x) \mapsto \sqrt{8}e_x$ defines an isomorphism of B_{24}^1 with $\mathbf{R} \otimes_{\mathbf{Z}} \Lambda$ in which $N_1\mu_2$ stabilizes the inverse image of Λ, so identifying this sublattice with Λ via the isomorphism, we have $N_1\mu_2 \leq C_2$ the stabilizer in $SL(B_{24}^1)$ of Λ. The map $\Lambda \to \tilde{\Lambda} \cong \tilde{Q}_1$ induces a surjection $\nu_2 : C_2 \to Co_1$.

We claim $\mu_1\nu_1 = \mu_2\nu_2$; this will show $N_1\mu_1 \leq C_1$ as mentioned earlier. Our claim amounts to the assertion that the map $\tilde{\xi}$ of 27.2 is N_1-equivariant. By construction during the proof of 27.2, $\tilde{\xi} = \tilde{\mu}\iota$, while by 23.10.4 and Exercise 4.6, the maps $\tilde{\mu}$ and ι are N_1-equivariant, establishing the claim. This proof depends on the general Exercise 4.6; we prove the equivariance directly in our special case now.

It suffices to show $g\mu_1\nu_1 = g\mu_2\nu_2$ for $g = \psi_1(\beta)$, $\beta \in M_{24}$, and $g = \psi_2(e)$, $\phi(e) \in C$, as these elements together with Q_1 generate N_1. Similarly as the elements $\tilde{\lambda}_x$, $x \in X$, generate $\tilde{\Lambda} \cong \tilde{Q}_1$, it suffices to check that $g\mu_1\nu_1$ and $g\mu_2\nu_2$ agree on $\tilde{\lambda}_x$ for each $x \in X$.

If $\beta \in M_{24}$ then by 27.4, $L_1^1(x)\psi_1(\beta) = L_1^1(x\beta)$, so as $L_1^1(x) \mapsto \sqrt{8}e_x$, we conclude $\tilde{\lambda}_x\psi_1(\beta)\mu_2\nu_2 = \tilde{\lambda}_{x\beta}$. On the other hand $\tilde{\xi}(\tilde{\psi}_1(\alpha_x)) = \tilde{\lambda}_x$ (cf. 27.8.3) and $\tilde{\psi}_1(\alpha_x)^{\psi_1(\beta)} = \tilde{\psi}_1(\alpha_{x\beta})$ by 14.2.4, so $\tilde{\lambda}_x\psi_1(\beta)\mu_1\nu_1 = \tilde{\lambda}_{x\beta}$ too. Next for $\phi(e) \in C$, $\tilde{\psi}_1(\alpha_x)^{\psi_2(e)} = \tilde{\psi}_1(e \cdot s^{\epsilon_x(e)})\tilde{\psi}_1(\alpha_x)$ (cf. 29.6.2) so $\tilde{\lambda}_x\psi_2(e)\mu_1\nu_1 = -\tilde{\lambda}_x + f_{\phi(e)+\epsilon_x(e)X} + 2\Lambda$. Finally by 27.4, $L_1^1(y)\psi_2(e) = L_1^{1+\epsilon_x(\phi(e))}(y)$, so $\psi_2(e)$ changes the signs of the entries a_y in $-\tilde{\lambda}_x$ for $y \in \phi(e)$, and hence indeed $\tilde{\lambda}_x\psi_2(e)\mu_2\nu_2 = \tilde{\lambda}_x\psi_2(e)\mu_1\nu_1$, establishing the claim.

Finally form the fiber product $C_0 = C_1 \times_{Co_1} C_2$ with respect to the diagram

$$C_1 \xrightarrow{\nu_1} Co_1 \xleftarrow{\nu_2} C_2$$

as in Section 6. As $\mu_1\nu_1 = \mu_2\nu_2$, 6.1 gives us a map $\mu : N_1 \to C_0$ with $\mu p_i = \mu_i$ for $i = 1, 2$, and 6.2 says $p_1 \otimes p_2 : C_0 \to O(B_-)$, where $p_i : C_0 \to C_i$ is the ith projection, and $\mu(p_1 \otimes p_2) = \mu_1 \otimes \mu_2 : N_1 \to O(B_-)$. Let C denote the image of C_0 under $p_1 \otimes p_2$, so that $\mu_1 \otimes \mu_2 : N_1 \to C \leq O(B_-)$.

Lemma 27.6: *(1)* $F^*(C) = Q_1(\mu_1 \otimes \mu_2) \cong D_8^{12}$.

(2) $C/F^*(C) \cong Co_1$.

(3) $\ker(\mu_1 \otimes \mu_2) = K$.

(4) $\tilde{Q}_1 = Q_1K/K\langle z_1 \rangle$ *is isomorphic to* $\tilde{\Lambda}$ *as a* $C/F^*(C)$*-module via the isomorphism* $\tilde{\xi}$ *of 27.2. Further* $\tilde{\xi}(\tilde{z}_2) = 8e_x + 2\Lambda$ *and* $N_1/Q_1K \cong M_{24}/E_{2^{11}}$ *is the stabilizer in* $C/F^*(C)$ *of* $\tilde{z}_2$ *with* E^+K/KZ *isomorphic to the 11-dimensional Todd module for* $N_1/O_2(N_1)$.

Proof: Part (3) is a restatement of 27.5.3. By (3) and 14.4, $Q_1(\mu_1 \otimes \mu_2) \cong Q_1/K \cong D_8^{12}$.

Next by 27.5.4, $\psi_2(s\pi) \in ker(\mu_1)$ with $\langle \psi_2(s\pi)\mu_2 \rangle = ker(\nu_2)$. Thus by 6.1, $ker(p_1) = \langle \psi_2(s\pi)\mu \rangle \le (Q_1 K)\mu$. Then as $(Q_1 K)\mu_1 = F^*(C_1)$ with $C_1/(Q_1 K)\mu_1 \cong Co_1$, $(Q_1 K)\mu = F^*(C_0)$ and $C_0/(Q_1 K)\mu \cong Co_1$. Hence (3) implies (1) and (2).

By construction, $\tilde{Q}_1 \cong \tilde{\Lambda}$ as a $C/\bar{Q}_1$-module via $\tilde{\xi}$, with $\tilde{\xi}(\tilde{z}_2) = f_X + 2\Lambda = 8e_x + 2\Lambda$. Then as N_1 stabilizes $\tilde{z}_2$ and $|N_1 : Q_1 K| = |M_{24}/E_{2^{11}}|$ with $M_{24}/E_{2^{11}}$ the stabilizer of $8e_x + 2\Lambda$ in Co_1, all but the final remark in (4) hold. Finally by 14.2.4, $\psi_1(\alpha_J)^{\psi_1(\beta)} = \psi_1(\alpha_{J\beta})$ for $J \in U$, $\beta \in M$, so the map

$$E^+ K/KZ \to T,$$

$$\psi_1(\alpha_J)KZ \mapsto J + V$$

is an N^+-equivariant isomorphism of $E^+ K/KZ$ with the 11-dimensional Todd module T.

Let $\bar{N} = N/K$ and for $g \in N$ set $\bar{g} = gK$. Then $\mu_1 \otimes \mu_2$ induces an injection $\bar{N}_1 \to C$ and we identify $\bar{N}_1$ with its image under this injection and regard $\bar{N}_1$ as a subgroup of C. In particular under this convention, $\bar{Q}_1 = F^*(C)$ by 27.6.1, while by 27.6.2, $C/\bar{Q}_1 \cong Co_1$.

Let $B_s = S^2(B_{24}^1)$ be the symmetric square of B_{24}^1 regarded as a C-module with quadratic form equal to the tensor product of the forms on B_{24}^1. Thus B_s has a basis $\{xy : x, y \in X\}$, where

$$xy = L_1^1(x) \otimes L_1^1(y) + L_1^1(y) \otimes L_1^1(x) \text{ if } x \neq y$$

and $x^2 = L_1^1(x) \otimes L_1^1(x)$. In particular $(xy, x'y') = 0$ if $xy \neq x'y'$, while $(x^2, x^2) = 1$ and $(xz, xz) = 2$ for $x \neq z$. Further

Lemma 27.7: *(1)* $\bar{Q}_1$ *is the kernel of the action of C on B_s.*

(2) B_s *is a monomial module for N_1 with $\bar{Q}_2^+$ fixing $\mathbf{R}xy$ for all $x, y \in X$.*

(3) $B_s = B_R \perp B_E^1$, where $B_R = \langle x^2 : x \in X \rangle$ and $B_E^1 = \langle xy : x, y \in X,\ x \neq y \rangle$.*

Proof: Part (1) follows from 27.4.4 and the fact that $\bar{C}/\bar{Q}_1$ is simple. Part (2) follows from 27.4.1 as $Q_2^+ = \langle \psi_2(d) : d \in L \rangle E^+$. Part (3) follows from the definition of the quadratic form on B.

Next identify $\tilde{Q}_1 = \bar{Q}_1/\langle \tilde{z}_1 \rangle$ with $\tilde{\Lambda}$ via the isometry $\tilde{\xi}$ of 27.2. By construction $C_1/(Q_1 K)\mu_1 \cong Co_1$ preserves this structure, so as $ker(p_1) \le (Q_1 K)\mu$, so does C. For $g \in Q_1$ or $\bar{g} \in \bar{Q}_1$, write $\tilde{g}$ for the image of g in $\tilde{Q}_1$. Write $\mathcal{R}$ for the set of $r \in \bar{Q}_1$ such that $\tilde{r} \in \tilde{\Lambda}_2$. Then C acts on $\mathcal{R}$ via conjugation, inducing a monomial $\mathbf{R}$-module B_r with basis $y(r)$,

$r \in \mathcal{R}$, subject to $y(rz_1) = -y(r)$, via $y(r)g = y(r^g)$ for $r \in \mathcal{R}$ and $g \in C$. Observe that C preserves the quadratic form on B_r for which our basis is orthonormal.

Remark. In the next lemma we associate to each $r \in \mathcal{R}$ an element $\xi(r) \in \Lambda_2$ such that $\xi(r) + 2\Lambda = \tilde{\xi}(r)$. Recall from 23.2 that there are exactly two $\lambda \in \Lambda_2$ with $\tilde{\lambda} = \tilde{\xi}(r)$. We caution that for $r \in \mathcal{R}^2$, $\xi(r)$ is not well defined in that $r(c, J + \phi(c)) = r(c, J)$ but $\xi(r(c, J + \phi(c))) = -\xi(r(c, J))$. However, we only encounter $\xi(r)$ in the guise of some function $h(\xi(r))$ with the property that $h(-\xi(r)) = h(\xi(r))$, so this abuse of notation causes no problems.

Lemma 27.8: *(1)* $\mathcal{R} = \mathcal{R}^2 \cup \mathcal{R}^3 \cup \mathcal{R}^4$, *where* $\mathcal{R}^i = \{r \in \mathcal{R} : \tilde{\xi}(r) \in \tilde{\Lambda}_2^i\}$ *and* $\tilde{\xi}(r) = \xi(r) + 2\Lambda$, *where* $\xi(r)$ *is defined below.*
  (2) $\mathcal{R}^2 = \{r(c, J) : c \in L, \ J \subseteq \phi(c) \in C, \ |J| \ even\}$, *where*

$$r(c, J) = \bar{\psi}_1(c)\bar{\psi}_1(\alpha_J)\bar{\psi}_1(s)^{|J|/2}$$

and $\xi(r(c, J)) = f_{\phi(c)} - 2f_J$.
  (3) $\mathcal{R}^3 = \{r(x, d) : x \in X, d \in L\}$, *where*

$$r(x, d) = \bar{\psi}_1(\alpha_x)\bar{\psi}_1(s)^{\epsilon_x(d)}\bar{\psi}_1(d)$$

and $\xi(r(x, d)) = f_{d + \epsilon_x(d)X} - \lambda_x$.
  (4) $\mathcal{R}^4 = \{r_u(xy)\bar{z}_1^\epsilon, r_v(xy)\bar{z}_1^\epsilon : x, y \in X, \ x \neq y, \ \epsilon = 0, 1\}$, *where* $r_v(xy) = \bar{\psi}_1(\alpha_{xy})$, $r_u(xy) = \bar{\psi}_1(\alpha_{xy})\bar{\psi}_1(s)$, $\xi(r_u(xy)) = 2f_{xy}$, *and* $\xi(r_v(xy)) = 2(f_x - f_y)$.
  (5) $B_r = B_Z \perp B_0^1 \perp B_E^2 \perp B_E^3$, *where* $B_Z = \langle y(r) : r \in \mathcal{R}^2 \rangle$, $B_0^1 = \langle y(r) : r \in \mathcal{R}^3 \rangle$, $B_E^i = \langle w_{xy}^i : x, y \in X \rangle$, $u_{xy} = y(r_u(xy))$, $v_{xy} = y(r_v(xy))$, $w_{xy}^2 = v_{xy} + u_{xy}$, *and* $w_{xy}^3 = v_{xy} - u_{xy}$.

Proof: Part (1) is 22.6. Also 22.6 and its proof tell when $\lambda \in \Lambda_2^i$; then we use this description and 27.2 to calculate (2)–(4). Part (5) follows from (2)–(4) and the definition of the quadratic form on B_r.

We close this section by defining B to be the F-space with quadratic form which is the orthogonal direct sum

$$B = B_- \perp B_s \perp B_r.$$

Thus C is a group of isometries of B.

Observe that z_1 centralizes B_s by 27.7.1. Also as $z_1 \in Z(Q_1)$ and C acts on $\mathcal{R}$ by conjugation, z_1 fixes $y(r)$ for each $r \in \mathcal{R}$. Thus z_1 centralizes B_r. Finally by 27.5.2, z_1 inverts B_-. Thus

Lemma 27.9: $B_- = [B, z_1]$ *and* $B_s + B_r = C_B(z_1)$.

28. The Griess algebra

In this section we use the results of Section 9 to define an algebra structure on the C-module B defined at the end of Section 27. Let $\gamma = (\ ,\ )$
be the bilinear form on B and recall

$$B = B_- \perp B_r \perp B_s$$

is the orthogonal direct sum of three subspaces. We define maps $\tau_i \in$
$L(B_i, B_i; B_i)$ and $\tau_{ij} \in L(B_i, B_j; B_j)$, which determine a symmetric
algebra map $\tau \in L^2(B; B)$ by

$$\tau = \tau_s + \tau_r + \tau_{sr} + \tau_{s-} + \tau_{r-}$$

subject to the conventions of Section 9, and τ and γ in turn determine
a symmetric trilinear form $b_\gamma(\tau) = \beta = (\ ,\ ,\ )$ by 9.4.

Recall $B_s = S^2(B_{24}^1)$ is the symmetric square of B_{24}^1 and has basis
xy, $x, y \in X$. It will sometimes be convenient to view B_s as the space
of all 24-by-24 symmetric matrices by identifying $\sum_{x,y} a_{xy} xy$ with the
matrix (a_{xy}). Subject to this identification we define $\tau_s \in L^2(B_s; B_s)$ by

(28.1) $A * A' = 2(AA' + A'A)$, for $A' \in B_s$,

where AA' is the usual matrix product of the matrices A and A'.

Next B_r has basis $\{y(r) : r \in \mathcal{R}\}$ and we define $\tau_{sr} \in L(B_s, B_r; B_r)$
by

(28.2) $A * y(r) = (A, \xi(r) \otimes \xi(r)) y(r)$, for $A \in B_s$, $r \in \mathcal{R}$.

Here we need to recall that $B_{24}^1 \cong \mathbf{R} \otimes_{\mathbf{Z}} \Lambda$ via $x = L_1^1(x) \mapsto \sqrt{8} e_x$
and $\xi(r) \in \Lambda$ is defined in 27.8. Thus $\xi(r) \otimes \xi(r)$ and A are members of
B_s and the inner product $(A, \xi(r) \otimes \xi(r))$ takes place in B_s. Indeed as
$e_x = x/\sqrt{8}$, we have:

Lemma 28.3: *If* $\xi(r) = \sum_x a_x e_x$ *then the inner product* $(\xi(r), x) =$
$a_x/\sqrt{8}$ *in* B_s.

Define $\tau_r \in L^2(B_r; B_r)$ by

(28.4)   *For* $r, r' \in \mathcal{R}$, $y(r) * y(r') = y(rr')$ *if* $rr' \in \mathcal{R}$ *and 0 otherwise.*

Next $B_- = B_{24}^1 \otimes B_-^1$ and we define $\tau_{s-} \in L(B_s, B_-; B_-)$ by

(28.5) $A * (a \otimes b) = aA \otimes b + (Tr(A)/8) \cdot (a \otimes b)$ *for* $A \in B_s$, $a \in B_{24}^1$,
and $b \in B_-^1$,

where aA is the product of the vector $a = \sum_x a_x x$ with the matrix A.

Finally we define $\tau_{r-} \in L(B_r, B_-; B_-)$ by

$$(28.6) \quad y(r) * (a \otimes b) = (a - 2(a, \xi(r))\xi(r)) \otimes (br)/8, \text{ for } r \in \mathcal{R}, a \in B_{24}^1,$$
$b \in B_-^1$.

Here br denotes the image of b under r.

Next by 9.4, τ_i and the restriction γ_i of γ to B_i determine a trilinear form $b_{\gamma_i}(\tau_i) = \beta_i$ on B_i via

$$\beta_i(a, b, c) = (\tau_i(a, b), c)$$

and τ_{ij} determines $\beta_{ij} \in L(B_i, B_j, B_j; F)$ via

$$\beta_{ij}(a, b, c) = (\tau_{ij}(a, b), c).$$

In the next few lemmas we determine the monomials of these forms with respect to our standard bases.

Lemma 28.7: β_s is symmetric with monomials $4 \cdot x^2 \cdot xy \cdot xy$, $4 \cdot x^2 \cdot x^2 \cdot x^2$, and $4 \cdot xy \cdot yz \cdot zx$ for distinct $x, y, z \in X$.

Proof: Let A, A' be the matrices corresponding to ab, $a'b'$, respectively, where $a, a', b, b' \in X$. Then unless $\{a, b\} \cap \{a', b'\} \neq \varnothing$, $AA' = A'A = 0$, so $A * A' = 0$ and hence there is no monomial $AA'A^*$. So assume $\{a, b\} \cap \{a', b'\} \neq \varnothing$.

If $a = a'$ and $b = b'$ then $A^2 = a^2 + b^2$, a^2, for $a \neq b$, $a = b$, respectively. Hence $A * A = 4(a^2 + b^2)$, $4a^2$, respectively. So as a^2 has norm 1, we get monomials $4 \cdot ab \cdot ab \cdot a^2$, $4 \cdot a^2 \cdot a^2 \cdot a^2$, respectively.

If $a' = b' = a \neq b$ then $A * A' = 2ab$, so as ab has norm 2 we get monomials $4 \cdot a^2 \cdot ab \cdot ab$. Finally if $a' = b$ and $b' = c$ with a, b, c distinct, then $A * A' = 2ca$, contributing a monomial $4 \cdot ab \cdot bc \cdot ca$. Thus the lemma is established.

Lemma 28.8: β_r is symmetric with monomials $y(r)y(r')y(rr')$, for $r, r', rr' \in \mathcal{R}$.

Proof: By definition, $y(r) * y(r') = 0$ unless $rr' \in \mathcal{R}$, where $y(r) * y(r') = y(rr')$, and we may assume the latter. As $y(rr')$ is of norm 1, we get monomials $y(r)y(r')y(rr')$, $r, r', rr' \in \mathcal{R}$.

So it remains to show the form is symmetric. But as $r, r', rr' \in \mathcal{R}$ and $\tilde{q}(\tilde{\xi}(s)) = 0$ for each $s \in \mathcal{R}$ with $\tilde{\xi}$ an isometry by 27.2, we conclude that $s^2 = 1$ and $0 = (\tilde{\xi}(r), \tilde{\xi}(r')) = [r, r']$. Further $r \cdot rr' = r'$. So the form is indeed symmetric.

Lemma 28.9: τ_{sr} *is commutative and* β_{sr} *has monomials*

$$(xy, \xi(r) \otimes \xi(r)) \cdot xy \cdot y(r) \cdot y(r),$$

for $x, y \in X$, $r \in \mathcal{R}$.

Proof: The list of monomials is immediate from 28.2 and the fact that $y(r)$ is of norm 1. In particular from the list of monomials we see β_{sr} is invariant under the permutation $(2,3)$, so its algebra map τ_{sr} is commutative.

Lemma 28.10: τ_{s-} *is commutative and* β_{s-} *has monomials*

$$y^2/8 \cdot (x \otimes \delta(d)) \cdot (x \otimes \delta(d)),$$

$$9x^2/8 \cdot (x \otimes \delta(d)) \cdot (x \otimes \delta(d)),$$

$$xy \cdot (x \otimes \delta(d)) \cdot (y \otimes \delta(d)),$$

for $x, y \in X$, $x \neq y$, $d \in L$, $i = 2, 3$.

Proof: Let A be the matrix of ab and $\delta = \delta(d)$. Then

$$A * (x \otimes \delta) = xA \otimes \delta + Tr(A)/8 \cdot (x \otimes \delta).$$

Now $Tr(A) = 0$ unless $a = b$, where $Tr(A) = 1$. Further $xA = 0$, b, a, for $x \notin \{a, b\}$, $x = a$, $x = b$, respectively. So as $(x \otimes \delta, y \otimes \delta) = (x, y) = \delta_{xy}$, the monomials are as claimed, and hence τ_{s-} is commutative.

Lemma 28.11: τ_{r-} *is commutative and* β_{r-} *has monomials*

$$(1 - a_x^2/4)/8 \cdot y(r) \cdot (x \otimes \delta(d)) \cdot (x \otimes \delta(d)r),$$

$$- a_x a_y/32 \cdot y(r) \cdot (x \otimes \delta(d)) \cdot (y \otimes \delta(d)r),$$

for $x, y \in X$, $x \neq y$, $r \in \mathcal{R}$, $d \in L$, *where* $\xi(r) = \sum_x a_x e_x$.

Proof: Write δ for $\delta(d)$. Then $y(r) * (x \otimes \delta) = (x - 2(x, \xi(r))\xi(r)) \otimes \delta r/8$, so

$$(y(r), x \otimes \delta, y \otimes \delta') = ((x, y) - 2(x, \xi(r))(y, \xi(r)))(\delta', \delta r)/8. \quad (*)$$

In particular $(*)$ is 0 unless $\delta' = \pm \delta r$, in which case we may normalize and take $\delta' = \delta r$. Then as δ' is of norm 1, appealing to 28.3 we obtain:

$$(y(r), x \otimes \delta, y \otimes \delta r) = ((x, y) - a_x a_y/4)/8. \quad (**)$$

Thus the lemma holds.

Lemma 28.12: *The algebra map*

$$\tau = \tau_r + \tau_s + \tau_{sr} + \tau_{s-} + \tau_{r-}$$

defined via the convention of Remark 9.6 is symmetric as is its trilinear form $\beta = b_\gamma(\tau)$ *defined via 9.4. Moreover*

$$\beta = \beta_r + \beta_s + \beta_{sr} + \beta_{r-} + \beta_{s-},$$

and τ and β have the same stabilizer in the isometry group of B.

Proof: The first remark follows from Remark 9.6, since from the lemmas above each of the summands τ_i are symmetric and the summands τ_{ij} are commutative. The second follows from 9.4.3.

Lemma 28.13: C *stabilizes* τ *and* β.

Proof: By 28.12 it suffices to show C stabilizes τ, and for this it suffices to show C stabilizes each τ_i and τ_{ij}. But this is almost immediate from the definitions.

29. The action of N on B

We have seen in 27.7 and 27.8 that:

Lemma 29.1: *As an N^+-module*

$$B = B_R \perp B_E \perp B_Z \perp B_0,$$

where

(1) $B_R = \langle x^2 : x \in X \rangle \le B_s$.

(2) $B_E = B_E^1 \perp B_E^2 \perp B_E^3$, *where* $B_E^1 = \langle xy : x, y \in X, x \ne y \rangle \le B_s$ *and* $B_E^i = \langle w_{xy}^i : x, y \in X \rangle \le B_r$ *for $i = 2, 3$ are described in 27.8.5.*

(3) $B_Z = \langle y(r) : r \in \mathcal{R}^2 \rangle \le B_r$.

(4) $B_0 = B_0^1 \perp B_0^2 \perp B_0^3$, *where* $B_0^1 = \langle y(r) : r \in \mathcal{R}^3 \rangle \le B_r$ *and* $B_- = B_0^2 \perp B_0^3$ *with* $B_0^i = \langle x \otimes \delta_i^1(d) : x \in X, d \in L \rangle$ *for $i = 2, 3$.*

Actually this is the first time we have defined B_0^2 and B_0^3. These subspaces are N^+-invariant by 27.3.

Recall from 14.5 that $R = (Q_1 \cap N_2)(Q_2 \cap N_1)$ is a normal subgroup of N with $\bar{R}$ of class at most 3, $\bar{Z} = Z(\bar{R})$, and $Z_2(\bar{R}) \ge \bar{E}^+ = \bar{Q}_1 \cap \bar{Q}_2$.

Lemma 29.2: *(1)* $B_R = C_B(R) \le C_B(Q_1)$.
   (2) $B_E = [C_B(E^+), R]$ *and* $B_E^i = C_{B_E}(Q_i^+)$ *with* $[B_E^1, Q_1] = 0$.
   (3) $B_Z = [C_B(Z), E^+]$.
   (4) $B_0 = [B, Z]$ *with* $B_0^i = C_{B_0}(Z_i)$.

Proof: As we have already observed in 27.9, $B_- = [B, z_1]$. By 27.3 and 27.4, $B_0^2 = C_{B_-}(z_2)$, so $B_0^3 = C_{B_-}(z_3)$ as $z_1 z_2 z_3 = 1$. For $r \in \mathcal{R}^3$, $y(r) \in Q_1 - C(z_2)$, so z_2 inverts B_0^1, and hence $B_0^1 = [B_0^1, Z]$. Then once we check that D centralizes B_D for $D = R$, E, Z, we have $B_0 = [B, Z]$ and (4) holds.

As $\mathcal{R}^2 \subseteq C(Z)$, $B_Z \leq C_B(Z)$. On the other hand

Lemma 29.3: $C_{E^+}(r(c, J)) = Z \cdot \langle \bar{\psi}_1(\alpha_K) : \epsilon_K(c) = 0 \rangle$,

since $[\bar{\psi}_1(\alpha_K), r(c, J)] = [\bar{\psi}_1(\alpha_K), \bar{\psi}_1(c)] = \bar{z}_1^{\epsilon_K(c)}$ by 27.8 and 14.2.8. Thus $B_Z = [B_Z, E^+]$, so (3) holds once we show $[D, B_D] = 0$ for $D = R$, E.

Next by 27.7.1, $B_R + B_E^1 = B_s \leq C(Q_1)$. Also Q_2^+ acts on $\mathbf{R}x \leq B_-^1$, $x \in X$, by 27.4.1, so Q_2^+ centralizes x^2 and hence $R \leq Q_1 Q_2^+ \leq C(B_R)$. Further by 27.4.1, $B_E^1 = [B_E^1, Q_2^+]$, so $B_E^1 = [B_E^1, R]$. Then 29.4 below completes the proof of 29.2.

Lemma 29.4: *(1)* $\psi_1(d) : w_{xy}^k \mapsto (-1)^{\epsilon_{xy}(d)} w_{xy}^k$ *for* $k = 2, 3$.

(2) $\psi_2(d) : w_{xy}^3 \mapsto (-1)^{\epsilon_{xy}(d)} w_{xy}^3$ *and fixes* w_{xy}^2.

(3) $\psi_1(\alpha) : w_{xy}^k \mapsto w_{x\alpha, y\alpha}^k$ *for* $k = 2, 3$ *and* α *even.*

(4) E^+ *centralizes* B_E^k *for* $k = 2, 3$, *and* $\psi_1(\alpha_z)$ *has cycle* (w_{xy}^2, w_{xy}^3) *for* $z \in X$.

Proof: By 14.2.8, $[\psi_i(d), \psi_1(\alpha_{xy})] = \psi_i(\epsilon_{xy}(d))$ and by 14.2.9 and 14.2.10, $\psi_i(d)$ centralizes $\psi_1(s)$ for all $d \in L$. Hence (1) holds and as $\psi_2(\epsilon) = (\psi_1(s)k_1)^\epsilon \equiv \psi_1(s)^\epsilon \mod K$ by 14.3, we also have (2). Part (3) is easy. Finally $\psi_1(E)$ is abelian and E^+ centralizes $\psi_1(s)$, so E^+ centralizes B_E^k for $k = 2, 3$. Also $[\psi_1(s), \psi_1(\alpha_z)] = z_1$ by 14.2.8, so $\psi_1(\alpha_z)$ centralizes v_{xy} and inverts u_{xy}, and hence has cycle (w_{xy}^2, w_{xy}^3).

We wish to extend the representation of N_1 on B to a representation of N that preserves our bilinear and trilinear forms, and hence also the Griess algebra. We see in general from 29.1 and 29.2 how this must be done. Namely B_D must be an N-submodule for $D = R$, E, Z, 0 (as each such D is normal in N), and N/N^+ must induce S_3 on the three summands of B_E and B_0. More precisely we have a permutation representation of N on $\{1, 2, 3\}$ with kernel N^+ such that for $g \in N$, $z_i^g = z_{ig}$, $k_i^g = k_{ig}$, $(\Delta_i)g = \Delta_{ig}$, etc. In particular, $\delta_i^j(d)g = \delta_{ig}^{jg}(d')$ and $L_i^\epsilon(x)g = L_{ig}^{\epsilon'}(x')$ for some $d' \in L$, $x' \in X$, and ϵ'.

Once extension is easy. If B_R is to be an N-submodule then as Q_1 centralizes B_R, so must $\langle Q_1, Q_2 \rangle$. But by 14.5.2, $N/R = \langle Q_1, Q_2 \rangle / R \times N^+/R$, so:

Lemma 29.5: B_R *extends uniquely to an* N-*module with* $[\langle Q_1, Q_2 \rangle, B_R] = 0$. *Further* N *preserves the restriction of our forms to* B_R.

Lemma 29.6: *Let $x \in X$, $b, e, d \in L$, $\alpha \in \Gamma$, and $r(x, b) \in \mathcal{R}^3$. Then*

(1) $[r(x, b), \bar{\psi}_1(d)] = \bar{\psi}_1(\epsilon_x(d) + P(d) + C(d, b))$.

(2) $r(x, b)^{\bar{\psi}_2(e)} = r(x, eb + \epsilon_x(e))$.

(3) $r(x, b)^{\bar{\psi}_2(\alpha)} = r(x\alpha, b\alpha)$, $r(x\alpha, b\alpha + \epsilon_x(b) + P(b))$ *for α even, odd, respectively.*

Proof: Straightforward calculations using 14.2 and the definition of $r(x, b)$ in 27.8. For example, by 14.2, $[\psi_1(d), \psi_1(\alpha_x)] = \psi_1(P(d) + \epsilon_x(d))$, $[\psi_1(d), \psi_1(s)] = 1$, and $[\psi_1(d), \psi_1(b)] = \psi_1(C(d, b))$, so (1) holds. In (2) use the fact that

$$\psi_3(\pi) = k_1 \psi_1(s) \psi_1(\pi) \equiv \psi_1(s) \psi_1(\pi).$$

Lemma 29.7: *(1) The map*

$$\chi_i : L_{i-1}^\epsilon(x) \otimes \delta_i^{i-1}(b) \mapsto L_{i+1}^\epsilon(x) \otimes \delta_i^{i+1}(b + \epsilon_x(b))$$

is an N^+-isomorphism of $B_{24}^{i-1} \otimes B^{i-1}(\Delta_i)$ with $B_{24}^{i+1} \otimes B^{i+1}(\Delta_i)$.
(2) The stabilizer in N^+ of $x \otimes \delta_2(1)$ consists of the elements

$$\psi_1(s^k \times \pi^j) \psi_2(e) \psi_1(\alpha)$$

such that $x\alpha = x$ and $\epsilon_x(e) + P(e) + j \equiv 0 \mod 2$.

Proof: Applying 14.2.1, we may assume $i = 2$. Observe first that the map is well defined. That is, $\delta_2^1(b) = \delta_2^1(b')$ if and only if $b' = b$ or bs, since $\delta_2(b)k_1 = \delta_2(bs)$. Similarly $\delta_2^3(b) = \delta_2^3(b')$ if and only if $b' = b$ or $bs\pi$, since $\delta_2(b)k_3 = \delta_2(bs\pi)$. But $bs + \epsilon_x(bs) = bs\pi + \epsilon_x(b)$, so $\delta_2^3(bs + \epsilon_x(bs)) = \delta_2^3(bs\pi + \epsilon_x(b)) = \delta_2^3(b + \epsilon_x(b))$, as desired. So (1) is established.

Next by 14.4 each member of N^+ is of the form $g = \psi_1(d)\psi_2(e)\psi_1(\alpha)$ for some $d, e \in L$ and even $\alpha \in \Gamma$. Then we calculate

$$g : L_1^1(x) \otimes \delta_2^1(b) \mapsto L_1^1(x\alpha) \otimes \delta_2^1((db)\alpha + \epsilon_x(e) + P(e) + C(db, e)), \quad (29.7.3)$$

$$g : L_3^1(x) \otimes \delta_2^3(b) \mapsto L_3^1(x\alpha) \otimes \delta_2^3((db)\alpha + \epsilon_x(ed) + P(e) + C(db, e)). \quad (29.7.4)$$

This shows the map χ_2 is N^+-equivariant and establishes (2).

We write $x \otimes \delta_i^j(b)$ for $L_j^1(x) \otimes \delta_i^j(b)$ and identify $B_{24}^{i-1} \otimes B^{i-1}(\Delta_i)$ with $B_{24}^{i+1} \otimes B^{i+1}(\Delta_i)$ via the map χ_i of 29.7 which identifies $x \otimes \delta_i^{i-1}(b)$ with $x \otimes \delta_i^{i+1}(b + \epsilon_x(b))$. We denote this space by $\bar{B}_0^i$ and let

$$\bar{B}_0 = \bar{B}_0^1 \perp \bar{B}_0^2 \perp \bar{B}_0^3 = (B_{24}^3 \otimes B^3(\Delta_1)) \perp (B_{24}^1 \otimes B^1(\Delta_2)) \perp (B_{24}^2 \otimes B^2(\Delta_3)).$$

Then N is naturally represented on $\bar{B}_0$ via

$$g : L_{i+1}^1(x) \otimes \delta_i^{i+1}(b) \mapsto L_{i+1}^1(x)g \otimes \delta_i^{i+1}(b)g, \quad (29.7.5)$$

and by 29.7 the map

$$\chi : B_0^2 \to \bar{B}_0^2,$$

$$x \otimes \delta_2^1(b) = L_1^1(x) \otimes \delta_2^1(b) \mapsto L_1^1(x) \otimes \delta_2^1(b) = x \otimes \delta_2^1(b);$$

$$\chi : B_0^3 \to \bar{B}_0^3,$$

$$x \otimes \delta_3^1(b) = L_1^1(x) \otimes \delta_3^1(b) \mapsto L_2^1(x) \otimes \delta_3^2(b + \epsilon_x(b)) = x \otimes \delta_3^1(b)$$

induces an N^+-isomorphism $B_0^i \to \bar{B}_0^i$ for $i = 2,3$, and of course this extends uniquely to an N_1-isomorphism. Next

Lemma 29.8: *The map*

$$y(r(x,b)) \mapsto L_3^1(x) \otimes \delta_1^3(b)$$

extends χ to an N_1-isomorphism of B_0^1 with $\bar{B}_0^1$.

Proof: We check first that the map is a well-defined bijection; that is, $r(x,b) = r(x',b')$ if and only if $x = x'$ and $b' = b$ or bs if and only if $L_1^1(x) \otimes \delta_1^3(b) = L_1^1(x') \otimes \delta_1^3(b')$ by 14.4.4. Then we observe that if $g = \psi_1(d)\psi_2(e)\psi_1(\alpha)$ with α even then using 29.6,

$$g : L_3^1(x) \otimes \delta_1^3(b) \mapsto L_3^1(x\alpha) \otimes \delta_1^3((eb)\alpha + \epsilon_x(ed) + P(d) + C(b,d)),$$

$$r(x,b)^g = r(x\alpha, (eb)\alpha + \epsilon_x(ed) + P(d) + C(d,b))$$

so χ is N^+-equivariant. Similarly using 29.6.3, we check that the map commutes with $\psi_1(\alpha_z)$ for $z \in X$ with $\psi_1(\alpha_z) : x \otimes \delta_1^3(b) = L_3^1(x) \otimes \delta_1^3(b) \mapsto L_2^1(x) \otimes \delta_1^2(b + P(b) + \epsilon_{xz}(b)) \sim L_3^1(x) \otimes (b + P(b) + \epsilon_{xz}(b)) = x \otimes \delta_1^3(b + P(b) + \epsilon_{xz}(b))$.

By 29.8, our N_1-isomorphism $\chi : B_0^i \to \bar{B}_0^i$, $i = 2,3$, extends to an N_1-isomorphism $\chi : B_0 \to \bar{B}_0$. We pull back the representation of N on $\bar{B}_0$ via χ to obtain a representation of N on B_0 which makes χ an equivalence. This is our action of N on B_0.

Our standard basis for B_0^i, $i = 2,3$, consists of the elements $x \otimes \delta_i(d)$, $x \in X$, $d \in L$, while the standard basis elements for B_0^1 are $y(r(x,b))$, $x \in X, b \in L$. We write $x \otimes \delta_1(b)$ for $y(r(x,b))$. This gives us a symmetric notation for the standard basis elements of B_0 and the notation makes sense thanks to 29.8.

Lemma 29.9: *N preserves the restriction of our forms to B_0.*

Proof: First N preserves γ as it permutes our standard basis up to sign, and that basis is orthonormal. To show N preserves β we appeal to 9.8, so we need information about monomials.

By 9.3 applied to Z in the role of the G of 9.3, and using 29.2.4, if $m = am_1m_2m_3$ is a nonzero monomial in β on B_0 then we may take $m_i \in B_0^i$. Hence by 28.11, m is one of

$$(1 - a_x^2/4)/8 \cdot y(r) \cdot (x \otimes \delta_2(d)) \cdot (x \otimes \delta_2(d)r),$$

$$-a_x a_y/32 \cdot y(r) \cdot (x \otimes \delta_2(d)) \cdot (y \otimes \delta_2(d)r),$$

with $r \in \mathcal{R}^3$ and with $x \neq y$ in the second case. We seek to determine m up to conjugation under N^+, so as N^+ is transitive on $\mathcal{R}^3$ we may take $m_1 = y(r)$, where $r = r(z,1) = \bar{\psi}_1(\alpha_z)$. So by our notational convention, $m_1 = z \otimes \delta_1(1)$. Next by 29.7.3, Q_1^+ is transitive on basis vectors $x \otimes \delta_2(d)$, so as $\bar{Q}_1^+ = \langle \bar{z}_3 \rangle C_{\bar{Q}_1^+}(\bar{\psi}_1(\alpha_z))$ and z_3 inverts B_0^2, each monomial is conjugate under N^+ to either

$$(1 - a_x^2/4)/8 \cdot (z \otimes \delta_1(1)) \cdot (x \otimes \delta_2(1)) \cdot (x \otimes \delta_2(1)r)$$

or

$$-a_x a_y/32 \cdot (z \otimes \delta_1(1)) \cdot (x \otimes \delta_2(1)) \cdot (y \otimes \delta_2(1)r)$$

with $x \neq y$ in the second case. But $x \otimes \delta_2(1)r = x \otimes \delta_3(1)$ by 27.3.4, and $\xi(r) = -\lambda_z$ by 27.8.3, so by 28.11, $a_x = 3, -1$, for $x = z, x \neq z$, respectively. Therefore m is conjugate to

$$a \cdot (z \otimes \delta_1(1)) \cdot (x \otimes \delta_2(1)) \cdot (y \otimes \delta_3(1)),$$

where $a = -1/32, 3/32, -5/32$ for x, y, z all distinct, two of x, y, z equal, $x = y = z$, respectively.

Let $\mathcal{M}_0$ be this set of monomials and $\mathcal{M}$ the set of all nonzero monomials of β on B_0. By 9.8.2, it remains to show $\langle \bar{\psi}_1(\alpha_w), \bar{\psi}_2(\alpha_w) \rangle$ permutes $\mathcal{M}_0$. But $\bar{\psi}_1(\alpha_w)$ fixes $z \otimes \delta_1(1)$ and interchanges $x \otimes \delta_2(1)$ and $x \otimes \delta_3(1)$, while $\bar{\psi}_2(\alpha_w)$ fixes $x \otimes \delta_2(1)$ and interchanges $z \otimes \delta_1(1)$ and $z \otimes \delta_3(1)$, completing the proof.

Write $\bar{B}_E^i$ for the subspace $\langle (xy)_i : x, y \in X, x \neq y \rangle$ of $S^2(B_{24}^i)$, where $(xy)_i = L_i^1(x) \otimes L_i^1(y) + L_i^1(y) \otimes L_i^1(x)$. Observe that the map $g : L_i^1(x) \mapsto L_i^1(x)g$ induces a representation of N on

$$\bar{B}_E = \bar{B}_E^1 \perp \bar{B}_E^2 \perp \bar{B}_E^3.$$

Of course $\bar{B}_E^1 \cong B_E^1$ as an N_1-module. Indeed

Lemma 29.10: Let $g = \psi_1(d)\psi_2(e)\psi_1(\alpha) \in N^+$. Then

    (1) $g : (xy)_i \mapsto (-1)^{\epsilon(g)}(x\alpha y\alpha)_i$ and $g : w_{xy}^i \mapsto (-1)^{\epsilon(g)} w_{xy}^i$ for $i = 2, 3$, where $\epsilon(g) = \epsilon_{xy}(d) + i\epsilon_{xy}(e)$.

(2) $\psi_1(\alpha_z)$ *has cycle* $((xy)_2, (xy)_3)$ *and fixes* $(xy)_1$ *while* $\psi_3(\alpha_z)$ *fixes* $(xy)_3$ *and has cycle* $((xy)_1, (xy)_2)$.

(3)   *The map* $xy \mapsto (xy)_1$, $w_{xy}^i \mapsto (xy)_i$, $i = 2, 3$, *is an* N_1*-isomorphism of* B_E *with* $\bar{B}_E$.

Proof: We use 27.4 to calculate the action of g on $\bar{B}_E^i$ and 29.4 to make the calculation on B_E^i for $i = 2, 3$. This gives (1), and (2) is obtained similarly. Then (1) and (2) and 29.4.4 imply (3).

We use the isomorphism of 29.10 to pull back the action of N on $\bar{B}_E$ to an action on B_E. Having defined our representation on B_E we prove:

Lemma 29.11: N *preserves the restriction of our forms to* B_E.

Proof: The argument is similar to 29.9. Again it is clear the bilinear form is preserved and we appeal to 9.8 to handle the trilinear form.

So let $m = am_1m_2m_3$ be a monomial in our basis $(xy)_i$, with $a \neq 0$. We apply 9.3 to R/E^+, recalling from 29.2 that $B_E \leq C(E^+)$ and $B_E^i = C_{B_E}(Q_i^+)$, and recalling also that R/E^+ is of exponent 2 by 14.5. Thus by 9.3 we may assume either

(a) $m_i \in B_E^i$ for $i = 1, 2, 3$, or

(b) $m_i \in B_E^j$ for some j and all i.

Let $Q_i(xy) = \langle E^+, \bar{\psi}_i(d) : \epsilon_{xy}(d) \rangle = 0$. Then $C_R((xy)_1) = Q_1^+ Q_2(xy)$ and $C_R((uz)_2) = Q_2^+ Q_1(uz)$. Thus in case (a) if $m_1 = (xy)_1$ and $m_2 = (uz)_2$ then $C_R(\langle m_1, m_2 \rangle) = Q_1(uz)Q_2(xy)$. But $C_{B_E}(Q_1(uz)Q_2(xy)) = \mathbf{R}(xy)_3 \cap \mathbf{R}(uz)_3$, so by 9.3, $uz = xy$ and $m_i = (xy)_i$.

Similarly in case (b) if $m_1 = (xy)_1$ and $m_2 = (uz)_1$ then

$$C_R(\langle m_1, m_2 \rangle) = Q_1^+(Q_2(xy) \cap Q_2(uz))$$

whose centralizer on B_E^1 is $\langle m_1, m_2 \rangle$ unless $y = u$ and $x \neq z$, where we get $\langle m_1, m_2, (zx)_1 \rangle$. Hence by 9.3 either $m_3 = m_1$, or m_2 or $m_1 = (xy)_1$, $m_2 = (yz)_1$, and $m_3 = (zx)_1$. But there exists $e \in L$ with $\psi_2(e)$ inverting m_1 (and centralizing m_2 if $\{x, y\} \cap \{u, z\} = \varnothing$) so 9.3.2 says $m_1 = (xy)_1$, $m_2 = (yz)_2$, and $m_3 = (zx)_1$.

Now in case (a) by 28.2 we have monomials

$$(xy, \xi(r_{v(\epsilon)}(xy)) \otimes \xi(r_{v(\epsilon)}(xy))) \cdot v(\epsilon)_{xy} \cdot v(\epsilon)_{xy},$$

where $v(1) = v$ and $v(-1) = u$. By 27.8, $\xi(r_{v(\epsilon)}(xy)) = 2(f_x - \epsilon f_y)$ so by 28.3,

$$(xy, \xi(r_{v(\epsilon)}(xy)) \otimes \xi(r_{v(\epsilon)}(xy)))$$
$$= 2(x, \xi(r_{v(\epsilon)}(xy)))(y, \xi(r_{v(\epsilon)}(xy))) = -4\epsilon.$$

Therefore $(xy, w_{xy}^2, w_{xy}^3) = -8$, so we have monomials

$$-8 \cdot xy \cdot w_{xy}^2 \cdot w_{xy}^3 \qquad (*)$$

in case (a).

Next in case (b) with $i = 1$ by 28.7 we have monomials

$$4 \cdot xy \cdot yz \cdot zx \qquad (**)$$

in B_E^1, while if $i = 2, 3$ the relevant monomials in B_r are $y(r)y(r')y(rr')$ for $r, r', rr' \in \mathcal{R}^4$. Further in our case $r = r_{v(\epsilon)}(xy)$ and $r' = r_{v(\delta)}(yz)$. Now $r_{v(\epsilon)}(xy)r_{v(\delta)}(yz) = r_{v(\epsilon\delta)}(zx)$, so we get monomials

$$4 \cdot w_{xy}^i \cdot w_{yz}^i \cdot w_{zx}^i, \quad i = 2, 3. \qquad (* * *)$$

Finally by 29.10.2, $\langle \psi_1(\alpha_z), \psi_2(\alpha_z) \rangle$ permutes the monomials in $(k*)$, $k = 1, 2, 3$, so 9.8 completes the proof.

Lemma 29.12: *Let $c \in L$ with $\phi(c) \in \mathcal{C}$, $J \subseteq \phi(c)$, $e, d \in L$, and $\alpha \in \Gamma^+$. Then*

(1) $r(c, J)^{\psi_1(\alpha)} = r(c\alpha, J\alpha)$.
(2) $r(c, J)^{\psi_1(d)} = r(c + \epsilon_J(d) + C(c, d), J)$.
(3) $r(c, J)^{\psi_2(e)} = r(c + C(e, c), J + (c \cap e))$.

Proof: Calculate using 14.2 and the definition of $r(c, J)$ in 27.8. In making the calculation in (3), recall $\bar{\psi}_2(\pi) = \bar{\psi}_1(s)$ and $\bar{\psi}_3(\pi) = \bar{\psi}_1(s\pi)$. This reduces us to a verification that $|J + (\phi(c) \cap \phi(e))|/2 \equiv |J|/2 + C(c, e) + |J \cap \phi(e)| \bmod 2$, which follows as $C(c, e) = |\phi(c) \cap \phi(e)|/2 \bmod 2$ and $|\phi(c) \cap \phi(e) \cap J| = |J \cap \phi(e)|$ since $J \subseteq \phi(c)$.

Lemma 29.13: *(1) $B_Z = \bigoplus_{c \in \mathcal{C}} B_{Z,c}$, where*

$$B_{Z,c} = \langle y(r(c', J)) : J \subseteq c = \phi(c') \rangle.$$

(2) $B_{Z,c} = C_{B_Z}(E_c^+)$, where $E_c^+ = \langle Z, \bar{\psi}_1(\alpha_J) : \epsilon_J(c) = 0 \rangle$.
(3) $C_\Gamma(B_{Z,c}) = C_\Gamma(c)$ and $N_{\Gamma^+}(c)/C_\Gamma(c) \cong A_8$.
(4) $C_R(B_{Z,c}) = E_c^+ \langle \bar{\psi}_1(d)\bar{\psi}_2(e) : e \cap c, d \cap c \in \{\varnothing, c\} \rangle$ and $R/C_R(B_{Z,c}) \cong 2^{1+12}$ is irreducible on $B_{Z,c}$.
(5) B_Z extends to an N-module in which $\bar{Q}_2$ acts on $B_{Z,c}$ for each $c \in \mathcal{C}$.

Proof: Part (1) is trivial while 29.12 implies (2) and the first statement of (3) and (4). Then 19.1 implies the second statement in (3). Similarly by 14.5 and the first part of (4), $R/C_R(B_{Z,c})$ is of order 2^{13} and the product of the elementary abelian groups $Q_i^+ C_R(B_{Z,c})/C_R(B_{Z,c})$ is of order 2^7, while by 14.2.10, $Z(R/C_R(B_{Z,c})) = E^+ C_R(B_{Z,c})/C_R(B_{Z,c})$, so $R/C_R(B_{Z,c}) \cong 2^{1+12}$. As $dim(B_{Z,c}) = 2^6$ is the minimal dimension of a faithful module for 2^{1+12}, (4) holds.

Finally by (1)–(4), B_Z is an induced module χ^{N_1}, where χ is the representation of $N_{1,c}$ whose kernel $K = C_\Gamma(c)C_R(B_{Z,c})$ is N_c-invariant and such that $N_{1,c}/K \cong A_8/2^{1+12}$ with χ the unique faithful irreducible representation for $F^*(N_{1,c}/K) = 2^{1+12}$. Then χ is N_c-invariant, so it extends to χ_{N_c} and then χ^{N_1} extends to $\chi^N_{N_c}$. Thus (5) is established.

We use 29.13.5 to define the action of N on B_Z. Thus we have defined N on each of the summands of B of 29.1, and hence have embedded N in $GL(B)$. We close this section by proving:

Lemma 29.14: *N preserves the restriction of our forms to B_Z. Hence N preserves γ on B.*

Proof: As B is the orthogonal direct sum of the four summands of 29.1 and we have shown N preserves the restriction of γ to the summands B_D, $D = R$, E, 0, it follows that N preserves γ on B if it preserves its restriction to B_Z. We prove N preserves this restriction by proving N^+ is irreducible on B_Z and hence, up to a scalar multiple, γ is the unique bilinear form on B_Z preserved by N^+. Hence N preserves γ up to a scalar, so as $N = \langle N_1^{N^+} \rangle$ and N_1 preserves γ, so does N. We use the same argument to show N preserves β on B_Z; that is, we show that, up to a scalar, β is the unique N^+-invariant trilinear form with a certain property and the image of β under each $g \in N$ has that property. So it remains to establish these claims.

Let $Y = \{y(r) : r \in \mathcal{R}^2\}$ so that Y is a signed basis for B_Z with $y(rz_1) = -y(r)$. Now for $r \in \mathcal{R}^2$, $Z(C_{Q_1^+}(r)) = \langle r, \bar{Z} \rangle$ and from 27.2, $\tilde{\xi}(\langle \tilde{r}, \tilde{Z} \rangle) \cap \tilde{\Lambda}_2^2 = \{\tilde{r}\}$, so the weight spaces of Q_1^+ on B_Z are 1-dimensional and are $\hat{Y} = \{\mathbf{R}y : y \in Y\}$. Thus as N^+ is transitive on $\mathcal{R}^2$, and hence also on $\hat{Y}$, N^+ is irreducible on B_Z as claimed.

Next let $\eta \neq 0$ be an N^+-invariant trilinear form on B_Z and $m = am_1m_2m_3$ be a nonzero monomial of η in the basis Y. Let $m_i = y(r_i)$. Then $m_i \in B_{Z,c_i}$ for some $c_i \in \mathcal{C}$, and by 29.13.2 and 9.3, $c_i \neq c_j$ for $i \neq j$. Further by 9.3, $r_3 \in Z(C_{Q_1^+}(\langle r_1, r_2 \rangle)) = \bar{Z}\langle r_1, r_2 \rangle$, so since we just showed that $\mathcal{R}^2 \cap \bar{Z}\langle r_i \rangle = \{r_i, r_i\bar{z}_1\}$ and $c_3 \neq c_1$ or c_2, it follows that $r_3 \in \bar{Z}r_1r_2$. Then $\tilde{\xi}(\tilde{r}_3) \in \tilde{\xi}(\bar{Z}\tilde{r}_1\tilde{r}_2) \cap \tilde{\Lambda}_2^2$, which implies $c_1 + c_2 = c_3$ and either

   (i) $\{c_1, c_2, c_3\}$ is a trio and $\tilde{r}_1\tilde{r}_2 = \tilde{r}_3\tilde{s}$, or

   (ii) $|c_1 \cap c_2| = 4$ and $\tilde{r}_1\tilde{r}_2 = \tilde{r}_3$.

By 28.8, $\beta_Z = \beta_{B_Z}$ has no monomials of type (i). Further N^+ is transitive on triples (r_1, r_2, r_3) with $r_1 r_2 = r_3$, so up to a scalar β_Z is the unique N^+-invariant trilinear form such that if $x_i \in B_{Z,c_i}$ with $\beta_Z(x_1, x_2, x_3) \neq 0$ then $c_1 + c_2 = c_3$ and $|c_1 \cap c_2| = 4$. But by 29.13.5, N permutes the subspaces $B_{Z,c}$, $c \in \mathcal{C}$, so N permutes such forms, completing the proof.

30. N preserves the Griess algebra

In the last section we constructed a representation of N on B preserving the decomposition of 29.1 and the bilinear form γ on B. In this section we prove N preserves the trilinear form β and hence also the Griess algebra.

Now by Remark 9.2,

$$\beta = \sum_{J \in I^3/S_3} \beta_J,$$

where $I = \{R, E, Z, 0\}$ and I^3/S_3 is some set of representatives for the orbits of S_3 on $I^3 = I \times I \times I$. Further N preserves β if and only if it preserves β_J for each $J \in I^3/S_3$. We have shown that N preserves the restriction of β to B_D for each $D \in I$; that is, N preserves β_J for $J = (D, D, D)$. It remains to treat the other projections β_J.

From the discussion in Section 29, we have a *standard basis* for B with respect to N. Namely the basis for B_R consists of the vectors x^2, $x \in X$, the basis for B_E^i consists of the vectors $(xy)_i$, $x, y \in X$ distinct, the basis for B_Z consists of the vectors $y(r)$, $r \in \mathcal{R}^2$, and the basis for B_0^i consists of the vectors $x \otimes \delta_i(d)$, $x \in X$, $d \in L$. Our subbases are monomial basis for N^+ and, except for B_Z, even monomial bases for N. We form the monomials for β from the standard basis.

Lemma 30.1: *For $c \in L$ with $\phi(c) \in \mathcal{C}$, $J \subseteq \phi(c)$ of even order, and $x \in X$:*

(1) $y(r(c, J))^{\psi_3(\alpha_x)} = 1/8 \cdot \sum_{K \subseteq c} t_K(J) y(r(c, K))$ *with $t_K(J) = \pm 1$ and the sum over any set $\mathcal{K}$ of 64 representatives for the even subsets K of $\phi(c)$ modulo the relation $K \sim K + \phi(c)$.*

(2) $t_K(\varnothing) = (1)^{\epsilon_x(c)}$ *for each $K \in \mathcal{K}$.*

Proof: Let $y(J) = y(r(c, J))$, $g = \bar{\psi}_3(\alpha_x)$, and $x(J) = y(J)g$. By 29.13.5, g acts on $B_{Z,\phi(c)}$, so as $\{Ry(J) : J \in \mathcal{K}\}$ is the set of weight spaces of Q_1^+ on $B_{Z,\phi(c)}$ and $(Q_1^+)^g = Q_2^+$, $\{x(J) : J \in \mathcal{K}\}$ is the set of weight spaces for Q_2^+ on $B_{Z,\phi(c)}$.

Next $x(J) = \sum_{K \in \mathcal{K}} a_K y(K)$ and as Q_2^+ is transitive on the Q_1^+ weight spaces and permutes $\{\pm y(J) : J \in \mathcal{K}\}$, we conclude $a_K = \pm a_{K'}$ for all

K, K'. As g preserves γ and $\{y(K) : K \in \mathcal{K}\}$ is orthonormal of order 64, it follows that $a_K = \pm 1/8$ for all $K \in \mathcal{K}$. That is, (1) is established.

Next by 29.12, $\mathbf{R}y(\varnothing) = C_{B_{Z,\phi(c)}}(Q_1(\varnothing))$, where

$$Q_1(\varnothing) = E_{\phi(c)}^+ \langle \psi_1(d) : C(c,d) = 0 \rangle,$$

so $Q_2^+(\varnothing) = (Q_1^+)^g = E_c^+ \langle \psi_2(e) : C(c,e) = 0 \rangle$ by 14.2.5, and $\mathbf{R}x(\varnothing) = C_{B_{Z,\phi(c)}}(Q_2(\varnothing))$. But by 29.12, $y(K)\psi_2(e) = (-1)^{C(e,c)} y(K + (\phi(c) \cap \phi(e)))$, so $t_K(\varnothing) = t_{K+(\phi(c) \cap \phi(e))}(\varnothing)$ for all $e \in L$ with $C(c,e) = 0$, since we just showed $\psi_2(e)$ fixes $x(\varnothing)$ for each such e. Now $C(c,e) = |\phi(c) \cap \phi(e)|/2$, so we conclude that the subgroup $\langle \phi(c) \cap \phi(e) : C(e,c) = 0 \rangle = \{K : K \subseteq \phi(c), |K| \text{ even}\}$, and hence that $t = t_K(\varnothing)$ is independent of K.

Let $h = \psi_1(\alpha_x)$ and $y = y(\varnothing)$. Then by 14.2 and the definition of $r(c, K)$ in 27.8, $y(K)h = (-1)^{|K|/2 + \epsilon_x(c)} y(K)$. Therefore $yh = by$, where $b = (-1)^{\epsilon_x(c)}$ and $ygh = t/8 \cdot \sum_K y(K)h = tb/8 \cdot \sum_K t_K y(K)$, where $t_K = (-1)^{|K|/2}$. Therefore $(y, yg) = t/8$ while

$$(yg, ygh) = \left(t/8 \cdot \sum_J y(J), tb/8 \cdot \sum_K t_K y(K) \right) = b/8$$

as there are 36, 28 choices for K with $|K| \equiv 0, 2 \mod 4$, respectively. Finally $h^g = g^h = \psi_2(\alpha_x)$, so $b/8 = (yg, ygh) = (y, yh^g) = (y, yg^h) = (yh, yhg) = (y, yg) = t/8$ as $yh = by$. That is, $t = b$.

Lemma 30.2: N preserves β_J for any $J \in I^3/S_3$ containing an entry R.

Proof: Assume $m = am_1m_2m_3$ is a nonzero monomial in β_J in our standard basis with $m_1 \in B_R$. Then $m_1 = x^2$ for some $x \in X$. As $[R, B_R] = 0$ we conclude from 9.3 that $C_R(m_2) = C_R(m_3)$ and then conclude from 29.2 that $J = (R, D, D)$ for some D. As N preserves the restriction of β to R, we may assume $y = m_2 \notin B_R$.

Assume $y \in B_Z$. Then $y = y(r)$ for some $r = r(c, J) \in \mathcal{R}^2$, and by 28.9 and 28.3, $y = m_3$ and $a = (x, \xi(r))^2 = a_x^2/8 = \epsilon_x(c)/2$. Let $g = \psi_3(\alpha_z)$. Then $x^2 g = x^2$ and by 30.1, $yg = 1/8 \cdot \sum_K t_K(J)y(K)$, so $(x^2 g, yg, yg) = 1/64 \cdot \sum_K (x^2, y(K), y(K)) = \epsilon_x(c)/2 = (x^2, y, y)$. That is, $N = \langle g, N_1 \rangle$ preserves β_J, as desired.

Assume next that $y \in B_E^i$. If $i = 1$ then $y = bc$ for distinct $b, c \in X$ and by 28.7, $y = m_3$ and $a = 4\epsilon_x(bc)$. On the other hand if $i \neq 1$ then $y = (bc)_i = w_{bc}^i$ and by 28.9 and 28.3, $\beta(x^2, v_{bc}, v_{bc}) = a_x^2/8 = 2\epsilon_x(bc) = \beta(x^2, u_{bc}, u_{bc})$, so $a = 4\epsilon_x(bc)$. Hence as g permutes $(bc)_i$ by 29.10.2, N preserves β_J in this case too.

This leaves the case $y \in B_0^i$. Then $y = u \otimes \delta_i(d)$ for some $u \in X$,

$d \in L$. If $i = 1$ then from 29.8, $y = y(r)$, where $r = r(u, d) \in \mathcal{R}^3$ and by 28.3 and 28.9, $y = m_3$ and $a = a_x^2/8 = 9/8$, $1/8$ for $x = u$, $x \neq u$, respectively. On the other hand if $i \neq 1$ then we obtain the same answer from 28.10. This time using 27.3 and 27.4, we calculate the action of g on $\bar{B}_0$ and using the N-equivariant isomorphism $\chi : B_0 \to \bar{B}_0$ of Section 29, we see that g permutes our monomials, completing the proof.

Lemma 30.3: N preserves β_J for any $J \in I^3/S_3$ containing an entry E.

Proof: Again take $m = am_1m_2m_3$ to be a nonzero monomial, this time with $m_1 \in B_E^i$. We have handled the case where each m_i is in B_E, so we may assume $m_2 \notin B_E$. As $m_1 \in B_E$, $[m_1, E^+] = 0$, so by 9.3, $C_{E^+}(m_2) = C_{E^+}(m_3)$. In particular if $m_2 \in B_E$ then so is m_3, contrary to the choice of m_3. Thus neither m_2 nor m_3 is in B_E. Similarly if $m_2 \in B_Z$ so is m_3. That is, $m_2, m_3 \in B_D$ for $D = Z$ or $D = 0$. So it remains to show N^+ preserves the restriction $\bar{\beta}$ of β to $B_E \times B_D \times B_D$.

Next $N = \langle N_1, g \rangle$, where $g = \psi_3(\alpha_z)$ for some $z \in X$, so as N_1 preserves $\bar{\beta}$, it remains to show g preserves $\bar{\beta}$. Write ζ_i for the restriction of β to $B_E^i \times B_D \times B_D$. It suffices to show g preserves ζ_3 and $\zeta = \zeta_1 + \zeta_2$, as $\bar{\beta} = \zeta_3 + \zeta$. We save the proof that g preserves ζ_3 till the end.

Notice $\zeta_1 g \in L(B_E^2, B_D, B_D; F)$ is N^+-invariant and to show g preserves ζ we must show $\zeta_1 g = \zeta_2$. In particular we may take $m_1 \in B_E^1$. Thus $m_1 = xy$ for some choice of distinct $x, y \in X$, and by 29.10.2, $m_1 g = (xy)_2 = w_{xy}^2$.

Let $\rho \in L(B_E^1, B_D, B_D; F)$ be N^+-invariant and $bm_1n_2n_3$ be a nonzero monomial in ρ. Suppose $D = Z$. Then $n_i = y(r_i)$ with $r_i = r(c_i, J_i) \in \mathcal{R}^2$, and as $C_{E^+}(n_2) = C_{E^+}(n_3)$, $\phi(c_2) = \phi(c_3)$ by 29.13.2. So replacing n_3 by $-n_3$ if necessary, we may assume $c_2 = c_3 = c$. Also $xy \in C_B(Q_1^+)$, so $C_{Q_1^+}(n_2) = C_{Q_1^+}(n_3)$ and hence by 29.12.2, $n_2 = n_3 = y(r)$, where $r = r(c, J)$. If $\{x, y\}$ is not contained in $\phi(c)$ we can pick $e \in L$ with $\phi(e) \cap \phi(c) = \varnothing$ and $\phi(e)$ containing exactly one of x, y. Then by 27.4, $\psi_2(e)$ inverts xy while by 29.12.3, $\psi_2(e)$ fixes n_2, contradicting 9.3. Thus $x, y \in \phi(c)$. We saw during the proof of 30.1 that Q_2^+ is transitive on the Q_1^+ weight spaces in $B_{Z,\phi(c)}$, so as Q_2^+ fixes x and y, we may take $J = \varnothing$ and ρ is determined up to a scalar. In particular $\zeta_1 g \in \mathbf{R}\zeta_2$, so it remains to show $a = (xy, n_2, n_2) = (w_{xy}^2, n_2 g, n_2 g)$.

As $r_i = r(c, \varnothing)$, $\tilde{\xi}(r_i) = f_{\phi(c)}$ by 27.8. So by 28.9 and 28.3, $a = 1$. By 30.1.2, $n_2 g = -1/8 \cdot \sum_K r(c, K)$. Further if we set $v(1) = v$, $v_{xy}(1) = v_{xy}$ and $v(0) = u$, $v_{xy}(0) = u_{xy}$, then using the definitions in 27.8, we calculate

$$r(c, K)r(c, K + xy) = r_{v(\epsilon_K(xy))}(xy)$$

so by 28.8,

$$(w_{xy}^2, n_2 g, n_2 g) = 1/64 \cdot \sum_K (w_{xy}^2, y(r(c, K)), y(r(c, K + xy))) = 1,$$

completing the proof in this case.

So take $D = 0$. By 9.3, $C_Z(m_2) = C_Z(m_3)$, so $m_2, m_3 \in B_0^k$ for the same k. Hence $\zeta_i = \zeta_i^1 + \zeta_i^2 + \zeta_i^3$, where ζ_i^j is the restriction of β to $B_E^i \times B_0^j \times B_0^j$. Therefore it suffices to show $\zeta_1^j g = \zeta_2^{3-j}$ for $j = 1, 2$, and $\zeta_1^3 g = \zeta_2^3$. Similarly we get maps $\rho_i^j \in L(B_E^i, B_0^j, B_0^j; F)$ associated to ρ.

Suppose $j = 1$. Then $n_i = y(r_i)$, $r_i \in \mathcal{R}^2$, and $C_{Q_1^+}(n_2) = C_{Q_1^+}(n_3)$, so $\tilde{r}_2 = \tilde{r}_3$ and hence we may take $n_2 = n_3$. Further by 29.6, $N_{N^+}(\mathbf{R}xy)$ has two orbits on the weight spaces of Q_1^+ on B_0^1 with representatives $q \otimes \delta_1(1) = y(r(q, 1))$, where $q = x$ or $q \neq x, y$, respectively. Now as $r = r(q, 1)$, $\tilde{\xi}(r) = -\lambda_q$ by 27.8. Then by 28.9 and 28.3, $b = -3/4$, $1/4$, for $q = x$, $q \neq x, y$, respectively. Thus $\zeta_1^1 g = \zeta_2^2$ if $(w_{xy}^2, (q \otimes \delta_1(1))g, (q \otimes \delta_1(1))g) = -3/4, 1/4$, for the respective q. Now using 27.3 and 27.4 we calculate the action of g on $\bar{B}_0$ and use the N-equivariant isomorphism $\chi : B_0 \to \bar{B}_0$ to check that $(q \otimes \delta_1(1))g = q \otimes \delta_2(1)$. Also by 28.11, $(v_{xy}(\epsilon), q \otimes \delta_2(1), q \otimes \delta_2(1)) = (1 - a_q^2/4)/8 = -3/8, 1/8$, for the respective q, and hence $(w_{xy}^2, n_2 g, n_2 g) = -3/4, 1/4$, as desired.

So take $j = 2$ or 3. By 27.4.1, $\psi_j(e)$ inverts xy if $\epsilon_{xy}(e) = 1$. Then from 29.7.3, there exists $e \in L$ such that $\psi_j(e)$ inverts xy and centralizes $q \otimes \delta_j(d)$ and $q' \otimes \delta_j(d')$ unless $\{q, q'\} = \{x, y\}$. Thus each nonzero monomial of ρ_1^j is conjugate under $N_{N^+}(\mathbf{R}xy)$ to $b_{xy} \cdot xy \cdot (x \otimes \delta_j(1)) \cdot (y \otimes \delta_j(1))$, so to show $\zeta_1^2 g = \zeta_2^1$ and $\zeta_1^3 g = \zeta_2^3$ it remains to show $(xy, x \otimes \delta_j(1), y \otimes \delta_j(1)) = (w_{xy}^2, x \otimes \delta_{k(j)}(1), y \otimes \delta_{k(j)}(1))$, where $k(2) = 1$ and $k(3) = 3$.

Now by 28.10, $(xy, x \otimes \delta_j(1), y \otimes \delta_j(1)) = 1$. Further $x \otimes \delta_2(1) = y(r(x, 1))$ and $r(x, 1)r(y, 1) = r_v(xy)$, so 28.8 gives the desired equality when $j = 2$. So take $j = 3$. Now from 27.3, $\delta_3(1)r_{v(\epsilon)}(xy) = \delta_3(\pi^{1-\epsilon})$, so $y \otimes \delta_3(1)r_{v(\epsilon)}(xy) = (-1)^{\epsilon+1}(y \otimes \delta_3(1))$. Then by 28.11, $(v_{xy}(\epsilon), x \otimes \delta_3(1), y \otimes \delta_3(1)) = 1/2$, so $(w_{xy}^2, x \otimes \delta_3(1), y \otimes \delta_3(1)) = 1$, completing the proof that $\zeta_1 g = \zeta_2$.

Finally we show g preserves ζ_3. Set $h = \psi_2(\alpha_z)$ and $k = \psi_1(\alpha_z)$. Then $h = g^k$. As ζ_1 is N_1-invariant and $N_1^h = N_3$, it suffices to show $\zeta_1 h = \zeta_3$. But $\zeta_1 h = \zeta_1 g^k = \zeta_1 g k = \zeta_2 k = \zeta_3$, completing the proof.

Lemma 30.4: N preserves the Griess algebra.

Proof: As we observed earlier, to show N preserves the Griess algebra, it suffices to show N preserves β_J for each $J \in I^3/S_3$. We may assume

$J \neq (D, D, D)$, so by 30.2 and 30.3, it remains to consider the cases $J = (Z, Z, 0)$ or $(Z, 0, 0)$. In particular we need only consider nonzero monomials $m = am_1 m_2 m_3$ with $m_1 \in B_Z$. Then by 9.3, $C_Z(m_2) = C_Z(m_3)$, so $J = (Z, 0, 0)$ and $m_2, m_3 \in B_0^i$ for some i. If we argue as in the proof of 30.3, it suffices to show $\zeta_1 g = \zeta_2$ and $\zeta_3 g = \zeta_3$, where ζ_i is the restriction of β to $B_Z \times B_0^i \times B_0^i$ and $g = \psi_3(\alpha_z)$. Let $\rho_i \in L(B_Z, B_0^i, B_0^i; F)$ be N^+-invariant and $n = bm_1 n_2 n_3$ a nonzero monomial in ρ_i. Conjugating in N^+ we may take $m_1 = y(r)$ for $r = r(c, \varnothing)$ and $z \notin \phi(c)$. Then $m_1 g = 1/8 \cdot \sum_K y(r(c, K))$ by 30.1.2.

Suppose $i = 1$. Then $n_j = y(r_j)$ for some $r_j = r(x_j, b_j) \in \mathcal{R}^3$ for $j = 2, 3$. Now unless $r \in \bar{Z}\langle r_2, r_3 \rangle$, there exists $h \in C_{Q_1^+}(\langle r_2, r_3 \rangle) - C_{Q_1^+}(r)$. But then h fixes n_2 and n_3 and inverts m_1, contrary to 9.3. Therefore $r \in \bar{Z}\langle r_2, r_3 \rangle$, which forces $r \in \bar{Z} r_1 r_2$. Recall that if $\rho = \zeta$ then by 28.8, $r = r_2 r_3$.

Now by Exercise 10.1.6, $C_{N^+}(m_1)$ has two orbits on pairs (r_2, r_3) such that $r r_2 = r_3 \in \mathcal{R}^3$, with representatives $(r(x, 1), r(x, c))$, $x \notin \phi(c)$, and $(r(x, d), r(y, cd))$, $d \in \mathcal{C}$ with $|\phi(c) \cap \phi(d)| = 2$ and $xy = \phi(c) \cap \phi(d)$. By 28.8, $(m_1, n_2, n_3) = 1$ when $r r_2 = r_3$. Similarly if $r_2 r_3 = r \bar{z}_2$ then by Exercise 10.1.7, up to conjugation in N^+, $r_2 = r(x, 1)$ and $r_3 = r(x, cs)$ with $x \in \phi(c)$.

Now $\zeta_1 g$ is an N^+-invariant form so by the previous paragraphs to show $\zeta_1 g = \zeta_2$, it suffices to show $\rho_2(m_1 g, n_2 g, n_3 g) = 1, 1, 0$ for the three choices of n_2, n_3 above. Now using 27.3 and 27.4, $y(r(x, d))g = (x \otimes \delta_2^3(d))g = x \otimes \delta_2^1(d + P(d) + \epsilon_{zx}(d))$. Thus if $r_2, r_3 = r(x, 1), r(x, c)$ with $x \notin \phi(c)$, then

$$(m_1 g, n_2 g, n_3 g) = 1/8 \cdot \sum_K (y(r(c, K)), x \otimes \delta_2^1(1), x \otimes \delta_2^1(c))$$

and as $\delta_2^1(b) r(c, K) = \delta_2^1(cb + \epsilon_K(cb))$ by 27.3, it follows from 28.11 that as $x \notin \phi(c)$, each term in this sum is $1/8$, so the sum is indeed 1 as desired.

Next take $r_2, r_3 = r(x, d), r(y, cd)$. Then this time we get a sum of 64 terms

$$1/8 \cdot (y(r(c, K)), x \otimes \delta(d + \epsilon_{zx}(d)), y \otimes \delta(cd + \epsilon_{zx}(cd)))$$

and as $xy = \phi(c) \cap \phi(d)$ such a term is equal to $(-1)^{s(K)}/64$ by 28.11, where $s(K) = 1 + \epsilon_{zx}(c) + \epsilon_K(cd) + \epsilon_K(xy)$. Further $s(K) = 0$, since $x \in \phi(c)$, $z \notin \phi(c)$, and $\epsilon_K(cd) = \epsilon_K(xy)$. So again we get 1 as desired.

Finally take $r_2, r_3 = r(x, 1), r(x, cs)$ with $x \in \phi(c)$. This time our terms are of the form

$$-1/8 \cdot (y(r(c, K)), x \otimes \delta(1), x \otimes \delta(c))$$

since $\delta_2(cs) = \delta_2(c)$. Now as $x \in \phi(c)$ each such term is 0 by 28.11.

Thus we have shown that $\zeta_1 g = \zeta_2$, so it remains only to show g preserves ζ_3. Set $h = \psi_2(\alpha_z)$ and $k = \psi_1(\alpha_z)$. Then $h = g^k$. So as ζ_1 is N_1-invariant and $N_1^h = N_3$, it suffice to show $\zeta_1 h = \zeta_3$. But $\zeta_1 h = \zeta_1 g^k = \zeta_1 g k = \zeta_2 k = \zeta_3$, completing the proof.

31. The automorphism group of the Griess algebra

In this section $G = O(B, \beta, \gamma)$ denotes the group of isometries of the bilinear form γ and the trilinear form β. In particular G is also a group of automorphisms of the Griess algebra.

Using the convention of Section 28 for regarding elements of B_s as symmetric matrices indexed by X, we write I for the element of B_s corresponding to the identity matrix. Observe first from the definition of τ in Section 28 that:

Lemma 31.1: *The element* $id = I/4$ *of* B_s *is the identity for the Griess algebra. In particular G fixes id and acts on the subspace $\hat{B} = id^\perp$ orthogonal to id.*

Observe that $\hat{B} = B_- \perp B_r \perp \hat{B}_s$, where

$$\hat{B}_s = \langle xy, x^2 - y^2 : x, y \in X, \ x \neq y \rangle.$$

Lemma 31.2: $C_G(z_1)$ *acts on* B_i *for each* $i \in \{r, s, -\}$.

Proof: Let $H = C_G(z_1)$. Certainly H acts on $B_- = [B, z_1]$ and $B_+ = C_B(z_1) = B_s \perp B_r$ (cf. 27.9). Then H acts on $\hat{B}_+ = \hat{B} \cap B_+$ and as C is absolutely irreducible on B_r and $\hat{B}_s$, with each space the orthogonal complement in $\hat{B}_+$ of the other, we may assume H is absolutely irreducible on $\hat{B}_+$. In particular, up to a scalar multiple, the restriction $\hat{\gamma}_+$ of γ to $\hat{B}_+$ is the unique quadratic form on $\hat{B}_+$ preserved by H.

Next by 9.9.4, H acts on $End(B_-)$ via conjugation; that is, for $g \in H$ and $\rho \in End(B_-)$, $g : \rho \mapsto \rho^g$, where $\rho^g(b) = \rho(bg^{-1})g$ for $b \in B_-$. Further for $a \in B_+$, it follows from 9.9.5 that $\lambda_a \in End(B_-)$, where $\lambda_a : b \mapsto b * a$, and the map $a \mapsto \lambda_a$ is in $Hom(B_+, End(B_-))$. Also by 9.9.4, $\lambda_a^g = \lambda_{ag}$.

Define $T \in L^2(\hat{B}_+)$ by $T(a,c) = Tr(\lambda_a \lambda_c)$. Then

$$T(ag, cg) = Tr(\lambda_{ag}\lambda_{cg}) = Tr(\lambda_a^g \lambda_c^g)$$
$$= Tr((\lambda_a \lambda_c)^g) = Tr(\lambda_a \lambda_c) = T(a,c),$$

so H preserves T. Hence by the uniqueness of $\hat{\gamma}_+$, $T = p\hat{\gamma}_+$ for some $p \in \mathbf{R}$.

Let x_i, $i = 1, 2$, be distinct elements of X and $\rho = \lambda_{x_1 x_2}$. Then by 28.10, $\rho(y \otimes \delta) = x_{3-i} \otimes \delta$ if $y = x_i$ and 0 otherwise. Hence

$$T(x_1 x_2, x_1 x_2) = 2dim(B^1_-) = 2^{13},$$

so as $\gamma(x_1 x_2, x_1 x_2) = 2$, we conclude $p = 2^{12}$.

On the other hand let $r \in \mathcal{R}$ and $\rho = \lambda_{y(r)}$. Then by 28.6:

$$\rho^2(x \otimes \delta) = 1/64 \cdot x \otimes \delta + 1/16 \cdot (x, \xi(r))((\xi(r), \xi(r)) - 1)(\xi(r) \otimes \delta). \quad (*)$$

Further by 28.3, $(x, \xi(r)) = a_x/\sqrt{8}$ and $(\xi(r), \xi(r)) = \sum_x a_x^2/8 = 4$, so the second term in $(*)$ has projection $3/2 \cdot a_x^2/64$ on $x \otimes \delta$. Hence summing the contributions to $Tr(\rho^2)$ from $(*)$ over our standard basis for B_- we get

$$T(\rho, \rho) = \sum_{x, \delta}(1 + 3/2 \cdot a_x^2)/64$$

$$= 2^{12}\sum_x(1 + 3/2 \cdot a_x^2)/64 = 2^6(24 + 48) = 9 \cdot 2^9.$$

So as $\gamma(x \otimes \delta, x \otimes \delta) = 1$, $p = 9 \cdot 2^9$, a contradiction.

Lemma 31.3: $C = C_G(z_1)$.

Proof: Let $H = C_G(z_1)$. For $(j, i) \in \{(r, -), (s, -), (s, r)\}$, and $b \in B_j$, let $\lambda_b^i = \lambda_b \in End(B_i)$ be defined by $\lambda_b : a \mapsto b * a$ as in 9.9.5, and recall $\lambda^i : b \mapsto \lambda_b^i$ is in $End(B_j, End(B_i))$. Further by 9.9, H acts on $\lambda^i(B_j)$ via conjugation by $(\lambda_b^i)^g = \lambda_{bg}^i$ for $g \in H$.

Now for $x, y \in X$, $r \in \mathcal{R}$, 28.9 and 28.3 say

$$\lambda_{xy}(y(r)) = (xy, \xi(r) \otimes \xi(r))y(r) = a_x a_y y(r)/4 \cdot 2^{\epsilon_x(y)}.$$

In particular we conclude that $\{\mathbf{R}y(r) : r \in \mathcal{R}\}$ is the set of weight spaces for $\lambda^r(B_s)$, and hence as H acts on $\lambda^r(B_s)$, H permutes these weight spaces. Moreover we see that $C_H(B_s)$ fixes each weight space.

Next we saw during the proof of the previous lemma that for distinct $x, y \in X$, λ_{xy}^- interchanges $x \otimes \delta$ and $y \otimes \delta$ and annihilates $u \otimes \delta$ for all $u \in X - \{x, y\}$. So as $C_H(B_s)$ commutes with all these maps, we conclude that $C_H(B_s)$ acts on B_- via $(x \otimes \delta)g = x \otimes \delta g$ for some representation $g : \delta \mapsto \delta g$ of $C_H(B_s)$ on B^1_-. Similarly by 28.6, for $r \in \mathcal{R}$:

$$\lambda^-_{y(r)}(x \otimes \delta) = (x - 2(x, \xi(r))\xi(r)) \otimes \delta r/8. \qquad (*)$$

But for $g \in C_H(B_s)$, we have observed that $y(r)g = c_r(g)y(r)$ for some $c_r(g) \in \mathbf{R}$, so $\lambda_{y(r)}(x\otimes\delta)g = ((x\otimes\delta)*y(r))g = (x\otimes\delta)g*y(r)g = (x\otimes\delta g)* c_r(g)y(r) = c_r(g)\lambda_{y(r)}(x \otimes \delta g)$. Pick x with $v = x - 2(x, \xi(r))\xi(r) \neq 0$. Then by $(*)$, $(v \otimes \delta rg)/8 = (v \otimes \delta r)g/8 = \lambda_{y(r)}(x \otimes \delta)g = c_r(g)\lambda_{y(r)}(x \otimes \delta g) = c_r(y)(v \otimes \delta gr)/8$. As $v \neq 0$ we conclude $[r, g]$ induces a scalar on B^1_- and then conclude the image $P(C_H(B_s))$ of $C_H(B_s)$ in $PGL(B^1_-)$ commutes with the image $P(r)$ of r. So as $\bar{Q}_1$ is generated by $\mathcal{R}$, $P(C_H(B_s)) \leq C_{PGL(B^1_-)}(P(\bar{Q}_1)) = P(\bar{Q}_1)$ since $\bar{Q}_1$ is irreducible on B^1_-. Therefore $C_H(B_s) \leq \bar{Q}_1 D$, where D is the subgroup inducing scalars on B_-. Finally as D preserves γ, $D = \langle z_1 \rangle C_H(B_-)$.

Thus we have shown $C_H(B_s) \leq \bar{Q}_1 C_H(B_- + B_s)$. Then as $C_H(B_s)$ acts on $\mathbf{R}y(r)$ for each $r \in \mathcal{R}$, $[C_H(B_s), \bar{Q}_1] \leq C_H(B_+)$, so $C_H(B_s) = \bar{Q}_1$. In particular $\bar{Q}_1 \trianglelefteq H$. Further as $C_{\bar{Q}_1}(r) = C_{\bar{Q}_1}(y(r))$, and H permutes $\{\mathbf{R}y(r) : r \in \mathcal{R}\}$, H permutes $\mathcal{R}$ by conjugation. So as $Co1$ is the subgroup of $Out(\bar{Q}_1)$ permuting $\mathcal{R}$, our proof is complete.

Lemma 31.4: *G is finite.*

Proof: Consider the stabilizer $\bar{G}$ of the forms β and γ on $\bar{B} = B \otimes_\mathbf{R} \mathbf{C}$. It suffices to show $\bar{G}$ is finite. The proof of the previous lemma applies equally well to $\bar{G}$ to show $C = C_{\bar{G}}(z_1)$. As the stabilizer of the forms β and γ, $\bar{G}$ is an algebraic group. As $\bar{G}$ is irreducible on $\hat{B}$, $\bar{G}$ is reductive. Thus if the connected component $\bar{G}^0 \neq 1$ then by a standard result from the theory of reductive algebraic groups, $C_{\bar{G}}(z_1)$ is infinite, a contradiction. Thus $\bar{G}^0 = 1$; that is, $\bar{G}$ is finite.

Lemma 31.5: *G is simple.*

Proof: By 31.3 and 27.6, $\bar{Q}_1 = F^*(C)$ is extraspecial. Further z_1 is fused to z_2 in N and into $C - \bar{Q}_1$ under $\langle C, N \rangle$, so the result follows from Exercise 2.4.

Remarks. We define the simple group G constructed in this section to be the Monster. By construction it satisfies Hypothesis $\mathcal{H}(Co_1, 12)$, and indeed is even of Monster type in the sense of the next chapter.

Griess constructed a 196,883-dimensional algebra and the Monster in 1980; see [Gr1]. Then in [Gr2], Griess adjoined an identity to obtain the 196,884-dimensional Griess algebra. Later Conway [Co3] and Tits [T2] supplied simplifications of some of Griess' arguments, which we have incorporated in our treatment. In the remainder of this subsection we briefly summarize the approaches of Griess, Conway, and Tits, and compare them with the approach adopted here.

Early in the study of the Monster it was determined that the smallest possible dimension for a nontrivial irreducible **C**-module for the Monster is 196,883. Moreover a calculation by Norton showed that such a module would admit an algebra structure invariant under the Monster. To obtain an identity for this algebra, we add an extra dimension.

Let C be the centralizer of an involution z in a potential Monster. Griess determined the 196,884-dimensional **R**C-module B which has a chance of extending to the Monster, and the algebra structures on B preserved by C. From these structures he picked one which can extend to the Monster and guessed the action of an extra automorphism σ of the algebra not contained in C. This treatment is a tour de force, but very long and complicated.

Let $G = \langle C, \sigma \rangle$. To identify G as the Monster, Griess observes that G preserves a $\mathbf{Z}[1/6]$ lattice in B closed under the algebra multiplication. He then reduces modulo p for each prime $p > 3$ to obtain a $GF(p)$-algebra $B(p)$ and a representation $\pi(p) : G \to G(p) \le O(B(p))$. He proves $C \cong C\pi(p)$ is the centralizer of $z\pi(p)$ in $G(p)$, so $G(p)$ satisfies Hypothesis $\mathcal{H}(12, Co_1)$ of the introduction. Then Griess appeals to a theorem of S. Smith [Sm] which determines the order of a group satisfying $\mathcal{H}(12, Co_1)$, to conclude $g = |G(p)|$ is independent of p. Hence $|G| = g$, so G is finite and satisfies $\mathcal{H}(12, Co_1)$.

Conway had the tremendous advantage of knowing the facts about B that Griess generated. This knowledge led Conway to a wonderful, inspired construction. He begins with the Parker loop L and uses L to construct the group N and its factor group $\bar{N}$ discussed here in Sections 14 and 27. He then extends $\bar{N}_1 = \bar{N} \cap C$ to C and uses $\bar{N}_1$ to construct an **R**C-module B and an algebra structure on B which is visibly preserved by C. Next he restricts this algebra to $\bar{N}_1$, chooses an $\bar{N}_1$-monomial basis X_1 for B, and writes down bases X_2 and X_3 of B and bijections $x_1 \mapsto x_2 \mapsto x_3$ among the three bases. Conway refers to this set up as a *dictionary*. Then he sketches a proof that the bijections define isometries of B induced in N. Therefore N also preserves B.

Finally Conway exhibits a vector $t \in B$ whose orbit under $O(B)$ is finite. Hence $O(B)$ is finite and Conway concludes by proving $C = C_{O(B)}(z)$.

Tits begins with C and the **C**C-module $\mathbf{C} \otimes_{\mathbf{R}} B$ of Griess, which we also denote by B. He writes down algebra structures on B preserved by

C and proves that for certain of these algebras, $C_{O(B)}(z) = C$. He then proves $O(B)$ is reductive as an algebraic group, and hence as $C_{O(B)}(z)$ is finite, $O(B)$ is finite. He does not prove C is proper in $O(B)$.

Our argument follows that of Conway initially, in that we use L to construct N, we use $\bar{N}_1$ to construct C and the action of C on B, and we take Conway's definition of the algebra structure τ on B.

At this point (in Section 28) we transfer emphasis to the trilinear form β determined by τ and the bilinear form γ, and calculate the monomials of β on our standard basis. Then in Section 29 we calculate the action of N_1 on our basis and extend the N_1-module B to an N-module in such a way that N preserves γ. Further in Section 30 we check that N preserves β and hence also τ. Finally in Section 31 we use Tits's approach to prove $C = C_G(z_1)$ and $G = O(B, \beta, \gamma)$ is finite. Our theory of large extraspecial subgroups then immediately says G is simple.

Exercises

1. Let $c_0 \in L$ with $\phi(c_0) = c \in \mathcal{C}$, $r = r(c_0, \varnothing) \in \mathcal{R}^2$, $v = f_c \in \Lambda$, $w = f_X - v$, and $u \in \Lambda_2^3$. Prove
   (1) $\tilde{u} + \tilde{v} \in \tilde{\Lambda}_2$ if and only if $u \in \Lambda_2(v, \pm 2)$.
   (2) $N_{\tilde{v}}$ has two orbits on $\{\tilde{u} \in \tilde{\Lambda}_2^3 : \tilde{u} + \tilde{v} \in \tilde{\Lambda}_2^3\}$, with representatives $u = -\lambda_x$, $x \notin c$, and $u = -\lambda_x + f_{d+X}$, $x \in c \cap d$, $d \in C_2(c)$.
   (3) $w \in \Lambda_4$ with coordinate frame $f_{X+c} - 2f_d$, $d \in V$, $d \cap c = \varnothing$, and $\pm(f_{X+c} - 2\lambda_x)$, $x \in c$.
   (4) $\tilde{u} + \tilde{w} \in \tilde{\Lambda}_2$ if and only if $(u, w) = 0$ or ± 32, and $(u, w_0) = -32$ for some w_0 in the coordinate frame of w.
   (5) $N_{\tilde{v}}$ is transitive on $\{\tilde{u} \in \tilde{\Lambda}_2^3 : \tilde{w} + \tilde{u} \in \tilde{\Lambda}_2^3\}$ with representative $u = -\lambda_x$, $x \in c$.
   (6) $C_{N^+}(r)$ has two orbits on pairs $(r_2, r_3) \in \mathcal{R}^3 \times \mathcal{R}^3$ such that $rr_2 = r_3$, with representatives $(r(x, 1), r(x, c_0))$, $x \notin c$, and $(r(x, d), r(y, c_0 d))$, $\phi(d) \in C_2(c)$, $\{x, y\} = c \cap \phi(d)$.
   (7) $C_{N^+}(r)$ is transitive on pairs (r_2, r_3) such that $r_2 r_3 = r\bar{z}_2$, with representatives $(r(x, 1), r(x, c_0 s))$, $x \in c$.

Chapter 11

Subgroups of Groups
of Monster Type

In Chapter 10 we constructed a finite simple group G possessing an involution z such that $F^*(C_G(z)) = Q$ is extraspecial of order 2^{1+24}, $C_G(z)/Q \cong Co_1$, $Q/\langle z \rangle$ is isomorphic to the Leech lattice modulo 2 as a $C_G(z)/Q$-module, and z is not weakly closed in Q with respect to G. We say that a group G satisfying these hypotheses is of *Monster type*.

In this short chapter we investigate groups of Monster type. In particular we see that such a group contains simple subgroups of *type* F_p, for $p = 2, 3, 5, 7$, and 24. See Section 32 for the definition of groups of type F_p.

Now from Chapter 5, there are twenty-six sporadic groups. In Chapter 6 we constructed the five Mathieu groups. In Chapters 8 and 9 we constructed the three Conway groups plus Suz, J_2, HS, and Mc. Hence each of these twelve sporadic groups is a section of the Monster. Similarly the sporadic groups F_1, F_2, F_3, and F_5 are of type F_p, and hence sections of the Monster. Held's group He is of type F_7 and the largest Fischer group $F_{24} = M(24)'$ is of type F_{24}, so these groups are sections of the Monster. Finally the Fischer groups $M(22) = F_{22}$ and $M(23) = F_{23}$ are sections of F_{24}, and hence also of the Monster. This is best seen by viewing $Aut(F_{24}) = M(24)$ as a 3-transposition group. Thus we have existence proofs for twenty of the twenty-six sporadic groups.

32. Subgroups of groups of Monster type

Define a finite group G to be of *Monster type* if G possesses an involution z such that $Q = F^*(C_G(z)) \cong 2^{1+24}$, $C_G(z)/Q \cong Co_1$, $Q/\langle z \rangle$ is isomorphic to the Leech lattice modulo 2 as a $C_G(z)/Q$-module, and z is not weakly closed in Q with respect to G.

Similarly define a finite group G to be of *type F_p* if G possesses an involution z such that $F^*(G) = Q$ is extraspecial, z is not weakly closed in Q with respect to G, and

(a) $p = 2$, $C_G(z)/Q \cong Co_2$, $Q \cong 2^{1+22}$, and $Q/\langle z \rangle$ is isomorphic to the subspace $v^\perp/\langle v \rangle$ of $\tilde{\Lambda}/\langle v \rangle$ as a $C_G(z)/Q$-module, where $v \in \tilde{\Lambda}_2$ and $C_G(z)/Q \cong Co_2$ is the stabilizer in Co_1 of v.

(b) $p = 3$, $C_G(z)/Q \cong A_9$, and $Q \cong 2^{1+8}$.

(c) $p = 5$, $C_G(z)/Q \cong A_5 wr \mathbf{Z}_2$, and $Q \cong 2^{1+8}$.

(d) $p = 7$, $C_G(z)/Q \cong L_3(2)$, $Q \cong 2^{1+6}$, and $|z^G \cap Q| = 29$.

(e) $p = 24$, $C_G(z)/Q \cong \mathbf{Z}_2/U_4(3)/\mathbf{Z}_3$, and $Q \cong 2^{1+12}$.

Throughout this section we assume G is of Monster type. We let $H = C_G(z)$, $\tilde{H} = H/\langle z \rangle$, $Q = F^*(H)$, and $H^* = H/Q$. By hypothesis we may identify $\tilde{Q}$ with $\tilde{\Lambda}$ as a $GF(2)H^*$-module. We adopt the notation and terminology of Chapter 8 in discussing this module. Let $\theta(z) = z^G \cap Q - \{z\}$. By hypothesis there is $z^g \in \theta(z)$. Let $s = z^g$ and $U = Q \cap Q^g$.

Lemma 32.1: *(1) H is transitive on $\theta(z)$ and $\tilde{\theta}(z) = \tilde{\Lambda}_4$.*

(2) $C_H(s)^*$ is the split extension of $E_{2^{11}}$ by M_{24}, $dim(\tilde{U}) = 12$, $(Q^g \cap H)^* = O_2(C_{H^*}(\tilde{s}))$ induces the group of transvections on $\tilde{U}$ with center $\langle \tilde{s} \rangle$, $Q \cong D_8^{12}$, and $\tilde{U}/\langle \tilde{s} \rangle$ is isomorphic to the 11-dimensional Todd module for $C_H(s)^*/O_2(C_H(s)^*)$.

Proof: By 23.2 and 8.3.4, H has two orbits on involutions in $Q - \langle z \rangle$ with representatives t_i, $i = 2, 4$, where $\tilde{t}_i \in \tilde{\Lambda}_i$. Now the stabilizer in H^* of $\tilde{t}_i$ is isomorphic to Co_2, $M_{24}/E_{2^{11}}$ for $i = 2, 4$, respectively, by 23.1 and 23.2. By 25.8, Co_2 is simple, so $O_2(C_{H^*}(\tilde{t}_2)) = 1$. Thus 8.15.8 and 8.15.7 imply that (1) holds, $dim(\tilde{U}) = 12$, $O_2(C_H(s))^* = (Q^g \cap H)^*$ induces the full group of transvections on $\tilde{U}$ with center $\langle \tilde{s} \rangle$, and $Q \cong D_8^{12}$. The final remark in (2) follows from 23.10.5.

Lemma 32.2: *(1) H is transitive on involutions in $Q - z^G$.*

(2) Let $t \in Q - z^G$ be an involution. Then $C_G(t)/\langle t \rangle$ is of type F_2.

Proof: We saw during the proof of the previous lemma that (1) holds and that $C_H(t) \cong Co_2/(\mathbf{Z}_2 \times 2^{1+22})$. Further as the Monster is of Monster type with (using 27.2 and its notation) $\bar{\psi}_1(s)$ in the role of s and $\bar{E}^+$ in the role of U, and as $\xi_1(\psi_1(s)\psi_1(\alpha_{xy})) = 2\tilde{f}_{xy} \in \tilde{\Lambda}_2$ with

$\psi_1(s)\psi_1(\alpha_{xy}) \in \bar{E}^+$, we may choose $t \in U$. Hence 8.14 completes the proof of (2).

Lemma 32.3: *Let* $B \in Syl_5(C_H(s))$. *Then* $C_G(B)/B$ *is of type* F_5.

Proof: We appeal to 8.13 and 26.5.

Lemma 32.4: *Let* $\mathbf{Z}_3 \cong Y \le C_H(s)$. *Then*

  (1) *If* Y *is not 3-central in* $C_H(s)$ *then* $C_G(Y)/Y$ *is of type* F_3.
  (2) *If* Y *is 3-central in* $C_H(s)$ *then* $C_G(Y)/Y$ *is of type* F_{24}.

Proof: Use 8.13, 26.4, and 26.6.

Lemma 32.5: *Let* $B \in Syl_7(C_H(s))$. *Then*

  (1) $C_G(B)/B$ *is of type* F_7.
  (2) $C_G(B)/B \cong He$.
  (3) $Aut_G(B) \cong \mathbf{Z}_6$ *and some element of order 2 inverting* B *induces an outer automorphism on* $E(C_G(B))$.

Proof: Use 8.13 and 26.2 to prove (1). Further by 26.2, $Aut_H(B) \cong \mathbf{Z}_6 \cong Aut(B)$, so $Aut_G(B) \cong \mathbf{Z}_6$. Also by 26.2, an involution $t \in H$ inverting B induces an outer automorphism on $C_H(B)/BO_2(C_H(B))$ and hence also on $C_G(B)$.

Next by Exercise 14.5 and by 44.4, $C_G(B)/B \cong L_5(2)$, M_{24}, or He. As t induces an outer automorphism on $C_G(B)/B$ and $Out(M_{24}) = 1$ (cf. Exercise 7.5), the second case is out. So to prove that (3) holds, we may assume $C_G(B)/B \cong L_5(2)$. Next there is an element $y \in H$ of order 3 faithful on B centralizing $C_H(B)/B$. Then as $L_5(2)$ contains no such element of order 3, y centralizes $E(C_G(B))$, so 31 divides $|C_G(y)|$. However, $C_Q(y) = C_Q(B)*[C_Q(y), B]$ with $C_Q(B) \cong 2^{1+6}$ and $[C_Q(y), B] \cong 2^{1+6k}$ as $[C_Q(y), B]$ is extraspecial (or $\langle z \rangle$) by Exercise 2.2, and as each faithful irreducible $GF(2)B$-module is of dimension 3. Therefore by 32.4, $C_G(y)/\langle y \rangle$ is of type F_{24}, and hence has order prime to 31.

Remarks. The general structure of the local subgroups of the Monster considered in this chapter was first studied in 1973 by the group at Cambridge, particularly Conway, Harada, Norton, and Thompson, and independently by Griess; see the discussion in Chapter 5. While I don't know the specific approach used by the Cambridge mathematicians or Griess (as most of that work remains unpublished), it must have been local group theoretic as no other techniques existed which could make a dent in the problem. Of course our analysis depends upon the theory of large extraspecial subgroups. That theory had not been systematically developed in 1973.

PART III

Chapter 12

Coverings of Graphs and Simplicial Complexes

A *covering* of a graph or simplicial complex K is a surjective morphism
$d : \tilde{K} \to K$ that is a local isomorphism. A precise definition appears
in Section 35. The graph or complex is *simply connected* if it admits
no proper connected coverings. In this chapter we establish basic results on coverings including criteria for deciding when a graph is simply
connected. Then in Chapter 13 we generate machinery for reducing the
question of the uniqueness of a group G, subject to suitable hypotheses, to the question of whether certain graphs associated to the group
are simply connected. This machinery is the basis for our uniqueness
treatment of the sporadic groups.

In Section 33 we consider certain equivalence relations on the set P
of all paths in a graph Δ. We find in Section 35 that these *invariant
relations* correspond to certain *fiberings* of the graph Δ; that is, surjective locally bijective morphisms onto Δ. Among these fiberings are the
coverings of Δ and of the simplicial complexes K with Δ as the graph
or 1-skeleton of K. We also find in Section 35 that each covering corresponds to a local system on the graph or complex. This discussion is
broader than is strictly necessary for our purposes, but it has the advantage of putting the concepts in a larger context which hopefully makes
them easier to understand. Further the expanded discussion is useful
in the study of various simplicial complexes associated to finite groups,
such as the Brown and Quillen complexes.

Finally in Section 34, we investigate criteria for proving that a graph or complex is simply connected. For example, in Section 35 we see that the graph is simply connected if and only if each cycle is a product in the fundamental groupoid of conjugates of triangles. We term a graph with this property to be *triangulable*. In Section 35 we record various results useful for proving that a graph is triangulable.

33. The fundamental groupoid

In this section Δ is a graph. We direct the reader to Section 3 for our notational conventions and terminology for graphs. In addition let $P = P(\Delta)$ be the set of paths in Δ. For each $p = x_0 \cdots x_r \in P$ we write $org(p)$, $end(p)$ for the *origin* x_0 and *end* x_r of p, respectively. Write $p \cdot q$ or simply pq for the concatenation of paths p and q such that $end(p) = org(q)$. Thus if $q = y_0 \cdots y_s$ with $y_0 = x_r$ then pq is the path $pq = z_0 \cdots z_{r+s}$ of length $r + s$ such that $z_i = x_i$ for $0 \le i \le r$ and $z_i = y_{i-r}$ for $r \le i \le r + s$.

Write p^{-1} for the path $x_r \cdots x_0$. A path $p = x_0 \cdots x_r$ is a *cycle* or *circuit* if $x_r = x_0$.

Define an equivalence relation $\sim$ on P to be *P-invariant* if the following four conditions are satisfied:

(PI1)   If $p \sim q$ then $org(p) = org(q)$ and $end(p) = end(q)$.
(PI2) $rr^{-1} \sim org(r)$ for all $r \in P$.
(PI3)   Whenever $p \sim p'$ and $q \sim q'$ with $end(p) = org(q)$, then also $pq \sim p'q'$.
(PI4) $x \sim xx$ for all $x \in \Delta$.

Define the *kernel* of an equivalence relation $\sim$ on P to be the set $ker(\sim)$ of all cycles s such that $s \sim org(s)$. Given a set S of cycles of Δ, define a relation $\sim_S$ on P by $p \sim_S q$ if p and q have the same origin and end and $pq^{-1} \in S$.

Lemma 33.1: *Let $\sim$ be a P-invariant equivalence relation. Then $\sim = \sim_{ker(\sim)}$.*

Proof: Let $p, q \in P$, $org(p) = x$, and $S = ker(\sim)$. If $p \sim q$ then $x = org(q)$ and $end(p) = end(q)$, so pq^{-1} is defined and $pq^{-1} \sim qq^{-1} \sim x$. Thus $pq^{-1} \in S$ and hence $p \sim_S q$.

Conversely suppose $p \sim_S q$. Then $pq^{-1} \in S$, $x = org(q)$, and $end(p) = end(q)$. Now $p = p \cdot end(p) \sim pq^{-1}q \sim x \cdot q = q$.

Define a subset S of P to be *closed* if S is a set of cycles satisfying the following six properties:

(C1) $rr^{-1} \in S$ for all $r \in P$.

(C2) If $p \in S$ then $p^{-1} \in S$.

(C3) If $p, q \in S$ and $org(p) = org(q)$, then $pq \in S$.

(C4) If $p \in S$ then $r^{-1}pr \in S$ for each $r \in P$ with $org(r) = end(p)$.

(C5) If p is a cycle and $r \in P$ with $org(p) = end(r)$ and $r^{-1}rp \in S$, then $p \in S$.

(C6) $xx \in S$ for all $x \in \Delta$.

Lemma 33.2: *A set S of cycles of Δ is closed if and only if $\sim_S$ is a P-invariant equivalence relation on P.*

Proof: Write $\sim$ for $\sim_S$. Observe that $\sim$ is reflexive if and only if (C1) holds and $\sim$ is symmetric if and only if (C2) holds.

Assume S is closed. We must show $\sim$ is transitive and satisfies (PI1)–(PI4). Notice (PI1) is satisfied by definition of $\sim$ while (C1) implies (PI2) and (C6) implies (PI4). Also

(i)   If $a, b \in P$ with $ab \in S$ then $ba \in S$.

For as $ab \in S$, $a^{-1}aba \in S$ by (C4), and then by (C5), $ba \in S$. Next

(ii) If $a, b, c \in P$ with $ab \in S$ and $c \in S$ then $acb \in S$.

Namely by (i), $ba \in S$, so by (C3), $bac \in S$, and then $acb \in S$ by another application of (i). Similarly

(iii) If $a, b, r \in P$ with $arr^{-1}b \in S$ then $ab \in S$.

For by (i), $rr^{-1}ba \in S$, so by (C5), $ba \in S$, and then $ab \in S$ by (i).

Now if $p \sim p_1$ and $q \sim q_1$ then $pp_1^{-1}, qq_1^{-1} \in S$, and then if $end(p) = org(q)$, $(pq)(p_1q_1)^{-1} = pqq_1^{-1}p_1^{-1} \in S$ by (ii), so $pq \sim p_1q_1$. That is, (PI3) holds.

Finally if $p \sim q \sim r$ then $pq^{-1}, qr^{-1} \in S$, so by (C3), $pq^{-1}qr^{-1} \in S$ and hence $pr^{-1} \in S$ by (iii). That is, $p \sim r$, so $\sim$ is transitive and we have completed the proof that if S is closed then $\sim$ is an invariant equivalence relation.

Conversely assume $\sim$ is an invariant equivalence relation. We have seen that (C1) and (C2) hold. Let $p, q \in P$ with $end(p) = org(q) = x$. Then if $p, q \in S$, we have $p \sim x \sim q$, so by (PI3), $pq \sim x \cdot x = x$, and hence $pq \in S$. That is, (C3) holds. Also when $p \in S$ then $q^{-1}pq \sim q^{-1} \cdot x \cdot q \sim x$, so (C4) holds. Notice (PI4) implies (C6), so it remains to establish (C5). But if $q^{-1}qp \in S$ then $x \sim q^{-1}qp \sim p$ by (PI2) and (PI3), so $p \in S$, completing the proof.

The *closure* $\langle T \rangle$ of a set T of cycles is the intersection of all closed subsets containing T. It is easy to check that the intersection of closed sets is closed using the axioms (C1)–(C6). Thus

Lemma 33.3: *The closure $\langle T \rangle$ of any set T of cycles is a closed set.*

Define the *basic relation* to be the relation $\sim_{Bas}$, where Bas is the smallest closed subset of P. Write $\equiv$ for $\sim_{Bas}$. Thus $\equiv$ is the P-invariant equivalence relation generated by the paths of length 0. Notice

Lemma 33.4: $\equiv$ *is characterized by the property that if $\sim$ is P-invariant and $a \equiv b$ then $a \sim b$.*

Given a set S of cycles of Δ and $p = x_0 \cdots x_r \in P$, write $I_S(p)$ for the paths obtained by replacing x_i with s or s^{-1} for some i and $s \in S$ with origin x_i. Write $D_S(p)$ for the paths obtained by replacing a subpath t of p such that t or t^{-1} is in S by $org(t)$.

Let Bas_1 denote the set of all paths xx and xyx with $x \in \Delta$ and $y \in \Delta(x)$. For $p = x_0 \cdots x_r$ let $I(p) = I_{Bas_1}(p)$ and $D(p) = D_{Bas_1}(p)$. The processes I and D are called *insertion* and *deletion*, respectively. Notice $D(p)$ is empty if and only if p has no subpaths in Bas_1. For $0 \leq l_i, k_i \in \mathbf{Z}$, define $I^{k_1} D^{l_1} \cdots I^{k_n} D^{l_n}(p)$ recursively in the obvious manner.

Regard P as a graph by decreeing that p is adjacent to q if $q \in I(p) \cup D(p)$. Write $[p]$ for the connected component of p in P. The following observation gives a characterization of the basic relation.

Lemma 33.5: $p \equiv q$ *if and only if* $[p] = [q]$.

Proof: Let $\sim$ be the relation $p \sim q$ if and only if $[p] = [q]$. Then $\sim$ is an equivalence relation satisfying (PI1), (PI2), and (PI4). If $p \sim q$ there exists a path $p = p_0, \ldots, p_n = q$ in P. Then for $a, b \in P$ with $end(a) = org(p)$ and $org(b) = end(p)$, we see that $ap = ap_0, \ldots, ap_n = aq$ and $pb = p_0 b, \ldots, p_n b = qb$ are paths in P, so $\sim$ is P-invariant.

To complete the proof we prove that if $p \sim q$ then $p \equiv q$, and appeal to 33.4. Proceeding by induction on the distance of p from q in P, it suffices to show that if $q \in I(p) \cup D(p)$ then $p \equiv q$. So as $\equiv$ is P-invariant it suffices to show $xvx \equiv x$ for all $x, v \in \Delta$ with $v \in \Delta(x)$, which follows from (C4).

Write $\Phi(\Delta)$ for the set of equivalence classes $[p]$ of the basic relation $\equiv$. Given a set S of cycles of Δ, define $I_S([p])$ to consist of the classes $[a]$ such that $a \in I_S(b)$ and $b \in [p]$. Define $D_S([p])$ analogously. Let P_S be the graph on $\Phi(\Delta)$ with $[p]$ adjacent to $[q]$ if and only if $[q] \in I_S([p]) \cup D_S([p])$. Write $[p]_S$ for the connected component of $[p]$ in P_S. For $p \in \langle S \rangle$, define the S-degree $deg_S(p)$ to be the distance from $[p]$ to $[org(p)]$ in P_S.

Arguing as in the proof of 33.5 we obtain:

Lemma 33.6: *Let S be a set of cycles of Δ. Then $p \sim_{\langle S \rangle} q$ if and only if $[p]_S = [q]_S$.*

Lemma 33.7: *Let S be a set of cycles and $p \in \langle S \rangle$. Then*

(1)   If $deg_S(p) = 0$ then $p \equiv org(p)$.
(2)   If $deg_S(p) = 1$ then $p \equiv r^{-1}tr$ for some $t \in S \cup S^{-1}$ and $r \in P$.
(3)   If $deg_S(p) > 1$ then $p \equiv cd$ for some $c, d \in \langle S \rangle$ with $deg_S(c) < deg_S(p) > deg_S(d)$.

Proof: Part (1) is trivial, so assume $n = deg_S(p) > 0$. Then there exists $q \in \langle S \rangle$ of degree $n-1$ such that either $q = ab$ and $p \equiv asb$ or $q = asb$ and $p \equiv ab$ for some $s \in S \cup S^{-1}$. In the first case $p \equiv asb \equiv abb^{-1}sb \equiv qb^{-1}sb$ and in the second $p \equiv qb^{-1}s^{-1}b$. But $deg_S(b^{-1}sb) = 1$ and if $n = 1$ then $q \equiv org(q)$, so the lemma holds.

Define an *inversion* on a category C to be a contravariant functor $Inv : C \to C$ such that $Inv(x) = x$ for each object x in C and $Inv^2 = id$. Thus for each morphism $f : x \to y$, $Inv(f) : y \to x$ and $Inv^2(f) = f$.

Define a *pregroupoid* to be a category C together with an inversion Inv on C. The composition map on C will be termed the *pregroupoid product*. A *morphism of pregroupoids* $\alpha : (C, Inv) \to (C', Inv')$ is a covariant functor $\alpha : C \to C'$ preserving inversion; that is, $\alpha(Inv(f)) = Inv'(\alpha(f))$ for each morphism f in C. The *kernel* of the morphism α consists of those morphisms $f : x \to x$ such that $\alpha(f) = id_x$ for some object x.

A *groupoid* is a category in which each morphism is an isomorphism. Notice a groupoid is a pregroupoid with the natural inversion map $Inv : f \mapsto f^{-1}$. A *morphism of groupoids* is just a covariant functor $\alpha : C \to C'$, since each such functor preserves the natural inversion.

We can make P into a category as follows. The objects of the category P are the vertices of Δ, the morphisms from a vertex x to a vertex y are the paths from x to y, and the composition of suitable paths p and q is the product $p \cdot q$. Further we have an inversion $Inv : p \to p^{-1}$ on P, so P is a pregroupoid.

Similarly given any invariant relation $\sim$ on P, write $\bar{p}$ for the equivalence class of $p \in P$ under $\sim$ and write $\tilde{P}$ for the set of equivalence classes of $\sim$. Given $\bar{p} \in \tilde{P}$ define $org(\bar{p}) = org(p)$ and $end(\bar{p}) = end(p)$. This is well defined by (PI1). Further if $\tilde{p}, \tilde{q} \in \tilde{P}$ with $end(\tilde{p}) = org(\tilde{q})$ define $\tilde{p} \cdot \tilde{q}$ to be the equivalence class of $p \cdot q$. This is well defined by (PI3). Define $(\tilde{p})^{-1}$ to be the equivalence class of p^{-1}. If $p \sim q$ then $pq^{-1} \in ker(\sim)$ so by (C4) and (C5), $q^{-1}p \in ker(\sim)$ and hence $q^{-1} \sim p^{-1}$. Thus $(\tilde{p})^{-1}$ is well defined. Finally observe $\tilde{P}$ is a groupoid. Namely the objects of $\tilde{P}$ are again the vertices of Δ, the morphisms from x to x are the equivalence classes $\tilde{p}$, for p a path from x to y, and composition is the product in $\tilde{P}$. Each morphism in $\tilde{P}$ is an isomorphism by (PI2). Thus we have essentially shown:

Lemma 33.8: *Let $\sim$ be an invariant equivalence relation on Δ and $P = P(\Delta)$. Then*

> *(1) P is a pregroupoid.*
>
> *(2) $\tilde{P}$ is a groupoid; in particular $\Phi(\Delta)$ is a groupoid called the fundamental groupoid.*
>
> *(3)   The map $p \mapsto \tilde{p}$ of P onto $\tilde{P}$ is a surjective morphism of pregroupoids which is the identity on the set Δ of objects and with kernel $ker(\sim)$.*
>
> *(4)   The map $[p] \to \tilde{p}$ of $\Phi(\Delta)$ onto $\tilde{P}$ is a surjective morphism of groupoids which is the identity on the set Δ of objects.*
>
> *(5)   If C is a groupoid with object set Δ and $\alpha : P \to C$ is a morphism of pregroupoids with kernel S which is the identity on Δ, then S is a closed subset of P and $p \sim_S q$ if and only if $\alpha(p) = \alpha(q)$.*

Given an invariant relation $\sim$ on Δ and $x \in \Delta$, denote by $\tilde{\pi}(\Delta, x)$ the subgroupoid of $\tilde{P}$ consisting of all $\tilde{p}$ such that p is a cycle with origin x. Observe that $\tilde{\pi}(\Delta, x)$ is a group under the groupoid product. Write $\pi_1(\Delta, x)$ for $\tilde{\pi}_1(\Delta, x)$ when $\sim$ is the basic relation $\equiv$. Then $\pi_1(\Delta, x)$ is the *fundamental group* of the graph Δ at x.

The last three lemmas in this section are not used elsewhere in *Sporadic Groups*, so they can be skipped if the reader chooses. However, they give a solution to the word problem in the fundamental groupoid, and hence seem worth including.

Lemma 33.9: *If $p \in P$ with $D(p) \neq \varnothing$ then $D(I(p)) = I(D(p))$.*

Proof: Let $p = x_0 \cdots x_r$ and for $s \in Bas_1$ with $x_i = org(s)$, write $D(p, i, s)$ for the path obtained by replacing the subpath s of p beginning at x_i by x_i, and write $I(p, i, s)$ for the path obtained by replacing x_i by s. Observe that if $s, s' \in Bas_1$ with $org(s) = x_i$ and $org(s') = x_t$, then

> (a) $D(I(p, i, s), t, s') = I(D(p, t, s'), i - l(s'), s)$ for $t + l(s') \leq i$.
>
> (b) $D(I(p, i, v), t + l(s), s') = I(D(p, t, s'), i, s)$ for $t \geq i$.

Further if $d = D(p, t, s')$ and $q = I(d, j, s)$, then either $j \leq t$ and by (b), $q = D(I(p, j, s), t + l(s), s') \in D(I(p))$, or $j \geq t$, in which case the jth entry d_j of d is x_i, where $i = j + l(s') \geq t + l(s')$, and thus by (a), $q = D(I(p, i, s), t, s') \in D(I(p))$. So in any event, $I(D(p)) \subseteq D(I(p))$, so to complete the proof we must show $D(I(p)) \subseteq I(D(p))$.

Let $q = I(p, i, s)$ and suppose $d = D(q, j, s') \in D(q)$. If $j \leq i - l(s')$ then by (a), $d = I(D(p, j, s'), i - l(s'), s) \in I(D(p))$. Similarly if $j \geq i + l(s)$ then $q_j = x_t$, where $t = j - l(s) \geq i$, so by (b), $d = I(D(p, t, s'), i, s) \in I(D(p))$. Thus we may assume $i - l(s') < j < i + l(s)$.

Observe first that if $j \leq i$ and $i + l(s) \leq j + l(s')$ then s is a subpath of s'. So as s and s' are of the form xx or xyx, we conclude $s = s'$ and then that $d = p \in I(D(p))$. So we may assume this is not the case.

Next s, s' have length 1 or 2. Suppose first $l(s) = 1$. Then $s = xx$ and as $j < i + l(s) = i + 1$, $j \leq i$. Similarly $i - l(s') < j$, so $i + l(s) = i + 1 \leq j + l(s')$, contrary to our reduction of the previous paragraph.

So $l(s) = 2$ and $s = xyx$. Then $j \leq i + 1$ and $i + l(s) = i + 2 \leq j + l(s') + 1$, and by our reduction, one of these inequalities is an equality. In either case $l(s') = 2$ and $d = p \in I(D(p))$, completing the proof.

Define the *basic degree* $bas(p)$ of $p \in ker(\equiv)$ to be the distance of p from $org(p)$ in the graph P.

Lemma 33.10: *For $p \in Bas$, $p \in I^{bas(p)}(org(p))$.*

Proof: Let $x = org(p)$. The proof is by induction on $n = bas(p)$. If $n = 0$, then $p = x$ and the lemma is clear, so take $n > 0$. Then the set Q of paths $q \in Bas$ with $bas(q) = n - 1$ and $p \in D(q) \cup I(q)$ is nonempty. Let $q \in Q$. By induction on n, $q \in I^{n-1}(x)$. Thus if $p \in I(q)$, then $p \in I^n(q)$, so we may assume

(∗) If $q \in Q$ then $p \notin I(q)$ so $p \in D(q)$.

Then $D(q) \neq \varnothing$, so $q \neq x$ and $n > 1$. If $n = 2$ then $q \in Bas_1$ so as $p \in D(q)$, $p = x$, contradicting $n = 2$. Hence $n > 2$.

Now $q \in I(r)$ for some $r \in I^{n-2}(x)$ and as $n > 2$, $D(r) \neq \varnothing$. Then $p \in D(q) \subseteq D(I(r)) = I(D(r))$ by 33.9. Thus $p \in I(s)$ for some $s \in D(r)$ and $bas(s) \leq bas(r) + 1 = n - 1$. That is, $s \in Q$, contradicting (∗).

Lemma 33.11: *For $p = x_0 \cdots x_r \in Bas$,*

$$bas(p) = (l(p) + r(p))/2,$$

where $r(p) = |\{i : x_i = x_{i+1}\}|$.

Proof: This follows from 33.10 by induction on $bas(p)$.

Remark.   Lemma 33.10 supplies an algorithm for deciding when two paths p and q are equivalent under the basic relation. Namely define p to be *reduced* if $D(p) = \varnothing$. Then it follows from 33.10 (cf. Exercise 12.3) that $[p]$ contains a *unique* reduced path r_p. Thus $p \equiv q$ if and only if $r_p = r_q$. Further $r_p \in D^n(p)$ for some n, so we can effectively calculate r_p.

34. Triangulation

In this section we continue the hypotheses and notation of Section 33. In particular Δ is a graph and $P = P(\Delta)$ is the set of paths in Δ.

   Write $\Delta^n(x)$ for the set of vertices at distance n from some vertex x in Δ. Write $\Delta^{\leq n}$, $\Delta^{<n}$ for the set of vertices at distance at most n, at distance less than n from x, respectively. For $X \subset \Delta$, let $\Delta(X) = \bigcap_{x \in X} \Delta(x)$. Thus, for example, $\Delta(x, y) = \Delta(x) \cap \Delta(y)$.

   Throughout this section S will be a closed subset of P and $\sim \, = \, \sim_S$ is the invariant relation determined by S. Define a cycle p to be *trivial* if $p \in S$ and *nontrivial* if $p \notin S$. Sometimes we write $p \sim 1$ to indicate p is a trivial cycle.

Lemma 34.1: S *contains all cycles of length at most 2.*

Proof: This is a consequence of (C1) and (C6).

   For n a positive integer define $C_n(\Delta)$ to be the closure of the set of all cycles of Δ of length at most n. Define Δ to be *n-generated* if $C_n(\Delta)$ is the set of all cycles. We say Δ is *triangulable* if Δ is 3-generated. In the remainder of this section we investigate conditions which will tend to force Δ to be triangulable.

Lemma 34.2: Δ *is triangulable if and only if for each $x \in \Delta$, $\pi_1(\Delta, x)$ is generated by classes $[rtr^{-1}]$, with $r \in P$, t a triangle, $org(r) = x$, $end(r) = org(t)$.*

Proof: This follows from 33.7 with the "S" of 33.7 the set of triangles.

Lemma 34.3: *(1) If pq, pr, and $r^{-1}q$ are cycles with pr, $r^{-1}q \in S$, then $pq \in S$.*

   (2) Let $a_i, b_i, c_j \in P$, $1 \leq i \leq n$, $1 \leq j < n$, such that $org(a_i) = x$, $end(b_i) = u$, and $end(a_i) = org(b_i) = org(c_i) = end(c_{i-1})$ for $1 \leq i \leq n$. Assume $a_i c_i a_{i+1}^{-1}$ and $b_i^{-1} c_i b_{i+1}$ are in S for $1 \leq i < n$. Then $a_n b_n b_1^{-1} a_1^{-1} \in S$.

Proof: Part (1) follows from the closure axioms (C1)–(C6). Or alternatively using the axioms (PI1)–(PI4), $pr \sim r^{-1}q \sim org(p)$, so

$pq \sim pr \cdot r^{-1}q \sim org(p)$ and hence $pq \in S$. Then (2) follows from (1) using either point of view by induction on n.

Given integers n, m with $n \geq 2$, define $|m|_n = r$, where $0 \leq r \leq n/2$ and $m \equiv r$ or $-r \mod n$. Then define a cycle $p = x_0 \cdots x_n$ of length n to be an *n-gon* if $d(x_i, x_j) = |i - j|_n$, for all i, j, $0 \leq i, j \leq n$.

Lemma 34.4: *Let p be a nontrivial cycle of minimal length r. Then p is an r-gon.*

Proof: Let $p = x_0 \cdots x_r$ with $x = x_0$. Pick n maximal subject to $d(x, x_n) = n$. Then there exists a path b from x_{n+1} to x_0 of length $m \leq n$. Let $a = x_0 \cdots x_{n+1}$ and $c = x_{n+1} \cdots x_r$. Thus $l(a) = n + 1$ and $l(c) = r - n - 1$. Then $l(ab) = l(a) + l(b) = n + m + 1$ and $l(b^{-1}c) = m + r - n - 1 < r$. However, if $l(ab), l(b^{-1}c) < r$ then by minimality of r, ab and $b^{-1}c$ are in S, so by 34.3.1, $p \in S$. Thus $m + n + 1 \geq r$, so as $m \leq n$, we have $n \geq (r - 1)/2$. Similarly applying the same argument to p^{-1} and letting N be maximal subject to $d(x, x_{r-N}) = N$, we have $N \geq (r - 1)/2$. It follows that $n = N$ and $r = 2n$ or $2n + 1$. That is, p is an r-gon.

Lemma 34.5: *Let d be the diameter of the graph Δ. Then S is the set of all cycles of Δ if and only if for all $r \leq 2d + 1$, each r-gon is in S.*

Proof: If p is an r-gon in Δ then $r \leq 2d + 1$, so the lemma follows from 34.4.

Lemma 34.6: *Let $p = x_0 \cdots x_4$ be a cycle such that $\Delta(x_0, x_2)$ is connected. Then $p \in C_3(\Delta)$.*

Proof: Let $S = C_3(\Delta)$ and suppose p is nontrivial. By hypothesis there is a path $x_1 = y_1, \ldots, y_n = x_3$ in $\Delta(x_0, x_2)$. Now appeal to 34.3.2 with $a_i = x_0 y_i$, $b_i = y_i x_2$, and $c_i = y_i y_{i+1}$.

Given a symmetric relation R on a set X, X becomes a graph with edge set R, and we say X is *R-connected* if this graph is connected and write $R(x)$ for the vertices of the graph adjacent to and distinct from a vertex x.

Lemma 34.7: *Define a sequence of symmetric relations on Δ recursively as follows: R_0 is the edge set of Δ. Given R_i, define R_{i+1} to be the set of pairs (x, y) such that $y \in \Delta^2(x)$ and $\Delta(x, y)$ is R_i-connected. Let $R_\infty = \bigcup_i R_i$ and $p = x_0 \cdots x_4$ a cycle with $x_2 \in R_\infty(x_0)$. Then $p \in C_3(\Delta)$.*

Proof: By hypothesis, $x_2 \in R_m(x_0)$ for some $m \geq 1$. We induct on m. If $m = 1$ the result is 34.6, so take $m > 1$. Then by hypothesis there is an R_{m-1}-path $x_1 = y_0, \ldots, y_k = x_3$ in $\Delta(x_0, x_2)$. As $y_{i+1} \in R_{m-1}(y_i)$, $p_i = x_0 y_i x_2 y_{i+1} x_0 \in S = C_3(\Delta)$ by induction on m. Thus as p is in the closure of the paths p_i, $0 \le i < k$ (cf. 34.3), also $p \in S$.

Lemma 34.8: *Let* $p = x_0 \cdots x_5$ *be a pentagon in* Δ *such that* $x_0^\perp \cap x_2^\perp \cap x_3^\perp \neq \varnothing$. *Then* $p \in C_4(\Delta)$.

Proof: Let $x \in x_0^\perp \cap x_2^\perp \cap x_3^\perp$. Then p is in the closure of the triangle $x x_2 x_3 x$ and the squares $x_0 x_1 x_2 x x_0$ and $x_0 x x_3 x_4 x_0$.

Lemma 34.9: *Let* $p = x_0 \cdots x_6$ *be a hexagon in* Δ *with* $x_0^\perp \cap x_2^\perp \cap x_4^\perp \neq \varnothing$. *Then* $p \in C_4(\Delta)$.

Proof: Let $x \in x_0^\perp \cap x_2^\perp \cap x_4^\perp$. Then p is in the closure of the squares $x x_i x_{i+1} x_{i+2} x$, $i = 0, 2, 4$.

Lemma 34.10: *Let* $\sim = \sim_S$, *where* $S = C_4(\Delta)$. *Let* $\mathcal{F}$ *denote the set of paths* p *of* Δ *of length 4 such that* $p \sim q$ *for some path* q *of length 3. Let* $p = x_0 \cdots x_n \in P$ *be such that* $p \not\sim q$ *for any* $q \in P$ *of length less than* n. *Then*

(1) $x_{i+2} \in \Delta^2(x_i)$ for each $i \le n - 2$.
(2)   If $n \ge 4$ then $d(\Delta(x_0, x_2), \Delta(x_2, x_4)) > 1$.
(3) p contains no subpath in $\mathcal{F}$.
(4)   If $x \in \Delta(x_0, x_2)$ then $p \sim x_0 x x_2 \cdots x_n$.

Proof: If (1) fails then $p \sim q$, where q is the path obtained by deleting x_{i+1} from p, contrary to the choice of p. Thus (1) is established. If $x_1 \neq x \in \Delta(x_0, x_2)$ then $x_0 x_1 x_2 x x_0 \in S$, so $x_0 x x_2 \sim x_0 x_1 x_2$ and hence (4) holds.

If p contains a subpath in $\mathcal{F}$ then without loss of generality it is $q = x_0 \cdots x_4$. Thus there is $r = y_0 \cdots y_3 \sim q$. Now $p = q \cdot x_4 \cdots x_n \sim r \cdot x_4 \cdots x_n$ which is of length less than n, contrary to the choice of p. So (3) holds.

Finally assume $n \ge 4$, $x \in \Delta(x_0, x_2)$, $y \in \Delta(x_2, x_4)$, and $y \in x^\perp$. By (4) we may take $x = x_1$ and $y = x_3$. But then (1) supplies a contradiction.

35. Coverings of graphs and simplicial complexes

We continue the hypotheses and notation of Section 33. In particular Δ is a graph and $P = P(\Delta)$ is the set of paths of Δ.

Recall that a *simplicial complex* K consists of a set X of *vertices* together with a distinguished set of nonempty subsets of X called the *simplices* of K such that each subset of a simplex is a simplex. The morphisms of simplicial complexes are the *simplicial maps*; that is, a simplicial map $f : K \to K'$ is a map $f : X \to X'$ of vertices such that $f(s)$ is a simplex of K' for each simplex s of K.

Example The *clique complex* $K(\Delta)$ of our graph Δ is the simplicial complex whose vertices are the vertices of Δ and whose simplices are the finite cliques of Δ. Recall a *clique* of Δ is a set Y of vertices such that $y \in x^{\perp}$ for each $x, y \in Y$. Conversely if K is a simplicial complex then the *graph* of K is the graph $\Delta(K)$ whose vertices are the vertices of K and with $x * y$ if $\{x, y\}$ is a simplex of K. Observe that K is a subcomplex of $K(\Delta(K))$.

Given a simplicial complex K and a simplex s of K, define the *star* of s to be the subcomplex $st_K(s)$ consisting of the simplices t of K such that $s \cup t$ is a simplex of K. Define the *link* $Link_K(s)$ to be the subcomplex of $st_K(s)$ consisting of the simplices t of $st_K(s)$ such that $t \cap s = \varnothing$.

The *dimension* of a simplex s is $|s| - 1$ and the dimension $dim(K)$ of K is the maximum of the dimensions of the simplices of K if those dimensions are bounded, with $dim(K) = \infty$ otherwise. An *n-simplex* is an n-dimensional simplex.

Define a morphism $d : \Lambda \to \Delta$ of graphs to be a *local bijection* if for all $\alpha \in \Gamma$,

$$d_\alpha = d|_{\alpha^{\perp}} : \alpha^{\perp} \to d(\alpha)^{\perp}$$

is a bijection. Define d to be a *fibering* if d is a surjective local bijection. The fibering is *connected* if its domain Γ is connected. The fibering is a *covering* if $d_\alpha : \alpha^{\perp} \to d(\alpha)^{\perp}$ is an isomorphism for all $\alpha \in \Gamma$. We say Δ is *simply connected* if Δ is connected and Δ possesses no proper connected coverings.

Similarly define a simplicial map $\phi : L \to D$ of simplicial complexes to be a *covering* if ϕ is surjective on vertices and a local isomorphism; that is, for each vertex x of L, the map $\phi_x : st_L(x) \to st_D(\phi(x))$ is an isomorphism.

Recall that the *n-skeleton* L^n of a simplicial complex L is the subcomplex consisting of all simplices of dimension at most n. Notice that the 1-skeleton L^1 of L and its graph $\Delta(L)$ are essentially the same, modulo the identification of the edges of $\Delta(L)$ with the 1-simplices of L.

Lemma 35.1: *(1) If $\phi : L \to D$ is a simplicial map then the induced map $\phi : \Delta(L) \to \Delta(D)$ of vertices is a morphism of graphs.*

(2) *If $d : \Lambda \to \Delta$ is a morphism of graphs then the induced map $d : K(\Lambda) \to K(\Delta)$ of vertices is a morphism of simplicial complexes.*

Lemma 35.2: *If $\phi : L \to K$ is a covering of simplicial complexes then*

(1) *ϕ is surjective on simplices.*

(2) *For each simplex s of L, $\phi : st_L(s) \to st_D(\phi(s))$ is an isomorphism.*

Lemma 35.3: *Let $d : \Lambda \to \Delta$ be a morphism of graphs. Then*

(1) *d is a fibering if and only if $d : K(\Lambda)^1 \to K(\Delta)^1$ is a covering of simplicial complexes.*

(2) *d is a covering of graphs if and only if $d : K(\Lambda) \to K(\Delta)$ is a covering of simplicial complexes.*

Given a P-invariant equivalence relation $\sim$, recall from Section 33 that $P/\sim = \tilde{P}$ is the set of equivalence classes $\tilde{p}$ of $\sim$. Each member of $\tilde{p}$ has the same origin and end, so we set $end(\tilde{p}) = end(p)$. Thus we have a map

$$end : \tilde{P} \to \Delta.$$

Make $\tilde{P}$ into a graph by decreeing that $\tilde{p}$ is adjacent to $\tilde{q}$ if $p \sim q \cdot (yx)$, where $y = end(q)$ and $x = end(p) \in \Delta(y)$. Notice that if $p \sim q \cdot (xy)$ then $q \sim p \cdot (yx)$, so our graph is undirected.

Denote by $\Sigma(\Delta, x)$ the set of paths with origin x. Write $\tilde{\Sigma}(\Delta, x) = \Sigma(\Delta, x)/\sim$ for the set of classes $\tilde{p}$ with $p \in \Sigma(\Delta, x)$. Recall that $\tilde{\pi}(\Delta, x)$ denotes the group of $\tilde{p} \in \tilde{\Sigma}(\Delta, x)$ with p a cycle and $\pi_1(\Delta, x) = \tilde{\pi}(\Delta, x)$ for $\sim$ as the basic relation $\equiv$; $\pi_1(\Delta, x)$ is the fundamental group of the graph Δ.

If D is a simplicial complex with graph Δ and $\sim$ an invariant relation, define $\tilde{D}$ to be the subcomplex of $K(\tilde{P})$ with simplices s such that $end(s)$ is a simplex of D. Let $\tilde{\Sigma}(D, x)$ be the corresponding subcomplex of $\tilde{\Sigma}(\Delta, x)$ and $\tilde{\pi}(D, x)$ the corresponding group. Let $\pi_1(D, x) = \tilde{\pi}(D, x)$, where $\sim = \sim_S$ for S the closure of the set of all 2-simplices of D. Then $\pi_1(D, x)$ is the *fundamental group* of D.

Lemma 35.4: *Assume Δ is connected and $\sim$ is a P-invariant equivalence relation on P. Then*

(1) *The connected components of $\tilde{P}$ are the sets $\tilde{\Sigma}(\Delta, x)$, $x \in \Delta$.*

(2) *The map end : $\tilde{p} \mapsto end(p)$ is a fibering of Δ and induces a connected fibering of $\tilde{\Sigma}(\Delta, x)$ onto Δ.*

(3) *The fibering end is a covering if and only if $\ker(\sim)$ contains all triangles of Δ.*

(4) Let D be a simplicial complex with graph Δ. Then $end : \tilde{D} \to D$ is a covering of simplicial complexes if and only if $ker(\sim)$ contains all 2-simplices of D.

(5) The fibering $end : \tilde{\Sigma}(\Delta, x) \to \Delta$ is an isomorphism if and only if $ker(\sim)$ is the set of all cycles in Δ.

(6) $\alpha : \tilde{\pi}_1(\Delta, x) \to Aut(\tilde{\Sigma}(\Delta, x))$ is a faithful representation of $\tilde{\pi}_1(\Delta, x)$ such that $\tilde{\pi}_1(\Delta, x)$ acts regularly on each fiber of the fibering of $end : \tilde{\Sigma}(\Delta, x) \to \Delta$, where for $\tilde{q} \in \tilde{\pi}_1(\Delta, x)$, $\alpha(\tilde{q}) : \tilde{p} \mapsto \tilde{q}^{-1} \cdot \tilde{p}$.

Proof: The proofs of parts (1), (2), and (6) are straightforward. Our fibering end is a covering if and only if whenever $\{x, y, z\}$ is a triangle in Δ and $p \in P$ with $end(p) = x$, then the equivalence classes of p, $p \cdot (xy)$, and $p \cdot (xz)$ form a triangle in $\tilde{P}$. But the latter holds if and only if $p \cdot (xz) \sim p \cdot (xy) \cdot (yz)$ if and only if $p \cdot (xzyx) \cdot p^{-1} \in S = ker(\sim)$ if and only if each triangle $xyzx$ is in S. That is, (3) holds. Similarly by definition, $end : \tilde{D} \to D$ is simplicial, while the map is a covering if and only if for each simplex $s = \{x, y, z\}$ of D, the unique preimage of s under end containing $\tilde{p}$ is a simplex of $\tilde{D}$ if and only if $xyzx$ is in S, so (4) holds.

As $end : \tilde{\Sigma}(\Delta, x) \to \Delta$ is a morphism of graphs surjective on vertices and edges, end is an isomorphism if and only if end is an injection. Now if end is injective and p is a cycle with origin x then $end(\tilde{p}) = end(\tilde{x}) = x$, so $p \sim x \in S$, and hence S contains all cycles. Conversely if S contains all cycles and $end(\tilde{p}) = end(\tilde{q})$ then pq^{-1} is a cycle so $pq^{-1} \in S$ and hence $p \sim q$. Thus (5) holds.

Define a *local system* on the graph Δ to be an assignment $F : x \mapsto F(x)$, $(x, y) \mapsto F_{xy}$ of each $x \in \Delta$ to a set $F(x)$ and each edge (x, y) to a bijection $F_{xy} : F(x) \to F(y)$, such that

(LS1) $F_{yx} \circ F_{xy} = id_{F(x)}$ for each edge (x, y) of Δ.

Example 35.5 If $d : \Lambda \to \Delta$ is a fibering we obtain a local system F^d on Δ defined by $F^d(x) = d^{-1}(x)$ and $F^d_{x,y}(\alpha) = d_\alpha^{-1}(y)$.

Recall from 33.8 that P is a pregroupoid; that is, P is a category possessing the inversion $Inv : p \mapsto p^{-1}$. Given a local system F on Λ, we can extend F to a pregroupoid homomorphism from the pregroupoid P to the groupoid of sets and bijections. Namely if $p = x_0 \cdots x_n \in P(\Delta)$ is a path, define $F_p : F(org(p)) \to F(end(p))$ recursively by $F_{x_0} = id_{F(x_0)}$ and $F_p = F_{x_{n-1}x_n} \circ F_q$, where $q = x_0 \cdots x_{n-1}$. We say paths p and q are *F-homotopic* if $F_p = F_q$ and write $p \sim_F q$.

Lemma 35.6: *(1) If p, q are paths with $end(p) = org(q)$ then $F_{pq} = F_q \circ F_p$.*

(2) $F_p \circ F_{p^{-1}} = id_{F(org(p))}$.

(3) F-homotopy is a $P(\Delta)$-invariant equivalence relation.

(4) F is a pregroupoid homomorphism from the path pregroupoid $P(\Delta)$ of Δ to the groupoid of sets and bijections.

Proof: Part (1) is immediate from the definitions while (2) follows from (LS1). Then (1) and (2) imply (4), while (4) and 33.8.5 imply (3).

Let D be a simplicial complex with graph Δ. A *local system* on D is a local system on Δ such that:

(LS2) $F_{xz} = F_{yz} \circ F_{xy}$ for each 2-simplex $\{x, y, z\}$ of D.

Example 35.7 If $d : L \to D$ is a covering of simplicial complexes then by 35.3.1, d induces a fibering $d : \Lambda \to \Delta$ of the graphs of L, D and hence a local system F^d on Δ by Example 35.5. Then F^d is a local system on D.

We can view D as a category whose objects are the simplices of D and morphisms are inclusions of simplices.

Lemma 35.8: *If F is a local system on D then F induces a functor F from D to sets via $F(s) = F(x_s)$ for some $x_s \in s$ and $F_{s,t} = F_p$ for some path p from x_s to x_t in t.*

Notice the definition in 35.8 is independent of p as t is simply connected.

Lemmas 35.6.4 and 35.8 show that a local system on Δ or D is a local system in the sense of Section 7 of [Q].

Given a local system F on Δ, define Δ_F to be the disjoint union of the sets $F(x)$, $x \in \Delta$, and define $d_F : \Delta_F \to \Delta$ to be the map with $d_F^{-1}(x) = F(x)$ for each $x \in \Delta$. Make Δ_F into a graph by decreeing that u is adjacent to v if there is an edge (x, y) of Δ with $u \in F(x)$ and $F_{xy}(u) = v$. Notice (LS1) says this relation is symmetric, so Δ_F is an undirected graph.

Similarly if F is a local system on D let D_F be the subcomplex of $K(\Delta_F)$ with simplices s such that $d_F(s)$ is a simplex of D.

Lemma 35.9: *Let F be a local system on D. Then*

(1) $d_F : D_F \to D$ is a covering of D.

(2) $F = F^{d_F}$ is the local system of d_F.

(3) If $\phi : L \to D$ is a covering then $D_{F^\phi} = L$ and $\phi = d_{F^\phi}$.

Proof: Let $d = d_F$, $x \in \Delta$, and $u \in F(x)$. Then $u^\perp = \{F_{xy}(u) : y \in x^\perp\}$ and the map $F_{xy}(u) \mapsto y$ is the restriction d_u of d to $u^\perp$. In particular $d_u : u^\perp \to x^\perp$ is a bijection so d is a fibering. Further if $t = \{x, y, z\}$ is a 2-simplex in D, then $F_{xz}(u) = F_{yz}(F_{xy}(u)) = F_{xyz}(u)$, so $F_{xz}(u)$ is adjacent to $F_{xy}(u)$ in Δ_F and hence $d_x^{-1}(xyzx)$ is a triangle in Δ_F. So $d_x^{-1}(t)$ is a 2-simplex of D_F and (1) holds. Parts (2) and (3) follow from the definitions.

Lemma 35.10: *The map $F \mapsto d_F$ is a bijection of local systems on D with coverings of D. The inverse of this bijection is $d \mapsto F^d$.*

Proof: This follows from 35.9. Indeed by Exercise 12.4, this bijection induces an equivalence of categories.

Lemma 35.11: *Let $d : L \to D$ be a connected covering of complexes, Δ the graph of D, $F = F^d$, and $\sim$ the relation on $P = P(\Delta)$ defined by $p \sim q$ if $F_p = F_q$. Then*

(1) *$\sim$ is a P-invariant relation such that $u \sim uvwu$ for each 2-simplex $\{u, v, w\}$ of D.*

(2) *For α a vertex of L and $x = d(\alpha)$, the map*

$$\psi : \tilde{\Sigma}(D, x) \to L$$

defined by $\psi(\bar{p}) = F_p(\alpha)$ is a covering with $d \circ \psi = \text{end}$.

(3) *For $p \in P$ with $org(p) = x$, there exists a unique path q in L with $org(q) = \alpha$ and $d(q) = p$. Further $F_p(\alpha) = end(q)$.*

(4) *For p a cycle of Δ, $p \in ker(\sim)$ if and only if $F_p = id$.*

Proof: The first statement of (1) is just 35.6.3, while (LS2) implies the second.

By definition of $\sim$, ψ is well defined with $d \circ \psi = end$, and as L is connected, ψ is a surjection. By (1) and 35.4.4, $end : \tilde{\Sigma}(D, x) \to D$ is a covering. Thus d and end are local isomorphisms of simplicial complexes so locally $\psi = d^{-1} \circ end$ is an isomorphism, and hence ψ is a covering.

Thus (2) holds. Part (3) follows by induction on the length of p. In (4), $F_p = F_x$ if and only if $p \sim x$ if and only if $p \in ker(\sim)$ by definition of $\sim$. So as $F_x = id$, (4) holds.

Lemma 35.12: *Let S be the closure of the 2-simplices of the connected simplicial complex D, $\sim = \sim_S$, and x a vertex of S. Then end : $\tilde{\Sigma}(D, x) \to D$ is the universal covering of D.*

Proof: By 35.4.4, $e = end : \tilde{\Sigma}(D, x) \to D$ is a connected covering. Conversely suppose $d : L \to D$ is a connected covering, and let $\simeq$ be the relation induced by d as in 35.11, $\psi : \Sigma(D,x)/\simeq \to L$ the map of 35.11, and $f = end : \Sigma(D,x) \to D$. Then by 35.7, F^d is a local system, so (LS2) implies $F_p^d = id$ for each triangle p determined by a 2-simplex of D. Thus $S \subseteq ker(\simeq)$ by 35.11.4. Hence $g : \tilde{\Sigma}(D,x) \to \Sigma(D,x)/\simeq$ is a covering, where $g(\tilde{p})$ is the $\simeq$-equivalence class of p, and $f \circ g = e$. Then by 35.11, $h = \psi \circ g$ is a covering with $d \circ h = e$.

Lemma 35.13: *For a connected simplicial complex D, the following are equivalent:*

  (1) *D possesses no nontrivial connected coverings.*
  (2) *$P(\Delta)$ is generated by the 2-simplices of D.*
  (3) *$F_p = id$ for each local system F on D and each cycle p of Δ.*
  (4) *$\pi_1(D,x) = 1$ for x a vertex of D.*

Proof: Parts (1) and (2) are equivalent by 35.12. Finally (3) and (4) are equivalent by 35.11.4.

Lemma 35.14: *Δ is simply connected if and only if Δ is triangulable.*

Proof: This is a restatement of the equivalence of parts (1) and (2) of 35.13 when $D = K(\Delta)$.

Lemma 35.15: *Let $d : \Gamma \to \Delta$ be a fibering of graphs with Δ connected and $\sim$ the invariant relation induced by d as in 35.11. Then*

  (1) *d is a covering if and only if $ker(\sim)$ contains all triangles of Δ.*
  (2) *d is an isomorphism if and only if $ker(\sim)$ contains all cycles of Δ.*

Proof: By 35.11 we have a fibering $\psi : \tilde{\Sigma}(\Delta,x) \to L$ with $d \circ \psi = end$. Thus for each $\alpha \in \Lambda$, $d_\alpha^{-1} = \psi \circ end_\alpha^{-1}$. Therefore if end is a covering then end_α^{-1} is a morphism so d_α^{-1} is the composition of morphisms and hence a morphism. That is, d is a covering. But by 35.4.3, if $ker(\sim)$ contains all triangles then end is a covering, so d is a covering.

Conversely if d is a covering then $F_p^d = id$ for each triangle p of Δ, so $ker(\sim)$ contains all triangles by 35.11.4. Thus (1) holds. The proof of (2) is similar, using 35.4.5 in place of 35.4.3.

Lemma 35.16: *Assume $d : \Gamma \to \Delta$ is a fibering of graphs such that Δ is connected and $(\Gamma_i : i \in I)$ and $(\Delta_i : i \in I)$ are families of graphs such that for all $i \in I$, $\Gamma_i \subseteq \Gamma$, $\Delta_i \subseteq \Delta$, $d(\Gamma_i) = \Delta_i$, the inclusions $\Gamma_i \to \Gamma$, $\Delta_i \to \Delta$ are morphisms, and $d : \Gamma_i \to \Delta_i$ is an isomorphism. Assume further that $X : G \to Aut(\Gamma)$ and $Y : G \to Aut(\Delta)$ are representations of a group G such that for all $\alpha \in \Gamma$ and $g \in G$, $d(\alpha(gX)) = d(\alpha)(gY)$ and $ker(Y)$ is transitive on the fibers of d. Then*

(1)   If p is a cycle G-conjugate to a path of Δ_i then $F_p^d = id$.

(2)   If each triangle of Δ is G-conjugate to a triangle of Δ_0 then d is a covering.

(3)   If Γ is connected and the closure of the set of G-conjugates of cycles of Δ_i, $i \in I$, is the set of cycles of Δ then d is an isomorphism.

Proof: Let $F = F^d$. Given an edge (x, y) of Δ, $\alpha \in d^{-1}(x)$, and $\beta = F_{x,y}(\alpha)$, we have $d(\alpha g) = d(\alpha(gX)) = d(\alpha)(gY) = xg$ and $d(\beta g) = yg$, so $F_{xg,yg}(\alpha g) = \beta g$. That is, $F_{xg,yg}(\alpha g) = F_{x,y}(\alpha)g$. Proceeding by induction on the length of a path p with origin x and using 35.6.1, we conclude $F_{pg}(\alpha g) = F_p(\alpha)g$.

Next if $F_p(\alpha) = \alpha$ and $\beta \in d^{-1}(x)$ then by transitivity of $ker(Y)$ on $d^{-1}(x)$, there is $g \in ker(Y)$ with $\beta = \alpha g$. Then $F_p(\beta) = F_{pg}(\alpha g) = F_p(\alpha)g = \alpha g = \beta$, so $F_p = id$.

Let p be a closed path with origin x which is G-conjugate to a path of Δ_i for some $i \in I$. We claim $F_p = id$, so by the previous paragraph it remains to show $F_p(\alpha) = \alpha$ for some $\alpha \in d^{-1}(x)$. By hypothesis pg is a path in Δ_i for some $g \in G$. Then if $F_{pg}(\delta) = \delta$ for some $\delta \in d^{-1}(xg)$ then $\delta = F_{pg}(\delta) = F_p(\delta g^{-1})g$, so $F_p(\alpha) = \alpha$ for $\alpha = \delta g^{-1}$.

Hence we may take p to be a path in Δ_i. But for each path $q = x_0 \cdots x_r$ in Δ_i, since the restriction e of d to Γ_i is an isomorphism of Γ_i with Δ_i, $e^{-1}(q) = e^{-1}(x_0) \cdots e^{-1}(x_r)$ is a path in Γ_i with $F_q(e^{-1}(x_0)) = e^{-1}(x_r)$ so in particular if q is a cycle then so is $e^{-1}(q)$, and therefore $F_q = id$. Therefore $F_p = id$, completing the proof of the claim. Therefore (1) is established.

Next by 35.11 the relation $p \sim q$ if and only if $F_p = F_q$ is an invariant relation and by (1), $ker(\sim)$ contains all cycles G-conjugate to a member of Δ_i. Thus under the hypothesis of (3), $ker(\sim)$ contains all cycles of Δ and hence d is an isomorphism by 35.15.2. Similarly (1) and 35.15.1 imply (2).

Remarks. The fundamental groupoid of a simplicial complex K is a standard construction in combinatorial topology; for example, the *edge path groupoid* of Chapter 3, Section 6 of [Sp] is essentially the fundamental groupoid. Similarly the fundamental group of K is essentially the *edge path group* of K as defined in [Sp]. Of course the fundamental group of K is isomorphic to the fundamental group of the topological space of K; compare Theorem 3.6.16 in [Sp].

Our notion of a covering of a graph or simplicial complex is suggested by the standard notion of a covering of topological spaces or by Tits's definition of a covering of geometries in [T1]. The reader is cautioned, however, that in the combinatorial group theoretic literature, the term "graph covering" is sometimes used as we use the term "fibering." The concept of a covering of a graph or simplicial complex was introduced in [AS1] and [AS3], and our treatment of these topics comes from those references.

Exercises

1. Let $d : L \to K$ be a universal covering of connected simplicial complexes, let $x \in L$ be a vertex, and let $Aut(L, d)$ be the subgroup of $Aut(L)$ permuting the fibers of d. Prove
   (1) The map $\psi : Aut(L, d) \to Aut(K)$ defined by $\psi(g) : d(x) \mapsto d(xg)$ for $x \in L$ and $g \in Aut(L, d)$ is a surjective group homomorphism with kernel $\pi_1(L, x)$.
   (2) If s is a simplex of L then the map $\psi : Aut(L, d)_t \to Aut(K)_{d(t)}$ is an isomorphism for each $t \subseteq s$.

2. Let $p = x_0 \cdots x_r$ be an r-gon in a graph Δ. Prove
   (1) If r is even and $\Delta^{(r-2)/2}(x_0) \cap \Delta(x_{r/2})$ is connected then $p \in C_{r-1}(\Delta)$.
   (2) If r is odd and $\Delta(x_{(r-1)/2}, x_{(r+1)/2}) \cap \Delta^{(r-3)/2}(x_0) \neq \varnothing$ then $p \in C_{r-1}(\Delta)$.

3. Define a path p in a graph Δ to be *reduced* if $D(p) = \varnothing$. Prove
   (1) The equivalence class $[p]$ of p under the basic relation contains a unique reduced path r_p.
   (2) $r_p \in D^n(p)$ for some nonnegative integer n.

4. Let D be a simplicial complex. Define a morphism $f : F \to E$ of local systems on D to be a family of maps $f_x : F(x) \to E(x)$, $x \in \Delta(D) = \Delta$, such that for each x and $y \in \Delta(x)$, $f_y \circ F_{xy} = E_{xy} \circ f_x$. Define a morphism $\phi : (L, d) \to (\bar{L}, \bar{d})$ of coverings of D to be a simplicial map $\phi : L \to \bar{L}$ with $\bar{d} \circ \phi = d$. Prove
   (1) If $\phi : (L, d) \to (\bar{L}, \bar{d})$ is a morphism of coverings then $f^\phi : F^d \to F^{\bar{d}}$ is a morphism of local systems, where $f^\phi_x = \phi_{|F^d(x)}$.
   (2) If $f : F \to E$ is a morphism of local systems then $\phi^f : (D^F, d_F) \to (D^E, d_E)$ is a morphism of coverings, where $\phi^f_{|F(x)} = f_x$.
   (3) $(F \xrightarrow{f} E) \mapsto ((D^F, d_F) \xrightarrow{\phi^f} (D^E, d_E))$ is an isomorphism of the category of local systems on D with the category of coverings of D. The inverse of this isomorphism is $d \mapsto F^d$, $\phi \mapsto f^\phi$.

5. Given a local system F on D let $\lim_{x \in \Delta}(F(x))$ be the disjoint union of the $F(x)$ modulo the equivalence relation $\simeq$ generated by the identifications $F_{xy} : F(x) \to F(y)$, (x, y) an edge of Δ. Prove
   (1) For $u, v \in \coprod_{x \in \Delta}(F(x))$, $u \simeq v$ in $\lim_{x \in \Delta}(F(x))$ if and only if there exists a path p from $x = d_F(u)$ to $y = d_F(v)$ with $F_p(u) = v$.
   (2) If D is simply connected then the map $u \mapsto \bar{u}$ is an isomorphism of $F(x)$ with $\lim_{x \in \Delta}(F(x))$.

Chapter 13

The Geometry of Amalgams

In this chapter we put in place the machinery we will use to establish the uniqueness of most of the sporadic groups. The general approach is as follows. Let $\mathcal{H}$ be some group theoretic hypothesis and G a group satisfying $\mathcal{H}$. We associate to each such group a coset graph Δ on a coset space G/G_1 defined by some self-paired orbital of G on G/G_1, as in 3.2. We pick some family $\mathcal{F} = (G_i : i \in I)$ of subgroups of G and show that the amalgam $\mathcal{A}(\mathcal{F})$ of this family is determined up to isomorphism by $\mathcal{H}$ independently of G. The family $\mathcal{F}$ determines a coset geometry $\Gamma = \Gamma(G, \mathcal{F})$ and a geometric complex $\mathcal{C}(G, \mathcal{F})$ as in Section 4, and indeed any completion $\beta : \mathcal{A}(\mathcal{F}) \to \tilde{G}$ of $\mathcal{A}(\mathcal{F})$ determines a corresponding geometry and complex. The family $\mathcal{F}$ is chosen so that Δ is isomorphic to a collinearity graph of Γ and so that the closure of the cycles in the collinearity graphs of proper residues of Γ is the set of all cycles of Δ. Further if $\iota : \mathcal{A}(\mathcal{F}) \to \tilde{G}$ is the universal completion of $\mathcal{A}(\mathcal{F})$ then we obtain a fibering $d : \tilde{\Delta} \to \Delta$ from the collinearity graph $\tilde{\Delta}$ of $\tilde{\Gamma}$ onto Δ, and as the cycles in residues of Γ lift to cycles in residues in $\tilde{\Gamma}$, the fibering d is even an isomorphism. Hence $G \cong \tilde{G} \cong \bar{G}$ for any pair of groups $G, \bar{G}$ satisfying $\mathcal{H}$, giving us our uniqueness. This is a broad outline of the approach.

In Section 36 we develop the theory of amalgams and geometries necessary for this procedure. Then in Section 37 we specialize to a particular class of rank 3 amalgams and geometries which will be sufficient to establish the uniqueness of most of the sporadic groups. Our starting point is a 4-tuple $\mathcal{U} = (G, H, \Delta, \Delta_H)$ consisting of our group G and graph Δ together with a suitable subgroup H of G and a graph Δ_H, which will be a residue in our geometry. We term such 4-tuples satisfying a few more axioms *uniqueness systems* and we study these objects in Section 37. We are usually able to choose our uniqueness system so that each triangle of Δ is fused into Δ_H under G. Thus it remains to show that Δ is triangulable in the sense of Chapter 12 in order to implement the procedure described above. So in the process of proving the uniqueness of a sporadic, we usually show some graph Δ associated to the sporadic is simply connected via the results in Section 34.

Finally in Section 38 we see how to obtain well-behaved uniqueness systems from a flag transitive group of automorphisms of a suitable rank 3 string geometry. In Chapter 14 we apply this point of view to the 2-local geometry of M_{24} and a truncation of the projective geometry of $L_5(2)$ to obtain uniqueness proofs for those groups. Similarly in Chapter 16 we use the approach to prove the uniqueness of J_2, Suz, and Co_1.

36. Amalgams

Let $I = \{1, \ldots, n\}$ be a set of finite order n. An *amalgam of rank n* is a family

$$A = (\alpha_{J,K} : P_J \to P_K : J \subset K \subset I)$$

of group homomorphisms such that for all $J \subset K \subset L$, $\alpha_{J,K}\alpha_{K,L} = \alpha_{J,L}$.

Example 36.1 Let $\mathcal{F} = (G_i : i \in I)$ be a family of subgroups of a group G. For $J \subseteq I$ let $J' = I - J$ be the complement to J in I and as in Section 4, define $G_J = \bigcap_{j \in J} G_j$. Define $P_J = G_{J'}$. Thus $P_J \cap P_K = G_{J'} \cap G_{K'} = G_{J' \cup K'} = G_{(J \cap K)'} = P_{J \cap K}$. Also for $J \subset K \subset I$, define $\alpha_{JK} : P_J \to P_K$ to be inclusion. Then

$$\mathcal{A}(\mathcal{F}) = (\alpha_{JK} : P_J \to P_K : J \subset K \subset I)$$

is an amalgam.

A *morphism* $\phi : A \to \bar{A}$ of rank n amalgams is a family

$$\phi = (\phi_J : P_J \to \bar{P}_J : J \subset I)$$

of group homomorphisms such that for all $J \subset K \subset I$ the obvious diagram commutes:

$$
\begin{array}{ccc}
P_J & \xrightarrow{\alpha_{J,K}} & P_K \\
\phi_J \downarrow & & \phi_K \downarrow \\
\bar{P}_J & \xrightarrow{\bar{\alpha}_{J,K}} & \bar{P}_K
\end{array}
$$

A *completion* $\beta : A \to G$ for A is a family $\beta = (\beta_J : P_J \to G)$ of group homomorphisms such that $G = \langle P_J \beta_J : J \subset I \rangle$ and for all $J \subset K \subset I$ the obvious diagram commutes:

$$P_J \xrightarrow{\alpha_{J,K}} P_K$$
$$\beta_J \searrow \quad \swarrow \beta_K$$
$$G$$

The completion $\beta : A \to G$ is said to be *faithful* if each β_J is an injection.

Example 36.2   Let $\mathcal{F} = (G_i : i \in I)$ be a family of subgroups of a group G with $G = \langle \mathcal{F} \rangle$. Form the amalgam $\mathcal{A}(\mathcal{F})$ of Example 36.1. Then the identity maps $id_J : P_J \to P_J$ form a faithful completion $id = (id_J : J \subset I)$ with $id : \mathcal{A}(\mathcal{F}) \to G$.

The *free product* $F(A)$ of the groups P_J, $J \subset I$, in an amalgam A is the free group on the disjoint union of the sets P_J modulo defining relations for the groups P_J. We have the following universal property:

Lemma 36.3: *If $(\varphi_J : P_J \to G)$ is a family of group homomorphisms then there exists a unique group homomorphism $\varphi : F(A) \to G$ with $g\varphi = g\varphi_J$ for all $g \in P_J$ and all $J \subset I$.*

Define the *free amalgamated product* $G(A)$ of the amalgam A to be the free product $F(A)$ of the groups P_J, $J \subset I$, modulo the relations $g^{-1}(g\alpha_{J,K}) = 1$ for $J \subset K \subset I$ and $g \in P_J$. Write $\tilde{g}$ for the image of $g \in F(A)$ in $G(A)$ and let $\iota = (\iota_J : P_J \to G(A))$ be defined by $g\iota_J = \tilde{g}$ for $g \in P_J$. Then

Lemma 36.4: $\iota : A \to G(A)$ *is a universal completion for A. That is, if $\beta : A \to G$ is a completion of A then there exists a unique group homomorphism $\psi : G(A) \to G$ such that $\iota_J \psi = \beta_J$ for all $J \subset I$.*

Proof: By construction $\iota : A \to G(A)$ is a completion of A. Suppose $\beta : A \to G$ is a completion. By the universal property of the free product recorded in 36.3, there exists a group homomorphism $\varphi : F(A) \to G$ defined by $g\varphi = g\beta_J$ for $g \in P_J$. As β is a completion of A, $g\alpha_{JK}\beta_K = g\beta_J$ for all $J \subset K \subset I$, so $g^{-1}(g\alpha_{JK}) \in ker(\varphi)$. for each $g \in P_J$. Thus φ induces a group homomorphism $\psi : G(A) \to G$ defined by $\tilde{g}\psi = g\varphi = g\beta_J$ for $g \in P_J$. This is the map of 36.4.

Lemma 36.5: *If A possesses a faithful completion then the universal completion $\iota : A \to G(A)$ is faithful.*

Proof: Suppose $\beta : A \to G$ is a faithful completion. Then $\beta_J = \iota_J \psi$ is injective, so ι_J is injective.

Lemma 36.6: *Isomorphic amalgams have isomorphic universal completions.*

Proof: If $\phi : A \to \bar{A}$ is an isomorphism of amalgams and $\iota : A \to G(A)$ and $\bar{\iota} : \bar{A} \to G(\bar{A})$ are universal completions, then $\phi\bar{\iota} : A \to G(\bar{A})$ is a completion, so there exists $\psi : G(A) \to G(\bar{A})$ with $\iota_J \psi = \phi \bar{\iota}_J$ for each J. Similarly we have $\bar{\psi} : G(\bar{A}) \to G(A)$ with $\bar{\iota}\bar{\psi} = \phi^{-1}\iota_J$. Then $\bar{\psi} = \psi^{-1}$.

Given a completion $\beta : A \to G$ of A let $\mathcal{F}(\beta) = (G_i : i \in I)$, where $G_i = P_{i'}\beta_{i'}$, $\Gamma(\beta) = \Gamma(G, \mathcal{F}(\beta))$, and $\mathcal{C}(\beta) = \mathcal{C}(G, \mathcal{F}(\beta))$ be the geometry and geometric complex of β, as defined in Examples 4 and 9 of Section 4. Further for $i \in I$, define the *collinearity graph* $\Delta(\beta, i)$ of $\mathcal{C}(\beta)$ at i to be the graph on the set G/G_i of objects of $\Gamma(\beta)$ of type i with x adjacent to y if there exist chambers C_u of $\mathcal{C}(\beta)$ for $u = x, y$ with $u \in C_u$ and $C_x \cap C_y$ a flag of type i'.

Lemma 36.7: *Let $\beta : A \to G$ be a faithful completion of an amalgam A and $\iota : A \to G(A)$ the universal completion of A. Let $\varphi : G(A) \to G$ be the surjection of 36.4. Then*

    (1) $\varphi : G(A)_i \to G_i$ is an isomorphism with $\varphi(G(A)_J) = G_J$ for each $i \in J \subseteq I$.

    (2) φ induces a morphism $\varphi : \mathcal{C}(\iota) \to \mathcal{C}(\beta)$ of geometric complexes which is a covering of simplicial complexes, via $\varphi : G(A)_i g \mapsto G_i \varphi(g)$.

    (3) Assume for some fixed $i \in I$ that:

    (*) $G_{i'} = G_I t G_I$ and $G_i \cap G_i^t = \langle G_{ij} \cap G_{ij}^t : j \in i' \rangle$ for some $t \in G_{i'} - G_I$.

Then the covering φ of (2) restricts to a fibering $\varphi : \Delta(\iota, i) \to \Delta(\beta, i)$ of collinearity graphs.

    (4) Assume (*) and

    (**) The closure of the set of cycles of $\Delta(\beta, i)$ conjugate under G to a cycle of the collinearity graph at i of $Link_{\mathcal{C}(\beta)}(G_j)$, as j ranges over i', is the set of all cycles of $\Delta(\beta, i)$.

Then $\varphi : G(A) \to G$ is an isomorphism.

Proof: Let $H = G(A)$. By 36.5, ι is faithful. Thus $\varphi : H_i \to G_i$ is the composition $\varphi = \iota_{i'}^{-1}\beta_{i'}$ of isomorphisms, so (1) holds.

As $\varphi(H_i) = G_i$, the map $\varphi : H_i g \mapsto G_i \varphi(g)$ of (2) is well defined and as $\varphi : H \to G$ is surjective, $\varphi : \mathcal{C}(\iota) \to \mathcal{C}(\beta)$ is surjective on vertices.

From the definition of $C(\iota)$ in Example 9 of Section 4, the chambers of $C(\iota)$ are of the form $S_{I,x} = \{H_i x : i \in I\}$, $x \in H$, and $\varphi(S_{I,x}) = S_{I,\varphi(x)}$ is a chamber of $C(\beta)$, so $\varphi : C(\iota) \to C(\beta)$ is a morphism of geometric complexes. By 4.5, $Link(H_i) \cong C(H_i, \mathcal{F}(\iota)_i)$ and $Link(G_i) \cong C(G_i, \mathcal{F}(\beta)_i)$, while by (1), $\varphi : C(H_i, \mathcal{F}(\iota)_i) \to C(G_i, \mathcal{F}(\beta)_i)$ is an isomorphism, so $\varphi : Link(H_i) \to Link(G_i)$ is an isomorphism and hence $\varphi : C(\iota) \to C(\beta)$ is a covering of simplicial complexes. That is, (2) is established.

Let $\Lambda = \Delta(\iota, i)$ and $\Delta = \Delta(\beta, i)$. By (2), φ restricts to a surjective morphism of graphs $\varphi : \Lambda \to \Delta$. Further as G is transitive on chambers of $C(\beta)$, each chamber through G_i is conjugate under G_i to $\mathcal{F}(\beta)$. Also $G_{i'}$ is the stabilizer of the wall $W = \mathcal{F}(\beta) - \{G_i\}$ of type i' and G_I is the stabilizer of $\mathcal{F}(\beta)$.

Assume (*). Then $G_{i'} = G_I t G_I$, so $G_{i'}$ is 2-transitive on the chambers through W and G_I is transitive on $Link(W) - \{G_i\}$. Hence G_i is transitive on $\Delta(G_i)$. Then by (1) and (2), H_I is transitive on $Link(U) - \{H_i\}$, where $U = \mathcal{F}(\iota) - \{H_i\}$, so H_i is transitive on $\Lambda(H_i)$.

Next $G_i t \in \Delta(G_i)$ and by (*), $G_i \cap G_i^t = \langle G_{ij} \cap G_{ij}^t : j \in i' \rangle$ is the stabilizer in G_i of $G_i t$. Let $s \in H_{i'}$ with $\varphi(s) = t$. By (1), $\varphi : H_i \to G_i$ is an isomorphism with $\varphi(H_{ij} \cap H_{ij}^s) = G_{ij} \cap G_{ij}^t$ so $\langle \varphi(H_{ij} \cap H_{ij}^s) : j \in i' \rangle = G_i \cap G_i^t$. Of course $\varphi(H_i \cap H_i^s) \le G_i \cap G_i^t$, so $\varphi(H_i \cap H_i^s) = G_i \cap G_i^t$. Thus by (1), $\varphi : H_i/(H_i \cap H_i^s) \to G_i/(G_i \cap G_i^t)$ is a bijection and hence $\varphi : \Lambda(H_i) \to \Delta(G_i)$ is a bijection and (3) is established.

Finally by (2), $\varphi : Link_{C(\iota)}(H_j) \to Link_{C(\beta)}(G_j)$ is an isomorphism of simplicial complexes and hence induces an isomorphism of the collinearity graphs at i of these links. Hence by 35.16.3, under the hypotheses of (4), $\varphi : \Lambda \to \Delta$ is an isomorphism. That is, the map $\varphi : H_i \to G_i$ is an isomorphism and the map $\varphi : H/H_i \to G/G_i$ is a bijection. Hence $\varphi : H \to G$ is an isomorphism and (4) is established.

37. Uniqueness systems

Define a *uniqueness system* to be a 4-tuple $\mathcal{U} = (G, H, \Delta, \Delta_H)$ such that Δ is a graph, G is an edge and vertex transitive group of automorphisms of Δ, $H \le G$, Δ_H is a graph with vertex set xH and edge set $(x, y)H$ for some $x \in \Delta$ and $y \in \Delta(x) \cap xH$, and:

(U) $G = \langle H, G_x \rangle$, $G_x = \langle G_{x,y}, H_x \rangle$, and $H = \langle H(\{x, y\}), H_x \rangle$.

In this section we assume $\mathcal{U} = (G, H, \Delta, \Delta_H)$ is a uniqueness system and (x, y) an edge in Δ_H. Notice by definition $y \in \Delta(x)$, so $x \ne y$.

Let $G_1 = G_x$, $G_2 = G(\{x, y\})$, $G_3 = H$, and $\mathcal{F} = \mathcal{F}(\mathcal{U}) = (G_i : i \in I)$, where $I = \{1, 2, 3\}$. The *amalgam of* $\mathcal{U}$ is the rank 3 amalgam

$\mathcal{A}(\mathcal{U}) = \mathcal{A}(\mathcal{F}(\mathcal{U}))$ as defined in Example 36.1. Notice $P_{12} = H$, $P_{23} = G_x$, $P_{13} = G(\{x,y\})$, $P_\varnothing = H_{x,y}$, $P_1 = H(\{x,y\})$, $P_2 = H_x$, and $P_3 = G_{x,y}$.

We say that a rank 3 amalgam A is *residually connected* if $P_{i,j} = \langle P_i, P_j \rangle$ for all distinct $i, j \in I$. Now:

Lemma 37.1: *Assume A is a residually connected rank 3 amalgam and $\beta : A \to M$ is a completion of A. Then*

(1) $\Gamma(\beta)$ is a residually connected geometry.

(2) If $P_{13} = P_1 P_3$ then $\Gamma(\beta)$ has a string diagram and M is flag transitive on $\Gamma(\beta)$.

Proof: Part (1) is 4.5.3. If $P_{13} = P_1 P_3$ then by 4.2, the diagram of $\mathcal{C}(\beta)$ is a string (in the sense of Section 4) so (2) follows from 4.11.1.

Lemma 37.2: *(1) There is $t \in H$ with cycle (x,y).*

(2) The amalgam $\mathcal{A}(\mathcal{U})$ is residually connected with $P_{13} = P_1 P_3$.

Proof: As $x \neq y$, (U) says $H(\{x,y\}) \neq H_x$, so (1) holds. Now $G(\{x,y\}) = G_{x,y}\langle t \rangle$. Thus $P_{13} = P_1 P_3$. This observation together with condition (U) in the definition of uniqueness system shows the amalgam is residually connected.

Observe that the inclusion map $\beta : \mathcal{A}(\mathcal{U}) \to G$ is a faithful completion of the amalgam $\mathcal{A}(\mathcal{U})$ by Example 36.2. Form the geometry $\Gamma = \Gamma(\beta)$ and its collinearity graph $\Delta(\beta, 1)$ as in Section 36. We call the objects of Γ of type 1 *points* and the objects of type 2 *lines*. By 37.1 and 37.2, Γ has a string diagram, so by Exercise 1.5, two points a, b of Γ are incident in $\Delta(\beta, 1)$ if and only if a and b are incident in Γ with a common line of Γ. Hence the term *collinearity graph*. Moreover as in 3.2:

Lemma 37.3: *The map $G_1 g \mapsto xg$ is an isomorphism of $\Delta(\beta, 1)$ with Δ. Further the line $G_2 g$ of Γ is identified with the edge $\{xg, yg\}$ of Δ via this isomorphism.*

Thus we identify Δ with $\Delta(\beta, 1)$ via the isomorphism of 37.3 and write Δ for both.

Define a *similarity* of uniqueness systems $\mathcal{U}, \bar{\mathcal{U}}$ to be a pair of isomorphisms $\alpha : G_x \to \bar{G}_{\bar{x}}$ and $\zeta : H \to \bar{H}$ such that $\alpha = \zeta$ on H_x, $H_x \zeta = \bar{H}_{\bar{x}}$, $G_{x,y} \alpha = \bar{G}_{\bar{x},\bar{y}}$, and $H(\{x,y\})\zeta = \bar{H}(\{\bar{x},\bar{y}\})$ for some edges $(x,y), (\bar{x},\bar{y})$ of $\Delta_H, \Delta_{\bar{H}}$, respectively. We say the similarity is *with respect to* $(x,y), (\bar{x},\bar{y})$ if we wish to emphasize the role of those edges. The similarity is an *equivalence* if there exists $t \in H$ with cycle (x,y) such that $(b^t)\alpha = (b\alpha)^{t\zeta}$ for all $b \in G_{x,y}$.

Define a *morphism* of uniqueness systems $\mathcal{U}, \bar{\mathcal{U}}$ to be a group homomorphism $d : G \to \bar{G}$ such that the restrictions $d : H \to \bar{H}$ and $d : G_x \to \bar{G}_{\bar{x}}$ are isomorphisms defining a similarity of $\mathcal{U}$ with $\bar{\mathcal{U}}$.

Lemma 37.4: *(1) The geometry of points and edges of Δ_H is isomorphic to the residue of H in Γ.*

(2) If $d : \bar{G} \to G$ is a morphism of uniqueness systems $\bar{\mathcal{U}}, \mathcal{U}$, then $\mathcal{A}(\bar{\mathcal{U}}) \cong \mathcal{A}(\mathcal{U})$ and d induces a covering $d : \bar{\Gamma} \to \Gamma$ of geometries and a fibering $d : \bar{\Delta} \to \Delta$ of graphs defined by $d : \bar{G}_i g \mapsto G_i(gd)$ and $(\bar{x}g)d = x(gd)$, respectively. The restriction $d : \bar{\Delta}_{\bar{H}} \to \Delta_H$ is an isomorphism and $\ker(d)$ is transitive on each fiber of d on $\bar{\Delta}$.

Proof: Part (1) is a consequence of 37.1, 37.2, and 4.11.2.

Assume the hypotheses of (2) and pick t as in 37.2.1. As $G(\{x, y\}) = \langle t \rangle G_{x,y}$, as $d : \bar{G}_{\bar{x}, \bar{y}} \to G_{x,y}$ is an isomorphism, and as $\bar{t}d = t$ for some $\bar{t} \in \bar{H}$ with cycle $(\bar{x}, \bar{y})$, $d : \bar{G}(\{\bar{x}, \bar{y}\}) \to G(\{x, y\})$ is an isomorphism. Thus as d is a similarity, d induces an isomorphism $\mathcal{A}(\bar{\mathcal{U}}) \cong \mathcal{A}(\mathcal{U})$ of amalgams.

As $\bar{G}_i d = G_i$ and $\bar{G}_{ij} d = G_{ij}$ for each i, j, d is a well-defined surjective morphism of geometries and graphs. Now G_i is incident with $G_j h$ in Γ when $h \in G_i$. But as $d : \bar{G}_i \to G_i$ is an isomorphism, $d : \bar{G}_i / \bar{G}_{ij} \to G_i / G_{ij}$ is a bijection and hence $d : \bar{\Gamma} \to \Gamma$ is a local bijection. Also $\Delta(x) = yG_x$ and $\Delta(\bar{x}) = \bar{y}\bar{G}_{\bar{x}}$ with $\bar{G}_{\bar{x}, \bar{y}} d = G_{x,y}$, so $d : \bar{x}g \mapsto x(gd)$ is a bijection of $\Delta(x)$ with $\Delta(\bar{x})$ and hence $d : \bar{\Delta} \to \Delta$ is a fibering of collinearity graphs. By 37.1 and 37.2, $\bar{G}$ and G are flag transitive on their respective geometries, so the same argument shows d is a local isomorphism of geometries.

Notice $\bar{G}_i g \in d^{-1}(G_i)$ if and only if $g \in \ker(d)\bar{G}_i$, so $\ker(d)$ is transitive on the fiber $d^{-1}(G_i)$. As d is a covering of geometries, (1) says $d : \bar{\Delta}_{\bar{H}} \to \Delta_H$ is an isomorphism.

Let $\iota : \mathcal{A}(\mathcal{U}) \to G(\mathcal{A}(\mathcal{U}))$ be the universal completion of $\mathcal{A}(\mathcal{U})$. Write $\tilde{G}$ for $G(\mathcal{A}(\mathcal{U}))$, $\tilde{H}$ for $H\iota$, $\tilde{G}_{\tilde{x}}$ for $G_x \iota$, etc. Let $\tilde{\Delta} = \Delta(\iota, 1)$ be the collinearity graph of $\tilde{G}$. Then $\tilde{G}_{\tilde{x}}$ is indeed the stabilizer of some $\tilde{x} \in \tilde{\Delta}$. Let $\tilde{\Delta}_{\tilde{H}}$ be the collinearity graph of the residue of $\tilde{H}$ in $\tilde{\Gamma}$ and $\tilde{\mathcal{U}} = (\tilde{G}, \tilde{H}, \tilde{\Delta}, \tilde{\Delta}_{\tilde{H}})$.

We say Δ_H is a *base* for $\mathcal{U}$ if the closure of the G-conjugates of all cycles of Δ_H is the set of all cycles of Δ.

Lemma 37.5: *(1) $\tilde{\mathcal{U}}$ is a uniqueness system equivalent to $\mathcal{U}$.*

(2) There exists a morphism $d : \tilde{\mathcal{U}} \to \mathcal{U}$ of uniqueness systems.

(3) If each triangle of Δ is G-conjugate to a triangle of Δ_H then $d : \tilde{\Delta} \to \Delta$ is a covering.

(4) If Δ_H is a base for Δ then $\tilde{\mathcal{U}} \cong \mathcal{U}$ and $\tilde{G} \cong G$.

Proof: Let $d : \tilde{G} \to G$ be the homomorphism supplied by the universal property of $\tilde{G}$. By construction $\tilde{G} = \langle \tilde{H}, \tilde{G}_{\tilde{x}} \rangle$. By 36.5, the universal completion is faithful, and by 37.1, $\tilde{G}$ is flag transitive on $\tilde{\Gamma}$. Then arguing as in the proof of 37.3 using Exercise 1.5, we have a natural $\tilde{G}$-equivariant bijection between the lines of $\tilde{\Gamma}$ and the edges of $\tilde{\Delta}$, so as $\tilde{G}$ is flag transitive on $\tilde{\Gamma}$, $\tilde{G}$ is also edge transitive on $\tilde{\Delta}$.

Next as β is faithful, $d : \tilde{H} \to H$ and $d : \tilde{G}_{\tilde{x}} \to G_x$ are isomorphisms. If K is the kernel of the action of $\tilde{G}$ on $\tilde{\Delta}$ then $K \leq \tilde{G}_{\tilde{x}}$ and $K \trianglelefteq \tilde{G}$. Therefore $Kd \leq G_x$ is normal in G so as G is faithful on Δ, $Kd = 1$. Then as $d : \tilde{G}_{\tilde{x}} \to G_x$ is an isomorphism, $K = 1$. That is, $\tilde{G}$ is faithful on $\tilde{\Delta}$. The other conditions in the definition of uniqueness system are properties of the amalgam $A(\mathcal{U})$ and hence are shared by $\tilde{\mathcal{U}}$. Thus $\tilde{\mathcal{U}}$ is a uniqueness system. Remarks above show $d : \tilde{\mathcal{U}} \to \mathcal{U}$ is a morphism of uniqueness systems. By construction, d defines an equivalence of $\tilde{\mathcal{U}}$ with $\mathcal{U}$.

We have established (1) and (2). Notice that if $d : \tilde{\Delta} \to \Delta$ is an isomorphism then $d : \tilde{G} \to G$ defines an equivalence of the actions of $\tilde{G}$ on $\tilde{\Delta}$ and G on Δ, so $d : \tilde{G} \to G$ is an isomorphism and hence $d : \tilde{\mathcal{U}} \to \mathcal{U}$ is an isomorphism. Thus to prove (4), it suffices to assume Δ_H is a base for Δ and prove $d : \tilde{\Delta} \to \Delta$ is an isomorphism. But 37.4.2 supplies the hypotheses of 35.16.3, so that lemma completes the proof of (4). Similarly 35.16.2 implies (3).

Lemma 37.6: *If $\mathcal{U}$ and $\bar{\mathcal{U}}$ are equivalent uniqueness systems then $A(\mathcal{U}) \cong A(\bar{\mathcal{U}})$.*

Proof: Assume $\alpha : G_x \to \bar{G}_{\bar{x}}$ and $\zeta : H \to \bar{H}$ define an equivalence of our systems and let t be the element of H supplied in the definition of equivalence. Define $\mu : G(\{x, y\}) \to \bar{G}(\{\bar{x}, \bar{y}\})$ via $\mu : bt^n \mapsto (b\alpha)(t\zeta)^n$ for $b \in G_{x,y}$ and $n \in \mathbf{Z}$. The map is well defined as α and ζ agree on H_x. The map is a homomorphism as $(b^t)\alpha = (b\alpha)^{t\zeta}$ for all $b \in G_{x,y}$. As α is injective on G_x the map is an isomorphism.

Now by construction $\zeta : H \to \bar{H}$, $\alpha : G_x \to \bar{G}_{\bar{x}}$, and $\mu : G(\{x, y\}) \to \bar{G}(\{\bar{x}, \bar{y}\})$ define our isomorphism of $A(\mathcal{U})$ with $A(\bar{\mathcal{U}})$.

We are now in a position to state one of the principal results of this section:

Theorem 37.7: *Assume $\mathcal{U}, \bar{\mathcal{U}}$ are equivalent uniqueness systems such* *that* $\Delta_H, \bar{\Delta}_{\bar{H}}$ *are bases for* $\Delta, \bar{\Delta}$, *respectively. Then* $\mathcal{U}$ *is isomorphic to* $\bar{\mathcal{U}}$.

Proof: By 37.6, $\mathcal{A}(\mathcal{U}) \cong \mathcal{A}(\bar{\mathcal{U}})$. Then by 36.6, $\tilde{G}$ is also the universal completion of $\mathcal{A}(\bar{\mathcal{U}})$. Then by 37.5.4, $\mathcal{U} \cong \tilde{\mathcal{U}} \cong \bar{\mathcal{U}}$.

Corollary 37.8: *Assume* $\mathcal{U}$ *and* $\bar{\mathcal{U}}$ *are equivalent uniqueness systems,* Δ *is triangulable, each triangle of* Δ *is G-conjugate to a triangle of* Δ_H, *and* $\bar{\mathcal{U}}$ *also satisfies these hypotheses. Then* $\mathcal{U} \cong \bar{\mathcal{U}}$.

In order to apply Theorem 37.7 and its corollary, we need effective means for verifying the equivalence of uniqueness systems. Theorems 37.9 through 37.12 supply such means.

Theorem 37.9: *Assume* $\mathcal{U}$ *and* $\bar{\mathcal{U}}$ *are similar uniqueness systems and for some edge* (x, y) *of* Δ_H, $Aut(G_{x,y}) \cap C(H_{x,y}) = 1$. *Then* $\mathcal{U}$ *is equivalent to* $\bar{\mathcal{U}}$.

Theorem 37.10: *Assume* $\alpha : G_x \to \bar{G}_{\bar{x}}$ *and* $\zeta : H \to \bar{H}$ *define a similarity of uniqueness systems* $\mathcal{U}$ *and* $\bar{\mathcal{U}}$ *with respect to edges* $(x, y), (\bar{x}, \bar{y})$, *and there exist* $K \leq G_x$ *and* $t, h \in N_H(K)$ *such that:*

(1) $C_{Aut(K)}(K \cap H) = Aut_{Z(H)}(K)$.

(2) t *has cycle* (x, y), $t^h \in G_x$, *and* $h\zeta \in N_{\bar{H}}(K\alpha)$.

(3) $G_{x,y} = \langle K_y, H_{x,y} \rangle$.

Then α, ζ, *define an equivalence of* $\mathcal{U}$ *and* $\bar{\mathcal{U}}$.

Hypothesis V: *The uniqueness system* $\mathcal{U} = (G, H, \Delta, \Delta_H)$ *satisfies the following four conditions for some edge* (x, y) *of* Δ_H:

(V1) $Aut(H_x) = Aut_{Aut(H)}(H_x) Aut_{Aut(G_x)}(H_x)$.

(V2) $N_{Aut(G_x)}(H_x) \leq N_{Aut(G_x)}(G_{x,y}^{H_x}) C_{Aut(G_x)}(H_x)$.

(V3) $N_{Aut(H)}(H_x) \leq N(H_x H(\{x, y\}) H_x) C(H_x)$.

(V4) $N_{H_x}(H_{x,y}) \leq N_{G_x}(G_{x,y})$.

Theorem 37.11: *Assume* $\mathcal{U}, \bar{\mathcal{U}}$ *are uniqueness systems satisfying Hypothesis V with respect to edges* $(x, y), (\bar{x}, \bar{y})$ *and* $\alpha : G_x \to \bar{G}_{\bar{x}}$ *and* $\zeta : H \to \bar{H}$ *are isomorphisms such that* $G_{x,y}\alpha = \bar{G}_{\bar{x}, \bar{y}}$, $H_x \zeta = \bar{H}_{\bar{x}} = H_x \alpha$, *and* $H(\{x, y\})\zeta = \bar{H}(\{\bar{x}, \bar{y}\})$. *Then* $\mathcal{U}$ *and* $\bar{\mathcal{U}}$ *are similar.*

Theorem 37.12: *Assume* $\mathcal{U}$ *and* $\bar{\mathcal{U}}$ *are uniqueness systems,* $(x, y), (\bar{x}, \bar{y})$ *are edges in* $\Delta_H, \bar{\Delta}_{\bar{H}}$, *and* $\alpha : G_x \to \bar{G}_{\bar{x}}$ *and* $\zeta : H \to \bar{H}$ *are isomorphisms such that* $H_x \alpha = \bar{H}_{\bar{x}}$ *and:*

(1)   *There exists* $Z(x)$ *char* G_x *with* $Z(x) \leq H_y$, *and either*

(a) $G_x = N_G(Z(x))$, *or*

(b) $H_x = N_H(Z(x))$ *and* $G_{x,y} = N_{G_x}(Z(y))$,

where for $h \in H$, $Z(xh)$ is defined by $Z(xh) = Z(x)^h$. Further the same conditions hold for $\bar{U}$.

(2) $Z(x)\alpha = Z(x)\zeta = Z(\bar{x})$ and $Z(y)\zeta = Z(\bar{y})$.

(3) $Aut(H_x) \cap N(Z(x)) = Aut_{Aut(H) \cap N(Z(x))}(H_x) Aut_{Aut(G_x)}(H_x)$.

(4) $N_{Aut(H) \cap N(Z(x))}(H_x) \le N(H_x H(\{x, y\}) H_x) C(H_x)$.

Then U and $\bar{U}$ are similar.

We next prove Theorems 37.9 and 37.10. We are supplied with a similarity $\alpha, \zeta : U \to \bar{U}$ with respect to $(x, y), (\bar{x}, \bar{y})$. In Theorem 37.10 we also have a subgroup K of G_x; in Theorem 37.9 let $K = G_{x,y}$. By 37.2 there exists $t \in H$ with cycle (x, y). Let $\bar{K} = K\alpha$ and $\phi : N_{\bar{G}}(\bar{K}) \to Aut(\bar{K})$ be the conjugation map, and define $\alpha^* : Aut(K) \to Aut(\bar{K})$ by $\alpha^* : \theta \mapsto \theta^\alpha$. For $g \in N_G(K)$ write $g\alpha^*$ for $g\psi\alpha^*$, where $\psi : N_G(K) \to Aut(K)$ is the conjugation map. Thus $(k\alpha)^{g\alpha^*} = (k^g)\alpha$ for each $k \in K$. Therefore for $g \in N_{G_x}(K)$, $g\alpha^* = g\alpha\phi$ as $(k^g)\alpha = (k\alpha)^{g\alpha} = (k\alpha)^{g\alpha\phi}$.

Now for $k \in K \cap H$ and $h \in N_H(K)$, $(k\alpha)^{h\alpha^*} = (k^h)\alpha = (k^h)\zeta = (k\zeta)^{h\zeta} = (k\alpha)^{h\zeta}$, so $h\alpha^* = h\zeta\phi$ on $\bar{K} \cap \bar{H}$, and hence $h\zeta\phi \cdot h^{-1}\alpha \in C_{Aut(\bar{K})}(\bar{H} \cap \bar{K})$. But in Theorem 37.9, $C_{Aut(K)}(K \cap H) = 1$, so $h\alpha^* = h\zeta\phi$. In particular this holds for $h = t$, so $(k^t)\alpha = (k\alpha)^{t\alpha^*} = (k\alpha)^{t\zeta}$ for all $k \in K = G_{x,y}$, completing the proof of Theorem 37.9.

Similarly in Theorem 37.10, hypothesis (1) of Theorem 37.10 says

$$h\zeta\phi \cdot h^{-1}\alpha^* \in Aut_{Z(\bar{H})}(\bar{K}) \le C(t\zeta\phi).$$

Then picking h as in hypothesis (2) of Theorem 37.10, $t^h \in H_x$ by hypothesis, so $(t^h)\alpha^* = (t^h)\alpha\phi$ by an earlier remark, and hence $(t\alpha^*)^{h\alpha^*} = (t^h)\alpha^* = (t^h)\alpha\phi = (t^h)\zeta\phi = (t\zeta\phi)^{h\zeta\phi}$, so $t\alpha^* = (t\zeta\phi)^{h\zeta\phi \cdot h^{-1}\alpha^*} = t\zeta\phi$.

We have shown $(k^t)\alpha = (k\alpha)^{t\zeta}$ for all $k \in K$. Also for $b \in H_{x,y}$, $(b^t)\alpha = (b^t)\zeta = (b\zeta)^{t\zeta} = (b\alpha)^{t\zeta}$, so by hypothesis (3) of Theorem 37.10, $(a^t)\alpha = (a\alpha)^{t\zeta}$ for all $a \in G_{x,y}$. This establishes Theorem 37.10.

Next the proof of Theorem 37.11. So assume U and $\bar{U}$ are uniqueness systems satisfying Hypothesis V and that $a : G_x \to \bar{G}_{\bar{x}}$ and $z : H \to \bar{H}$ are isomorphisms satisfying the hypotheses of Theorem 37.11. Then $az_{|H_x}^{-1} \in Aut(H_x)$, so by (V1), there exist $\delta \in N_{Aut(H)}(H_x)$ and $\epsilon \in N_{Aut(G_x)}(H_x)$ with $az^{-1} = \epsilon\delta$ on H_x. Let $\alpha = \epsilon^{-1}a$ and $\zeta = \delta z$; then $\alpha = \zeta$ on H_x. Further by (V2), we may pick $\epsilon \in N(G_{x,y}^{H_x})$. Therefore $G_{x,y}\epsilon^{-1} = G_{x,y}^g$ for some $g \in H_x$ and hence $G_{x,y}\alpha = (G_{x,y}^g)a = \bar{G}_{\bar{x},\bar{y}}^{ga} = \bar{G}_{\bar{x},\bar{y}(ga)}$ with $\bar{y}(ga) \in \Delta(\bar{x})$.

Similarly by 37.2 there is $t \in H$ with cycle (x, y) and by (V3) we may pick $\delta \in N(H_x t H_x)$. Then $xt\delta = yh$ for some $h \in H_x$, so $t\delta$ has cycle (x, yh) and $H_{x,y}\delta = (H_x \cap H_x^t)\delta = H_x \cap H_x^{t\delta} = H_{x,y}^h$. Thus

$H_{x,y}\zeta = H_{x,y}\delta z = H_{x,y}^h z = \bar{H}_{\bar{x},\bar{y}}^{hz} = H_{\bar{x},\bar{y}hz}$, while

$$t\zeta = t\delta z \in (H(\{x,y\})z)^{hz} - (H_{x,y}z)^{hz}$$
$$= \bar{H}(\{\bar{x},\bar{y}hz\}) - \bar{H}_{\bar{x},\bar{y}hz},$$

so $H(\{x,y\})\zeta = \bar{H}(\{\bar{x},\bar{y}(hz)\})$. Finally $H_{\bar{x},\bar{y}hz} = H_{x,y}\zeta = H_{x,y}\alpha \le \bar{G}_{\bar{x},\bar{y}(ga)}$, so $\bar{y}(hz)$ and $\bar{y}(ga)$ are fixed points of $\bar{H}_{\bar{x},\bar{y}(hz)}$ on $\Delta_{\bar{H}}(\bar{x})$. But as $\bar{H}_{\bar{x}}$ is transitive on $\bar{\Delta}_{\bar{H}}(\bar{x})$, $N_{\bar{H}_{\bar{x}}}(\bar{H}_{\bar{x},\bar{y}hz})$ is transitive on the fixed points of $\bar{H}_{\bar{x},\bar{y}hz}$ on $\Delta_{\bar{H}}(\bar{x})$, and then by (V4), $\bar{G}_{\bar{x},\bar{y}hz} = \bar{G}_{\bar{x},f}$ for each such fixed point f. In particular $\bar{G}_{\bar{x},\bar{y}ga} = \bar{G}_{\bar{x},\bar{y}hz}$, so α and ζ define a similarity of our systems. This completes the proof of Theorem 37.11.

Finally we prove Theorem 37.12. Assume the hypotheses of that theorem and let $t \in H$ have cycle (x,y). Observe that hypothesis (a) of Theorem 37.12 implies hypothesis (b), so we may assume (b) holds. Notice that $H_x\zeta = N_H(Z(x))\zeta = N_{\bar{H}}(Z(\bar{x})) = \bar{H}_{\bar{x}}$. Then arguing as in the proof of Theorem 37.11, using hypotheses (3) and (4) of Theorem 37.12 in place of (V1) and (V3), respectively, we may adjust α and ζ so that $\alpha = \zeta$ on H_x while still retaining hypotheses (1) and (2). For example, for $\delta \in Aut(H) \cap N(H_x) \cap N(Z(x))$, $xt\delta = yh$ for some $h \in H_x$, so $Z(y)\delta = Z(x)^t\delta = Z(x)^{t\delta} = Z(xt\delta) = Z(yh) = Z(y)^h$. So for $\zeta = \delta z$ with $Z(y)z = Z(\bar{y})$, we have $Z(y)\zeta = Z(y)^h z = Z(\bar{y}hz)$ with $\bar{y}hz \in \bar{\Delta}_{\bar{H}}(\bar{x})$.

Next by hypothesis (1), $Z(x) \le H_y$, so $Z(y) = Z(xt) = Z(x)^t \le (H_y)^t = H_{yt} = H_x$. Thus as $\alpha = \zeta$ on H_x, $Z(y)\alpha = Z(y)\zeta = Z(\bar{y})$. Further $Z(\bar{x}(t\zeta)) = Z(\bar{x})^{t\zeta} = (Z(x)\zeta)^{t\zeta} = (Z(x)^t)\zeta = Z(y)\zeta = Z(\bar{y})$. Also $Z(\bar{x}t\zeta) = Z(\bar{y}) = Z(\bar{x}\bar{t})$ for some $\bar{t} \in \bar{H}$, so $t\zeta \in N_{\bar{H}}(Z(\bar{x}))\bar{t} = \bar{H}_{\bar{x}}\bar{t}$ and hence $\bar{x}t\zeta = \bar{x}\bar{t} = \bar{y}$. That is, $t\zeta$ has cycle $(\bar{x},\bar{y})$.

Next $G_{x,y}\alpha = (N_{G_x}(Z(y)))\alpha = N_{\bar{G}_{\bar{x}}}(Z(\bar{y})) = \bar{G}_{\bar{x},\bar{y}}$. Finally $H_{x,y} = N_H(Z(x)) \cap N_H(Z(y))$ so $H_{x,y}\zeta = N_{\bar{H}}(Z(\bar{x})) \cap N_{\bar{H}}(Z(\bar{y})) = H_{\bar{x},\bar{y}}$. Therefore $H(\{x,y\})\zeta = (H_{x,y}\langle t \rangle)\zeta = \bar{H}_{\bar{x},\bar{y}}\langle t\zeta \rangle = \bar{H}(\{\bar{x},\bar{y}\})$. Therefore α and ζ define a similarity of our systems, and the proof of Theorem 37.12 is complete.

38. The uniqueness system of a string geometry

In this section we assume:

(Γ0)　　G is a flag transitive group of automorphisms of a residually connected rank 3 string geometry Γ and (x,l,π) is a flag in Γ such that the residues of x and π are *not* generalized digons.

See Section 4 for definitions and terminology. We recall that by 4.2 and 4.5, group theoretically these hypotheses are equivalent to the assertions that $G = \langle G_x, G_l, G_\pi \rangle$, $G_x = \langle G_{x,l}, G_{x,\pi} \rangle$, $G_\pi = \langle G_{\pi,x}, G_{\pi,l} \rangle$, and

$$G_l = G_{l,x} G_{l,\pi} \text{ but } G_x \neq G_{x,l} G_{x,\pi} \text{ and } G_\pi \neq G_{\pi,x} G_{\pi,l}. \qquad (*)$$

The three classes of objects of type 1, 2, and 3 will as usual be called *points*, *lines*, and *planes*, respectively. Given an object $a \in \Gamma$, write $\Gamma(a)$ for the residue of a, $\Gamma_i(a)$ for the objects of type i in $\Gamma(a)$, and $Q(a)$ for the kernel of the action of G_a on $\Gamma(a)$. Let p be a prime and $P(a) = O_p(Q(a))$.

Define the *collinearity graph* of Γ to be the graph whose vertices are the points of Γ and with points adjacent if and only if they are incident with a common line of Γ. Notice that by 38.1, this agrees with the notion of "collinearity graph" in Section 36. Let Δ be the collinearity graph of Γ.

Lemma 38.1: $\Gamma_1(l) \subseteq \Gamma_1(\pi)$ *and* $\Gamma_3(l) \subseteq \Gamma_3(x)$.

Proof: This is Exercise 1.5.

Lemma 38.2: Δ *is connected.*

Proof: This follows from 4.6.

Consider the following hypotheses:

(Γ1)   Each pair of distinct collinear points x, y is on a unique line $x + y$.

(Γ2)   If $x, y \in \Gamma_1(\pi)$ are collinear then $x + y \in \Gamma_2(\pi)$.

(Γ3)   Each triangle of Δ is incident with a plane.

(Γ4) $G_{\pi,l}$ is 2-transitive on $\Gamma_1(l)$.

(Γ5) $G_{x,l} = \langle G_{x,y,l}, G_{x,l,\pi} \rangle$ for $x \neq y \in \Gamma_1(l)$.

Theorem 38.3: *Assume* (G, Γ, x, l, π) *satisfies hypotheses* (Γi) *for* $i = 0, 4, 5$. *Define* Δ *to be the collinearity graph of* Γ, $H = G_\pi$, *and* Δ_H *the collinearity graph of the residue of* π. *Then*

(1) $\mathcal{U} = (G, H, \Delta, \Delta_H)$ *is a uniqueness system.*

(2)   *If hypotheses* (Γi), $0 \leq i \leq 5$, *hold then each triangle of* Δ *is* G-*conjugate to a triangle of* Δ_H.

Proof: Let $y \in \Gamma_1(l) - \{x\}$. By flag transitivity and (Γ4), G is edge transitive on Δ. By (Γ4) there exist $t \in H_l$ with cycle (x, y) and $H_l = \langle H_{x,l}, t \rangle$. By (Γ5), $G_{x,l} = \langle G_{x,y,l}, H_{x,l} \rangle$. These observations together with (Γ0) show that $\mathcal{U}$ satisfies condition (U) for uniqueness systems. Therefore (1) is established.

Next assume hypotheses (Γi), $0 \le i \le 5$. If abc is a triangle in Δ then by $(\Gamma 3)$, abc is G-conjugate to a subset of $\Gamma_1(\pi) = \Delta_H$, so without loss of generality $abc \subset \Delta_H$. By $(\Gamma 1)$ and $(\Gamma 2)$, the lines $a+b, a+c$, $b+c$ are incident with π, so abc is a triangle in Δ_H and (2) is established.

Example 38.4   Let V be a 5-dimensional vector space over the field F of order 2. From Example 1 in Section 4, the projective geometry $PG(V)$ of V is a rank 4 geometry whose objects are the points, lines, planes, and hyperplanes of V with incidence equal to inclusion. Let $\Gamma = PG(V)^t$ be the *truncation* of $PG(V)$ obtained by suppressing the hyperplanes of V. Thus Γ is a rank 3 geometry over $I = \{1, 2, 3\}$ whose objects are the points, lines, and planes of V. Pick a flag (x, l, π) from Γ; that is, $x \subset l \subset \pi$ is a chain of subspaces of V of dimension 1, 2, 3, respectively.

Observe that Γ is a rank 3 string geometry, since by Example 6 in Section 4, the residue $\Gamma(l) \cong PG(l) \oplus PG(V/l)^t$. Similarly $\Gamma(x) \cong PG(V/x)^t$ and $\Gamma(\pi) \cong PG(\pi)$.

Let $G = GL(V) \cong L_5(2)$. Then G is flag transitive on Γ, so Hypothesis $(\Gamma 0)$ is satisfied. Indeed visibly Hypotheses $(\Gamma 1)$ through $(\Gamma 5)$ are satisfied, so Theorem 38.3 supplies us with a uniqueness system $\mathcal{U} = (G, H, \Delta, \Delta_H)$. The vertices of Δ are the points of V and as each pair of points of V are collinear, Δ is a complete graph. Hence Δ is of diameter 1, so by 34.5, Δ is triangulable. Therefore by 35.14, Δ is simply connected while by Theorem 38.3 and 37.5.4, G is the free amalgamated product of $\mathcal{A}(\mathcal{U})$. Finally by Exercise 13.1, if $\bar{\mathcal{U}}$ is a uniqueness system equivalent to $\mathcal{U}$ then $G \cong \bar{G}$.

Lemma 38.5: *Assume M is a finite group with $F^*(M) = O_p(M) = R$ and $\Phi(Z(R)) = 1$, and $\alpha \in Aut(M)$ centralizes a Sylow p-subgroup of M. Then $\alpha \in Inn(M)$.*

Proof: Form the semidirect product S of M with $\langle \alpha \rangle$, and let $Z = Z(R)$ and $E = C_S(R)$. Then $E = Z \times \langle \alpha \rangle$.

Observe that if $A \le \langle \alpha \rangle$ with $A \trianglelefteq S$ then $[M, A] \le M \cap A = 1$, so as $\langle \alpha \rangle$ is faithful on M, $A = 1$. It follows that α is a p-element and as $\Phi(Z) = 1$, α is of order p. Now as α centralizes a Sylow p-subgroup of M, E splits over Z as an S-module, (cf. 12.8 in [FGT]) so $S = MC_S(M)$ and hence $\alpha \in Inn(M)$.

Write $d(x, y)$ for the distance between vertices x and y in Δ and for $S, T \subseteq \Delta$ let $d(S, T) = min\{d(s, t) : s \in S, \ t \in T\}$.

The following two lemmas can be used to study M_{24} and its 2-local geometry; see Exercise 13.2.

Lemma 38.6: *Assume (Γ1) and*

   (a) *For each line k of Γ there exists a unique $y \in \Gamma_1(k)$ with $d(x,y) = d(x,k)$.*

   (b) *Each pair of lines of $\Gamma(x)$ is incident with at least one plane.*

   (c) *If $d(x,z) = 2$ then $|\Gamma_3(x,z)| \leq 1$.*

Then

   (1) *Each triangle of Δ is contained in a line of Γ.*

   (2) *If $y \in \Delta^2(x)$ then there exists a unique plane π incident with x and y. Further $\Delta(x) \cap \Delta^{\leq 2}(y) \subseteq \Gamma_1(\pi)$.*

   (3) *Each square and pentagon of Δ is incident with a plane.*

   (4) *Assume σ is a plane with $d(x,\Gamma_1(\sigma)) = 1$ and the collinearity graph of $\Gamma(\sigma)$ is of diameter 2. Then $\Delta(x) \cap \Gamma_1(\sigma) = \{u\}$ for some point u, and $\Delta(x)^{\leq x} \cap \Gamma_1(\sigma) = u^{\perp} \cap \Gamma_1(\sigma)$.*

Proof: First if axb is a triangle in Δ then a and b are distinct points at distance 1 from x on $a+b$, so by (a), $x \in a+b$. Thus (1) holds.

Let $y \in \Delta^2(x)$. Then there is $u \in \Delta(x,y)$ and by (b), there is $\pi \in \Gamma_3(u+x,u+y)$. Hence by (c), $\pi = \pi(x,y) = \pi(u+x,u+y)$ is the unique plane incident with x and y and $u+x$ and $u+y$. Similarly if $v \in \Delta(x,y)$ then $\pi(v+x,v+y) = \pi(x,y) = \pi$, so $v \in \Gamma_1(\pi)$. Finally if $w \in \Delta(x) \cap \Delta^2(y)$ then $x, w \in \Delta^2(y) \cap (x+w)$, so by (a), there is $v \in \Delta(y) \cap (x+w)$. Then $v \in \Delta(x,y) \subseteq \pi$, so $w \in x+v \subseteq \pi$. Therefore (2) is established.

Notice (2) implies (3). Suppose σ is a plane with $d(x,\sigma) = 1$ and the collinearity graph of $\Gamma(\sigma)$ is of diameter 2. Then there is $u \in \Delta(x) \cap \sigma$. If $u \neq v \in \Delta(x) \cap \sigma$ then by (1), $d(u,v) = 2$, so by (2), $x \in \Delta(u,v) \subseteq \sigma$, contradicting $d(x,\sigma) = 1$. So u is unique. Clearly $u^{\perp} \cap \sigma \subseteq \Delta^{\leq 2}(x) \cap \sigma$. Conversely suppose y is a point on σ at distance 2 from x, but $y \notin \Delta(u)$. Then $d(u,y) = 2$ so $x \in \Delta^2(y) \cap \Delta(u) \subseteq \sigma$ by (2), a contradiction.

Lemma 38.7: *Assume the hypotheses of 38.6 and in addition assume*

   (d) *The collinearity graph of $\Gamma(\pi)$ is of diameter 2 and 4-generated.*

   (e) *Δ has diameter 3.*

   (f) *For each $z \in \Delta^3(x)$ the graph on $\Gamma_2(x)$ defined by $l * k$ if and only if $d(\Gamma_1(\pi(l,k)), z) = 1$ is connected, where $\pi(l,k)$ is the unique plane through l and k.*

Then Δ is 4-generated.

Proof: Let $S = \mathcal{C}_4(\Delta)$ be the closure of the set of all cycles of Δ of length at most 4; we must show S contains all cycles of Δ. By hypothesis (e) and 34.5 we must show S contains each r-gon for $r \leq 7$. By definition, S

contains all squares and triangles. By 38.6.3, each pentagon is incident with a plane, so by hypothesis (d), S contains all pentagons.

Let $p = x_0 \cdots x_r$ be an r-gon. Assume first that $r = 6$ and let $\pi = \pi(x_2, x_4)$. Now if $d(x_0, \pi) = 1$, then by 38.13.4, $\Delta(x_0) \cap \pi = \{x\}$ with $x \in \Delta(x_2, x_4)$. But then 34.9 says $p \in S$.

In general by (f) there is a sequence of lines $x_3 + y_i$, $0 \leq i \leq n$, with $y_0 = x_4$, $y_n = x_2$, and $\pi(x_3 + y_i, x_3 + y_{i+1})$ of distance 1 from x_0. Let $u_i \in \Delta(x_0, y_i)$ with $u_0 = x_5$ and $u_n = x_1$. Proceeding by induction on i, $x_0 x_5 x_4 x_3 y_i u_i x_0 \in S$, while by the previous paragraph, $x_0 u_i y_i x_3 y_{i+1} u_{i+1} x_0 \in S$. Thus by 34.3, the product $x_0 x_5 x_4 x_3 y_{i+1} u_{i+1} x_0$ of these two hexagons is in S, so $p \in S$ by induction on i.

Therefore all hexagons are in S, so it remains to consider the case $r = 7$. But $x_3, x_4 \in \Delta^3(x_0) \cap (x_3 + x_4)$, so by hypothesis (a), there is $x \in \Delta^2(x_0) \cap (x_3 + x_4)$, and hence p is in the closure of two hexagons and a triangle.

The remaining lemmas in this section give criteria for establishing the hypotheses (Γi), $1 \leq i \leq 5$. With the exception of 38.8, these lemmas will not be used elsewhere in *Sporadic Groups*, so the reader can skip them. They will be used to treat the uniqueness of other sporadic groups not considered here.

Lemma 38.8: *Assume*

 $(\Gamma 6)$ G_x *is primitive on planes and lines through* x.
 $(\Gamma 7)$ $Q(x)Q(\pi)$ *is transitive on* $\Gamma_1(l) - \{x\}$.

 Then

 (1) $Q(x)$ *is transitive on* $\Gamma_1(l) - \{x\}$ *and* $G_{x,l} = G_{x,y,l}Q(x)$ *for* $y \in \Gamma_1(l) - \{x\}$.
 (2) $(\Gamma 1)$, $(\Gamma 4)$, *and* $(\Gamma 5)$ *hold*.

Proof: By 38.1, $Q(\pi)$ fixes $\Gamma_1(l)$ pointwise. Further by $(\Gamma 7)$, $Q(x)Q(\pi)$ is transitive on $\Gamma_1(l) - \{x\}$, so $Q(x)$ is too. Thus (1) holds. Notice that as $Q(x) \leq G_{\pi, x, l}$, (1) implies $(\Gamma 5)$.

Next if y is on a line $k \in \Gamma_2(x)$ distinct from l, then by $(\Gamma 6)$, $G_x = \langle G_{x,l}, G_{x,k} \rangle = \langle G_{x,y,l}, G_{x,y,k}, Q(x) \rangle = G_{x,y}Q(x)$. Then $\Delta(x) = yQ(x) = l - \{x\}$. Thus by flag transitivity and 38.2, l is incident with all points of Γ, so by flag transitivity and 38.1, $G_\pi = G_{\pi,x}G_{\pi,l}$, contrary to (*). Thus $(\Gamma 1)$ holds. As $Q(x) \leq G_{\pi,l}$, (1) and flag transitivity imply $(\Gamma 4)$.

Lemma 38.9: *Assume the lines and planes in* $\Gamma(x)$ *form a linear space; that is, each pair of lines in* $\Gamma(x)$ *is incident with a unique plane of* $\Gamma(x)$.

Assume also that distinct points of $\Gamma(\pi)$ are incident with at most one line in $\Gamma(\pi)$. Then

(1) ($\Gamma 1$) and ($\Gamma 3$) hold.

(2) If $l, k \in \Gamma_2(x)$ then $|\Gamma_3(l, k)| = 1$.

Proof: Let $l, k \in \Gamma_2(x)$. As $\Gamma(x)$ is a linear space there is a unique plane $\pi \in \Gamma_3(x, l, k)$. Therefore by 38.1, π is the unique member of $\Gamma_3(l, k)$; that is, (2) holds. Further if $y \in \Gamma_1(l, k)$ is distinct from x then by 38.1, $x, y \in \Gamma_1(\pi)$ are distinct points incident with the lines l, k of $\Gamma(\pi)$, so by hypothesis, $l = k$. That is, ($\Gamma 1$) holds. If $xyzx$ is a triangle then by (2) there is $\pi \in \Gamma_3(x + y, x + z)$ so xyz is incident with π by 38.1. Hence ($\Gamma 3$) holds. Notice that we are not claiming, however, that $xyzx$ is a triangle in the collinearity graph Δ_H of $\Gamma_3(\pi)$; that is, y and z may not be collinear in $\Gamma(\pi)$.

Lemma 38.10: *Assume ($\Gamma 1$)and*

($\Gamma 8$) If k is a line with $Q(\pi) \leq G_k$ then $k \in \Gamma_2(\pi)$.

Then ($\Gamma 2$) holds.

Proof: If $x, y \in \Gamma_1(\pi)$ are collinear then $Q(\pi) \leq G_{x,y} \leq G_{x+y}$ by ($\Gamma 1$). Thus $x + y \in \Gamma_2(\pi)$ by ($\Gamma 8$).

Lemma 38.11: *Assume ($\Gamma 1$) and $P(\pi)$ is weakly closed in $N_G(P(\pi)) = N_G(G_\pi) = G_\pi$. Then ($\Gamma 2$) holds.*

Proof: By 38.10 it suffices to prove ($\Gamma 8$). Let $Q(\pi) \leq G_k$ and $\sigma \in \Gamma_3(k)$. Then by the weak closure of $P(\pi)$ and Sylow's Theorem, there is $g \in G_k$ with $P(\pi)^g = P(\sigma)$. Then as $N_G(P(\pi)) = N_G(G_\pi) = G_\pi$, $\pi g = \sigma$, so $k \in \Gamma_2(\pi)$.

Lemma 38.12: *Assume ($\Gamma 1$) and ($\Gamma 4$), and assume for collinear points x, y and $l = x + y$ that*

(a) $P(l) = (P(l) \cap P(x))(P(l) \cap P(y))$, and

(b) $|P(x) : P(x)_y| < |P(l)Q(x) : P(l)_k Q(x)|$ for all $k \in \Gamma_2(x)$ such that $\Gamma_3(l, k)$ is empty.

Then ($\Gamma 3$) holds.

Proof: Suppose $xyzx$ is a triangle in Δ and let $k = x + z$. By ($\Gamma 4$) and flag transitivity, $|P(y) : P(y)_z| = |P(x) : P(x)_y|$.

Now $|P(l)Q(x) : P(l)_k Q(x)|$ is the length of the orbit $kP(l)$ of k under $P(l)$ and by (a), $kP(l) = k(P(l) \cap P(y))$. So

$$|P(l)Q(x) : P(l)_k Q(x)| = |P(l) \cap P(y) : (P(l) \cap P(y))_k|$$

$$\leq |P(l) \cap P(y) : P(l) \cap P(y)_z|$$

(as $P(l) \cap P(y)_z \leq (P(l) \cap P(y))_k) \leq |P(y) : P(y)_z| = |P(x) : P(x)_y|$. Hence by (b) there is $\pi \in \Gamma_3(l, k)$. Then by 38.1, xyz is incident with π.

Lemma 38.13: *If $G \neq G_\pi G_x$ and $G_{x,y}$ is transitive on the planes of $\Gamma(x)$ not through l for $y \in \Gamma_1(l)$, then $(\Gamma 2)$ holds.*

Proof: If $x, y \in \Gamma_1(\sigma)$ for some plane σ not incident with l then as $G_{x,y}$ is transitive on planes in $\Gamma(x)$ not incident with l, $\Gamma_3(x) = \Gamma_3(y)$. Then by 38.2, $\Gamma_3(x) = \Gamma_3(v)$ for all points v, contradicting $G \neq G_x G_\pi$. Thus $(\Gamma 2)$ holds.

Remarks. The idea of establishing the uniqueness of a group G with respect to some hypothesis $\mathcal{H}$ by first proving

> (a)   a certain amalgam A associated to G is determined up to isomorphism by $\mathcal{H}$ independently of G,

and then proving

> (b)   the collinearity graph of the completion of A via G is simply connected,

was introduced in [AS1] by Yoav Segev and the author. Most of the material in this chapter comes from that reference.

Exercises

1. Let $\mathcal{U}$ and $\bar{\mathcal{U}}$ be equivalent uniqueness systems such that $\Delta_{\bar{H}}$ is a base for $\bar{\mathcal{U}}$. Assume either
   - (a) $|G| = |\bar{G}|$, or
   - (b) $\bar{G}$ is simple.

   Then $G \cong \bar{G}$ is the free amalgamated product of $A(\mathcal{U})$.
2. Let $G = M_{24}$, Γ the 2-local geometry for G, and Δ its collinearity graph. Prove
   - (1) G, Γ satisfy hypotheses $(\Gamma 0)$–$(\Gamma 5)$ of Section 38, and hence determined a uniqueness system $\mathcal{U}$ via Theorem 38.3.
   - (2) Δ is 4-generated.
   - (3) Each triangle and square of Δ is incident with a plane of Γ.
   - (4) G is the free amalgamated product of $A(\mathcal{U})$.
   - (5) If $\bar{U}$ is a uniqueness system equivalent to $\mathcal{U}$ then $G \cong \bar{G}$.
     (Hint: Use Exercise 7.1 together with 38.6 and 38.7.)
3. Assume hypotheses $(\Gamma 0)$, $(\Gamma 1)$, $(\Gamma 2)$, and $(\Gamma 4)$ of Section 38 with G_l finite. Prove
   - (1) G is transitive on triples x, y, π with $x, y \in \Gamma_1(\pi)$ and x, y collinear.
   - (2) If for each triple in (1), $(|\Gamma_3(l)|, p) = 1$, $F^*(G_{x,y}) = O_p(G_{x,y})$, $\Phi(Z(O_p(G_{x,y}))) = 1$, and $Z(G_{x,y,\pi}) = 1$, then the uniqueness

system $\mathcal{U}$ of Theorem 38.3 is equivalent to any uniqueness system similar to $\mathcal{U}$.

(3) Let $K = G_{\Gamma_1(l)}$ and assume $|\Gamma_i(l)|$ *vert* is odd for $i = 1, 3$, $G_l = KG_{l,\pi}$, $F^*(K) = O_2(K)$, $\Phi(Z(O_2(K))) = 1$, and $Z(K_\pi) = Z(G_\pi)$. Let $\mathcal{U}$ be the uniqueness system defined by Theorem 38.3, suppose $\bar{\mathcal{U}}$ is a uniqueness system defined via Theorem 38.3 by a pair $\bar{G}, \bar{\Gamma}$ satisfying hypothesis (Γi), $i = 0, 1, 4, 5$, and α, ζ is a similarity of $\mathcal{U}$ with $\bar{\mathcal{U}}$ with respect to $(x, y), (\bar{x}, \bar{y})$ such that $G_{\pi, x+y}\zeta = \bar{G}_{\bar{\pi}, \bar{x}+\bar{y}}$. Then $\mathcal{U}$ is equivalent to $\bar{\mathcal{U}}$.

(Hint: Use Theorem 37.9 to prove (2) and Theorem 37.10 to prove (3).)

Chapter 14

The Uniqueness of Groups of Type M_{24}, He, and $L_5(2)$

In this chapter we consider groups G possessing an involution z such that $H = C_G(z)$ is isomorphic to the centralizer of a transvection in $L_5(2)$. We saw in 8.9 that $F^*(H) = Q \cong D_8^3$ is an extraspecial 2-group. Thus we can use the theory of large extraspecial subgroups developed in Section 8 to study G. To eliminate the trivial case $G = H$, we assume in addition that z is not weakly closed in Q with respect to G. Recall that by 8.10, M_{24} satisfies these hypotheses, and of course so does $L_5(2)$. From Chapter 5, there is one more simple group satisfying the hypotheses: the sporadic group He discovered by Held.

We begin our study of G in Sections 39 and 40 by generating various facts about the 2-local subgroups of G which will define a coset geometry Γ for G. This culminates in lemma 40.5, which singles out the three possibilities for these subgroups and the geometry Γ. We then begin to use the machinery of Chapters 12 and 13 to establish that each case gives rise to at most one group. In particular in Section 41 we associate a uniqueness system to G when G is of type M_{24} or $L_5(2)$ and use the results in Section 38 to prove G is unique. Then in the remaining sections we analyze the final case leading to He, eventually using Lemma 36.7 to prove uniqueness here. In addition to our uniqueness result we also derive the order of He and the structure of various local subgroups of He.

39. Some 2-local subgroups
in $L_5(2)$, M_{24}, and He

In this section z is an involution in a finite group G such that $H = C_G(z)$ is isomorphic to the centralizer of a transvection in $L_5(2)$. Let $Q = O_2(H)$ and $Z = \langle z \rangle$. Let $T \in Syl_2(H)$. Assume z is not weakly closed in Q with respect to G.

For $X \leq G$ write $\mathcal{A}(X)$ for the set of elementary abelian 2-subgroups of X of maximal order and set $J(X) = \langle \mathcal{A}(X) \rangle$.

We regard H as the stabilizer in $GL(V)$ of an incident-point–hyperplane pair V_1, V_4 of a 5-dimensional $GF(2)$-space V and let

$$ 0 = V_0 \leq \cdots \leq V_5 = V $$

be a T-invariant chain of subspaces with $dim(V_i) = i$. Let U_1 be the subgroup of H of all transvections with center V_1, U_2 the subgroup of transvections with axis V_4, A_1 the subgroup centralizing V_2 and V/V_2, and A_2 the subgroup centralizing V_3 and V/V_3.

Lemma 39.1: *(1) $Q \cong D_8^3$ and H is the split extension of Q by $L_3(2)$.*

(2) The subgroups U_1 and U_2 are normal subgroups of H isomorphic to E_{16}. Moreover $Q = U_1 U_2$, $U_1 \cap U_2 = Z$, $C_H(U_i) = U_i$, and H splits over U_i.

(3) $\mathcal{A}(T) = \{A_1, A_2\}$ with $A_i \trianglelefteq T$, $A_i \cong E_{64}$, and $A_1 \cap A_2 \cong E_{16}$. Further A_1 and A_2 are the maximal elementary abelian subgroups of $J(T) = A_1 A_2$, T splits over A_i, $A_i \cap U_i$ is of rank 3, and $A_i \cap U_{3-i}$ is of rank 2.

(4) The transpose-inverse automorphism of $L_5(2)$ centralizing z and acting on T interchanges U_1 and U_2 and A_1 and A_2.

(5) H has four orbits on involutions of Q: $\{z\}$, $U_1 - Z$, $U_2 - Z$, and the set $\mathcal{T}_Q$ of involutions in $Q - (U_1 \cup U_2)$.

(6) If $t \in \mathcal{T}_Q$ then $|C_H(t)| = 2^9$ and $Z = C_H(t)^{(2)}$ is the second member of the derived series for $C_H(t)$.

(7) $N_H(A_1)$ is of index 7 in H and has seven orbits Δ_i, $1 \leq i \leq 7$, on $A_1^{\#}$ of lengths 1, 6, 2, 6, 12, 12, 24, respectively. Indeed $\Delta_1 - \{z\}$, $\Delta_2 = U_1 \cap A_1 - Z$, $\Delta_3 = U_2 \cap A_1 - Z$, and $\Delta_4 = \mathcal{T}_Q \cap A_1$. For $a \in \Delta_7$ and $h \in C_H(a) - A_1$, $C_{A_1}(h) \cong E_8$ contains z, two members of Δ_2, and four members of Δ_7.

(8) $C_H(A_1 \cap U_1) = A_1 U_1$ with A_1 and U_1 the maximal elementary abelian subgroups of $A_1 U_1$.

(9) $N_H(J(T)) = T$ has six orbits Γ_i, $0 \le i \le 5$, on $Z(J(T))^\#$ of lengths 1, 2, 2, 2, 4, 4, respectively. Further $\Gamma_0 = \{z\}$, $\Gamma_i = (U_i \cap Z(J(T))) - Z$ for $i = 1, 2$, and $\Gamma_3 = T_Q \cap Z(J((T)))$.

Proof: These facts can be checked inside of $GL(V) \cong L_5(2)$. The check is left as an exercise, but see 8.8 for (1).

Lemma 39.2: $T \in Syl_2(G)$.

Proof: As $Z = Z(T)$, $N_G(T) \le H$.

Lemma 39.3: *Either $N_G(U_i) = H$ or $N_G(U_i)$ is the split extension of U_i by $L_4(2)$ acting faithfully on U_i.*

Proof: As $z \in U_i$, $C_G(U_i) = C_H(U_i) = U_i$. Therefore $N_G(U_i)/U_i \le GL(U_i) = L_4(2)$. But H/U_i is the stabilizer of z in $GL(U_i)$, and hence maximal in $GL(U_i)$. Thus $N_G(U_i) = H$ or $N_G(U_i)/U_i \cong L_4(2)$. In the latter case as H splits over U_i, so does T, and hence $N_G(U_i)$ splits over U_i by Gaschutz's Theorem (cf. 10.4 in [FGT]).

Lemma 39.4: *Let $\mathcal{T}$ be the set of involutions G-conjugate to an element of T_Q. Then $z \notin \mathcal{T}$. Indeed if $t \in T_Q$ then $C_H(t) \in Syl_2(C_G(t))$.*

Proof: By 39.1.6, $S = C_H(t)$ is of order 2^9 and $Z = S^{(2)}$. Thus $N_G(S) \le N_G(Z) = H$, so $S \in Syl_2(C_G(t))$.

Lemma 39.5: *Either $U_1^\# \subseteq z^G$ or $U_2^\# \subseteq z^G$.*

Proof: This follows from 39.1.5, 39.4, and the hypothesis that z is not weakly closed in Q with respect to G.

Lemma 39.6: *(1) A_1 and A_2 are not conjugate in G.*
(2) A_i is weakly closed in G for $i = 1$ and 2.
(3) $N_G(A_i)$ controls fusion in A_i.
(4) $N_G(U_i \cap A_i) = N_G(A_i) \cap N_G(U_i)$.

Proof: By 39.1.3, $\mathcal{A}(T) = \{A_1, A_2\}$ is of order 2 and by 39.1.7, A_1 and A_2 are normal in T. So (1) and (2) follow from 7.7.2, and (3) follows from (2) and 7.7.1.

Evidently $N_G(A_i) \cap N_G(U_i) \le N_G(A_i \cap U_i)$. The opposite containment follows from 39.1.8.

Lemma 39.7: *If $U_1^\# \subseteq z^G$ then either*

(1) $N_G(U_1)/U_1 \cong L_4(2)$, $N_G(A_1) \le N_G(U_1)$, and $z^G \cap A_1 \subseteq U_1$, or
(2) $U_2^\# \subseteq z^G$.

Proof: Adopt the notation of 39.1.7, and assume $U_2^{\#}$ is not contained in z^G. Then z is weakly closed in U_2 by 39.1.5. Thus Δ_3 contains no member of z^G. By 39.4, Δ_4 contains no conjugate of z. Let $n = |z^G \cap A_1|$. Then $n = \sum_{i \in I} d_i$, where $I = \{i : \Delta_i \subseteq z^G\}$ and d_i is the order of Δ_i. We have $1, 2 \in I$ but $3, 4 \notin I$. Thus $n = 7, 19, 31, 43$, or 55 by 39.1.7. But by 39.6.3, $N_G(A_1)$ is transitive on $z^G \cap A_1$, so n divides the order of $L_6(2)$. It follows that $n = 7$ and $z^G \cap A_1 = A_1 \cap U_1^{\#}$. Thus $N_G(A_1) \le N_G(A_1 \cap U_1)$. So $N_G(A_1) \le N_G(U_1)$ by 39.6.4. In particular as $N_G(A_1)$ is transitive on $A_1 \cap U_1^{\#}$, $N_G(U_1) \not\le H$. Now 39.3 completes the proof.

Lemma 39.8: *Assume $N_G(U_1)/U_1 \cong L_4(2)$. Then either*

(1) $U_1^{\#} = z^G \cap N_G(U_1)$, or

(2) $U_2^{\#} \subseteq z^G$ and $N_G(A_1)/A_1 \cong L_3(2) \times S_3$.

Proof: Let $M = N_G(U_1)$. Then $N_M(A_1)/A_1 \cong L_3(2) \times \mathbf{Z}_2$ with $A_1 = B_1 \oplus B_2$ the sum of two equivalent natural modules for $N_M(A_1)^{\infty}/A_1 \cong L_3(2)$, and with $B_1 = U_1 \cap A_1$ and $B_2^u \neq B_2$ for $u \in U_1 - B_1$. Thus $M_1 = N_M(A_1)$ has three orbits Γ_i, $1 \le i \le 3$, on $A_1^{\#}$ of order 7, 14, and 42, respectively, where $\Gamma_1 = B_1^{\#}$, $\Gamma_2 = B_2^{\#} \cup B_2^{u\#}$, and $\Gamma_3 \subseteq T$. Thus by 39.4, $z^G \cap A_1$ is Γ_1 or $\Gamma_1 \cup \Gamma_2$ of order 7 or 21.

Let $\Gamma = z^G \cap A_1$, $L = N_G(A_1)$, $K = ker_{M_1}(L)$, and assume Γ is of order 21. Then $|L : M_1| = 3$, so $L/K \le S_3$. In particular $M_1^{\infty} \le K$, so $M_1^{\infty} = K^{\infty} \trianglelefteq L$. If $M_1 \trianglelefteq L$, then $B_1 = Z(O_2(M_1)) \trianglelefteq L$, contradicting 39.6.3 and our hypothesis on fusion of z in A_1. So $L/K \cong S_3$. Then as $[u, M_1] \le A_1$, $L = C_L(M_1^{\infty}/A_1)M_1$, so $L/A_1 \cong L_3(2) \times S_3$. Also $U_2^{\#} \cap A_1 \subseteq \Gamma$, so $U_2^{\#} \subseteq z^G$ by 39.1.5. That is, (2) holds.

So assume $\Gamma \subseteq U_1$. Then L acts on $\langle \Gamma \rangle = U_1 \cap A_1$, so $L = M_1$ by 39.6.4. Now if v is an involution in $M - U_1$ then as $M/U_1 = GL(U_1)$, either v induces a transvection on U_1 and v is fused into A_1 under M, or $m([U_1, v]) = 2$ and v is fused into A_2 under M. Thus if $v \in z^G$, v is of the second type and we may take $v \in A_2$. Then $[v, U_1] = C_{U_1}(v) = U_1 \cap A_2$, so each involution in vU_1 is in $v(U_1 \cap A_2) \subseteq A_2$, and is conjugate to v under U_1. Notice we may assume such a v exists, or else (1) holds.

Let $M^* = M/U_1$. Then A_2^* contains six conjugates of v^* and nine transvections. We have seen that if r^* is a conjugate of v^* then all four involutions in rU_1 are in z^G, while if r^* is a transvection then no member of rU_1 is in z^G. Thus $z^G \cap A_2$ is of order $3 + 6 \cdot 4 = 27$.

Further if r^* is a transvection with $r \in U_2 \cap A_2$ then $r(A_2 \cap U_1)$ contains two members of T and two members of $U_2 - Z$. So there are eighteen elements of A_2 fused into $U_2 - Z$ and eighteen in T. As elements in $U_2 \cap A_1 - Z$ are not fused to those in $T_Q \cap A_1$ in $M_1 = N_G(A_1)$, these elements are not fused in G by 39.6.3. Hence these eighteen element sets are the orbits of $N_G(A_2)$ on $A_2^{\#}$ distinct from $z^G \cap A_2$ by 39.6.

By 39.1.4 and 39.1.7, there is an element $a \in z^G \cap A_2$ with $C_H(a) = A_2\langle h \rangle = R$ of order 2^7 and $C_{A_2}(R) = B$ contains five elements from z^G and two elements b and c from $U_2 - Z$. Then as $|N_G(A_2) : N_G(A_2) \cap C_G(a)| = 27$, R is of index 2 in some subgroup S of $N_G(A_2) \cap C_G(a)$ and S acts on $\{b, c\}$ and hence fixes $z = bc$. But then $S \leq C_H(a) = R$, a contradiction.

Lemma 39.9: *If $z^G \cap N_G(U_1) = U_1^{\#}$ then $U_1 \trianglelefteq G$.*

Proof: Let $M = N_G(U_1)$. By our hypothesis and 7.3, M is the unique point of G/M fixed by z. Let $x \in A_1 - U_1$. Then $D = C_{U_1}(x) = U_1 \cap A_1$ is a hyperplane of U_1 but $N_M(\langle x, U_1 \rangle)$ is not transitive on $D^{\#}$; instead it fixes the center $[U_1, x]$ of the transvection x. Hence $G = M$ by Exercise 2.7.

Lemma 39.10: *Either*

(1) $z^G \cap Q = U_1^{\#} \cup U_2^{\#}$, *or*
(2) $G = N_G(U_i)$ *is the split extension of U_i by $L_4(2)$ for $i = 1$ or 2.*

Proof: By 39.5 and the symmetry supplied by 39.1.4, we may assume $U_1^{\#} \subseteq z^G$. Then by 39.1.5 and 39.4, either (1) holds or $z^G \cap Q = U_1^{\#}$, and we may assume the latter. Now 39.3, 39.7, 39.8, and 39.9 complete the proof.

40. Groups of type $L_5(2)$, M_{24}, and He

In this section we continue the hypotheses and notation of the previous section. In addition we assume $z^G \cap Q = U_1^{\#} \cup U_2^{\#}$. This hypothesis is justified in view of lemma 39.10. Recall the definition of $J(T)$ in Section 39 and observe that by 39.1.3, $J(T) = A_1 A_2$ is of order 2^8 and index 4 in T, with $Z(J(T)) = \Phi(J(T)) = A_1 \cap A_2 \cong E_{16}$.

Lemma 40.1: *(1)* $N_G(J(T))/J(T) \cong S_3 \times S_3$ *has two orbits $z^G \cap Z(J(T))$ and $T \cap Z(J(T))$ on $Z(J(T))^{\#}$ of order 9 and 6, respectively.*
(2) $z^G \cap A_i$ *contains a $N_H(A_i)$-orbit of length 12 for $i = 1$ and 2.*

Proof: Let $B = Z(J(T))$. By 39.1.9, $T = N_H(J(T))$ has orbits Γ_i, $0 \leq i \leq 5$, on $B^{\#}$. Notice $J(T)$ is weakly closed in T, so by 7.7, $N_G(J(T))$ controls fusion in B. Also by 39.1.9, $\Gamma_i \subseteq z^G$ for $i \leq 2$, while $\Gamma_3 \subseteq T$, so from the orders of Γ_i listed in 39.1.9 we conclude $\Gamma = z^G \cap B$ is of order 5, 9, or 13.

The last case is out as 13 does not divide the order of $L_4(2)$. In the first case $Aut_G(B)$ is of order $20 = |\Gamma||Aut_H(B)|$ with Sylow 2-group $Aut_H(B) \cong E_4$. But then a Sylow 5-group P of $Aut_G(B)$ is normal in $Aut_G(B)$ and centralizes an involution in $Aut_G(B)$. This is impossible as P is irreducible on B.

So $|\Gamma| = 9$ and $|T \cap B| = 6$. Further $Aut_G(B)$ is of order 36 with Sylow 2-group E_4. We conclude $Aut_G(B) \cong S_3 \times S_3$ and for $b \in \Gamma - Q$, $|b^T| = 4$. Next the $N_H(A_i)$-orbits on $A_i^{\#}$ are listed in 39.1.7. As $|T : C_T(b)| = 4$, the length of the orbit Δ_j of b under $N_H(A_i)$ is not divisible by 8. But by 39.1.7, orbits of $N_H(A_i)$ on $A_i - Q$ have length 12 or 24. Hence $d = 12$.

Lemma 40.2: *Either*

(1) $|z^G \cap A_1| = 21$, *or*

(2) $|z^G \cap A_1| = 45$ *and* $|T \cap A_1| = 18$.

Proof: Adopt the notation of 39.1.7. By the hypotheses of this section, $\Delta_i \subseteq z^G$ for $i \leq 3$, while $\Delta_4 \subseteq T$ and by 40.1 we may choose notation so that $\Delta_6 \subseteq z^G$. So $\Delta = z^G \cap A_1$ is of order 21, 33, 45, or $57 = 3 \cdot 19$. But by 39.6.3, $N_G(A_i)$ is transitive on Δ, so the order of Δ divides that of $L_6(2)$. Hence $|\Delta| = 21$ or 45.

Suppose $|\Delta| = 45$. Then $N_H(A_1)$ has two orbits Δ_4 and Δ_5 on $A_1^{\#} - \Delta$ of order 6 and 12, with $\Delta_4 \subseteq T$. Thus either $T \cap A_1 = A_1^{\#} - \Delta$ is of order 18, or $T \cap A_1 = \Delta_4$ and Δ_5 are orbits of $N_G(A_1)$. We may assume the latter. But then $\langle \Delta_4 \rangle = Q \cap A_1 \trianglelefteq N_G(A_1)$, contradicting that Δ is of order 45.

Lemma 40.3: *If* $z^G \cap A_1$ *is of order 21 then* $N_G(U_1)/U_1 \cong L_4(2)$, $N_G(A_1)/A_1 \cong L_3(2) \times S_3$, *and* $A_1^{\#} = (z^G \cap A_1) \cup (T \cap A_1)$.

Proof: Let $M = N_G(A_1)$ and $M^* = M/A_1$. Then

$$|M^*| = |z^M||(H \cap M)^*| = 2^4 \cdot 3^2 \cdot 7.$$

Let $I = N_G(J(T))$ and $K = ker_I(M)$. By 40.1, $|M : I| = 7$, so $M^*/K^* \cong M/K \leq S_7$.

Now $I^* \cong S_3 \times S_4$. Further if $K^* = 1$ then $M^* \leq S_7$. But S_7 has a unique subgroup isomorphic to $S_3 \times S_4$ and that subgroup is maximal but not of index 7. So $K^* \neq 1$. Thus either $O_3(I^*) \trianglelefteq M^*$ or $J(T) \leq K$, since $O_3(I^*)$ and $J(T)^* = O_2(I^*)$ are the minimal normal subgroups of I^*. However if $J(T) \leq K$ then $I = M$ by a Frattini argument.

So $O_3(I^*) \trianglelefteq M^*$. Let X be of order 3 in I with $X^* = O_3(I^*)$ and let $B_0 = C_M(X)$. From the action of X on $J(T)$, $C_{A_1}(X) = 1$, so $N_M(X)$ is a complement to A_1 in M and $|N_M(X) : B_0| = 2$. By Gaschütz's Theorem (cf. 10.4 in [FGT]) there is a complement B to X in B_0. Claim $|B : H \cap B| = 7$. As $|B_0 : B_0 \cap H| = 21$, it suffices to show B contains a Sylow 3-group P of $B_0 \cap H$. This follows as P is inverted by some $b \in B \cap H$ and $B \cap H \trianglelefteq B_0 \cap H$.

So $7 = |B : H \cap B| = |z^B|$ and hence by 39.1.7, z^B is composed of two orbits under $H \cap B$ of length 1 and 6, and $\langle z^B \rangle = U_1 \cap A_1$. Thus $B \leq N_G(U_1 \cap A_1) \leq N_G(U_1)$ by 39.6.4. Thus $N_G(U_1) \not\leq H$, so 39.3 and 39.8 complete the proof.

Lemma 40.4: *Assume $z^G \cap A_1$ is of order 45 and let $M = N_G(A_1)$, $a \in T \cap Z(J(T))$, and $I = C_M(a)$. Then*

(1) $M/A_1 \cong S_6/Z_3$.

(2) $I/A_1 \cong S_5$ and $C_{A_1}(I^\infty) = B \cong E_4$ with $N_M(B) = XI$ for some X of order 3 with $XA_1 \trianglelefteq M$.

Proof: Let $M^* = M/A_1$. Then

$$|M^*| = |z^M||(H \cap M)^*| = 2^4 \cdot 3^3 \cdot 5.$$

By 40.2, $|T \cap A_1| = 18$, so

$$|I^*| = |M^*|/|T \cap A_1| = 2^3 \cdot 3 \cdot 5.$$

Now by 40.1, $N_I(J(T))^* \cong S_4$ is of index 5 in I^* and as in the proof of the previous lemma, $J(T)^*$ is not normal in I^*, so $I^* \leq S_5$. Thus $I^* \cong S_5$.

Let $B = C_{Z(J(T))}(O^2(N_I(J(T))))$. Then $B \cong E_4$ and $\langle (B/\langle a \rangle)^I \rangle \leq A_1/\langle a \rangle$ is an image of the 5-dimensional permutation module for I^*. Hence as an element of order 3 in I is fixed point free on A_1/B, $B \trianglelefteq I$. Further by 40.1, $O^2(N_G(J(T)))^* \cong Z_3 \times A_4$ is transitive on $B^\#$. Therefore $L = O^2(N_G(J(T)))I = N_M(B)$ is of index 6 in M, so $M^*/K^* \leq S_6$, where $K = \ker_L(M)$. As $F^*(I^*)$ is simple and $O^2(N_G(J(T)))^* \cong Z_3 \times A_4$, it follows that $|K^*| \leq 3$. As $|M^*| = 3|S_6|$, it follows that $K^* \cong Z_3$ and $M^*/K^* \cong S_6$, as desired.

Lemma 40.5: *Up to a permutation of 1 and 2, one of the following holds:*

(1) $N_G(U_i)/U_i \cong L_4(2)$ and $N_G(A_i)/A_i \cong L_3(2) \times S_3$ for $i = 1$ and 2.

(2) $N_G(U_1)/U_1 \cong L_4(2)$, $N_G(U_2) = H$, $N_G(A_1)/A_1 \cong L_3(2) \times S_3$, and $N_G(A_2)/A_2 \cong S_6/Z_3$.

(3) $N_G(U_i) = H$ and $N_G(A_i)/A_i \cong S_6/Z_3$ for $i = 1$ and 2.

Proof: By 40.2, $n_i = |z^G \cap A_i| = 21$ or 45. Further if $n_i = 21$, then by 40.3, $N_G(U_i)/U_i \cong L_4(2)$ and $N_G(A_i)/A_i \cong L_3(2) \times S_3$. On the other hand if $n_i = 45$ then $N_G(A_i)/A_i \cong S_6/Z_3$ by 40.4. Thus in this case $H = N_G(U_i)$ by 39.3 and 39.8. So the lemma holds.

We will say that G is of *type $L_5(2)$* in case 40.5.1 holds, G is of *type M_{24}* if 40.5.2 holds, and G is of *type He* if 40.5.3 holds. We will consider each type in succeeding sections.

Lemma 40.6: *G has two conjugacy classes of involutions: z^G and T.*

Proof: First by 40.2 and 40.3, $z^G \cap A_i$ and $T \cap A_i$ are the orbits of $N_G(A_i)$ on $A_i^{\#}$. Thus it remains to show that each involution in G is fused into A_1 or A_2.

Now A_1 and A_2 are the maximal elementary abelian subgroups of $J(T)$, so each involution in $J(T)$ is in A_1 or A_2. Next if $N_G(U_i)/U_i$ is isomorphic to $L_4(2)$, then each involution in $N_G(U_i)$ is conjugate under $N_G(U_i)$ to an element of $J(T)$ (cf. the proof of 39.8). So we may assume G is of type He. Let $L_i = N_G(A_i)^{\infty}$. Then each involution in L_i is L_i-conjugate to a member of $J(T)$ and there are two L_i-classes of involutions in $N_G(A_i) - L_i$, each of which has a representative in L_{3-i}. So the proof is complete.

41. Groups of type $L_5(2)$ and M_{24}

We continue the hypotheses, notation, and terminology of Sections 39 and 40. In addition we assume G is of type $L_5(2)$ or M_{24} as defined after 40.5. Let $G_1 = N_G(U_1)$, $G_2 = N_G(A_1)$, and $G_3 = N_G(A_2)$. Let $\mathcal{F} = (G_1, G_2, G_3)$ and form the coset geometry $\Gamma = \Gamma(G, \mathcal{F})$ as in Example 4 of Section 4. We adopt the notation and terminology of Sections 4 and 38 in discussing Γ.

Lemma 41.1: *Γ is a residually connected string geometry and G is flag transitive on Γ.*

Proof: Let $L = \langle \mathcal{F} \rangle$ and suppose for the moment that $L = G$. From 4.5 and 4.11 it suffices to prove $G_{1s} = \langle G_{1s,2s}, G_{1s,3s} \rangle$ for each $s \in S_3$ and $G_2 = G_{12}G_{23}$. But $T \leq G_{123}$ and from 40.5, for each choice of distinct i, j there is $K_i \leq T$, $K_i \trianglelefteq G_i$ with G_{ij}/K_i maximal in G_i/K_i, so indeed $G_i = \langle G_{ij}, G_{ik} \rangle$. For example, $K_1 = U_1$, $G_1/U_1 = GL(U_1) \cong L_4(2)$, G_{12}/U_1 is the maximal parabolic of G_1/U_1 stabilizing a 3-dimensional subspace of U_1, and G_{13}/U_1 is the maximal parabolic stabilizing a 2-dimensional subspace of U_1. Similarly $K_2 \cong E_{64}$, $|G_2 : G_{12}| = 3$, and $|G_2 : G_{23}| = 7$, so $G_2 = G_{12}G_{23}$.

Thus it remains to show $L = G$. Notice that L satisfies the hypothesis of G so all results in Sections 39 and 40 can be applied to both G and L. In particular by 40.6, G and L have two classes of involutions with representatives z and $t \in T$ and $z^G \cap L = z^L$ and $t^G \cap L = t^L$. Further by 39.1.2, $C_G(z) = H \le G_1 \le L$, so by 7.3, L is the unique fixed point of z on G/L. Then by 7.5 it suffices to show t fixes a unique point of G/L, and hence by 7.3 it suffices to show $C_G(t) \le L$.

By 39.1.9 we may pick $t \in E = Z(J(T))$. Let $M = C_G(t)$ and $I = M \cap L$. We may assume $M \ne I$. For $v \in z^G \cap I$, L is the unique fixed point of v on G/L, so I is the unique fixed point of v on M/I.

Proceeding by induction on the order of G, we may assume $L \cong L_5(2)$ or M_{24}. Suppose first that $L \cong L_5(2)$. We view L as $GL(V)$ (where V is the 5-dimensional space over the field of order 2 of Section 39) and observe $V_2 = [V,t]$ and $V_3 = C_V(t)$, so $I \le G_{23} = N_L(J(T))$. By 40.1.1, $|z^G \cap E| = 9$ and $I/J(T) \cong S_3$ has orbits of length 3,6 on $z^G \cap J(T)$. We pick z in the orbit of length 3 and observe that by 7.5, there is $v \in z^G \cap J(T)$ with $K = C_M(zv) \not\le I$.

Let $B = \langle t, zv \rangle$, $D = I \cap K$, and $K^* = K/B$. Observe that as $C_G(\langle zv, t \rangle) \not\le L$, $z^G \cap \langle zv, t \rangle = \varnothing$ and hence $zv \in C_E(O^2(I)) \cong E_4$. So $B = C_E(O^2(I))$ and $D = O^2(I)$. Next D^* is the unique point of K^*/D^* fixed by x^* for each $x \in z^G \cap D$. But $A_1 \trianglelefteq D$ while $D = O^2(I)$ has two orbits on involutions of A_1^* of length 3,12, and each orbit has a representative x^* with $x \in z^G$. Therefore 7.5 supplies a contradiction and completes the proof.

Therefore $L = M_{24}$. Thus it suffices to prove:

Lemma 41.2: *Let $t \in T \cap Z(J(T))$ and assume G is of type M_{24}. Then*

(1) $C_G(t) \le G_3$.

(2) $C_G(t)$ has three orbits on $z^G \cap C_G(t)$ with representatives $z, z_2 \in Z(J(T))$ and $z_1 \in A_1 - A_2$.

Proof: Let $I = C_{G_3}(t)$. By 40.4.2, $I/A_2 \cong S_5$, $B = B(t) = C_{A_2}(I^\infty) \cong E_4$, and there is X of order 3 in G_3 with $N_{G_3}(B) = XI$ and $G_3 = A_2 N_{G_3}(X)$. Then XI acts faithfully on A_2/B as $GL_2(4)$ extended by a field automorphism, so in particular each involution in $A_2 - B$ is fused into zB under $C_I(X)$. From 40.1.1, $zB \cap T = \{a\}$. Let $s \in I \cap T - I^\infty$ be an involution and $b \in B - \langle t \rangle$. As $z \in Z(T)$, s centralizes z and $zB \cap T = \{a\}$, and hence also za. Therefore as $b^s = bt$, $za = t$, and then

$(ab)^s = abt$. Therefore I has two orbits on $z^G \cap A_2$ with representatives z and $z_2 = ab$, respectively.

Next each involution in $I^\infty - A_2$ is fused under I into $A_1 - A_2$. By 40.2, $|z^G \cap A_1| = 21$ and by 40.1, $|z^G \cap Z(J(T))| = 9$, so $|z^G \cap A_1 - A_2| = 12$. But $12 = |N_I(J(T)) : C_I(z_1)|$ for $z_1 \in z^G \cap A_1 - A_2$, so I is transitive on $z^G \cap I^\infty - A_2$.

Now $C_T(t)$ is transitive on the involutions in $C_T(t) - C_T(B)$ by Exercise 2.8, and hence each is fused to s. Further $|C_I(s)|$ is divisible by 3. But all involutions in H centralizing an element of H of order 3 or 7 are in Q so by 39.1.5 and 39.1.6, if $t \in T$, and $r \in z^G \cap C_G(t)$, then 21 is prime to $|C_G(\langle t, r\rangle)|$. Therefore $z^G \cap \langle s, t\rangle = \varnothing$, completing the proof of (2), modulo (1).

So it remains to prove (1). Let $M = C_G(t)$. Suppose $g \in M$ and $u, u^g \in z^G \cap T \cap M$. We claim $g \in I$. By (2), $u, u^g \in J(T)$, so $u, u^g \in A_1 \cup A_2$ by 39.1.3. If $u^g \in A_2$ then conjugating in I we may take $u^g \in Z(J(T)) = A_1 \cap A_2$. Now $u \in A_i$ for $i = 1$ or 2, and we saw in the previous paragraph that $C_M(u)$ is a 2-group, so A_i is weakly closed in $C_M(u)$. Similarly A_i is weakly closed in $C_M(u^g)$, so $g \in N_M(A_i)$. So as $N_M(A_i) \leq I$, our claim holds in this case. Thus we may assume $u, u^g \in A_1 - A_2$. But then again $g \in N_M(A_1) \leq I$.

So the claim is established. Then the claim and 7.3 show I is the unique point of M/I fixed by $u \in z^G \cap I$.

Let $K = C_I(B)$ and $C_G(B)^* = C_G(B)/B$. Then K^* is the unique point of $C_G(B)^*/K^*$ fixed by u^*. However, each involution in K^* is fused to z^* or z_1^*, so by 7.5, $K = C_G(B)$.

Finally let $M^* = M/\langle t\rangle$. Then I^* is the unique point of M^*/I^* fixed by $b^* = B^*$, z^*, and z_1^*, and each involution in A_2^* is fused to one of these involutions, so again 7.5 shows $M = I$ and completes the proof.

Let (x, l, π) be the flag of Γ whose member of type i is stabilized by G_i and more generally adopt the notation and terminology of Section 38. Thus, for example, Δ is the collinearity graph of Γ. Recall this is the graph whose vertices are the points of Γ and with points adjacent in Δ if they are incident with a common line of Γ.

By 41.1, Hypothesis ($\Gamma 0$) of Section 39 is satisfied by the pair G, Γ. We observed during the proof of 41.1 that hypothesis ($\Gamma 6$) of 38.8 holds.

By 41.1 and 4.11, $\Gamma(x_i) \cong \Gamma(G_i, \mathcal{F}_i)$ for x_i of type i. But we essentially saw the structure of G_i and $\mathcal{F}_i$ during the proof of 41.1. Namely we saw that $G_1/U_1 \cong L_4(2)$ with G_{12}/U_1 and G_{13}/U_1 the maximal parabolics stabilizing a point and line in some natural module for G_1/U_1. In particular $U_1 = Q(x)$. Similarly l has $|G_2 : G_{1,2}| = 3$ points and U_1 is transitive on the two points of l distinct from x, so hypothesis ($\Gamma 7$) of 38.8 holds. Hence by 38.8, G and Γ satisfy hypotheses ($\Gamma 1$), ($\Gamma 4$), and ($\Gamma 5$), so we can appeal to Theorem 38.3 to obtain a uniqueness system $\mathcal{U}$ for G.

Further as hypothesis ($\Gamma 1$) of Section 38 is satisfied, each pair of points is incident with at most one line. In particular if $y \neq x$ is a point incident with l then $G_{x,y} \leq G_l$.

Lemma 41.3: Let $I = G_{13}$. Then $I = Aut(I)$.

Proof: Let $A = Aut(I)$ and $I^* = I/A_2$. As $Z(I) = 1$ we can identify I with $Inn(I) \trianglelefteq A$. Notice $I^* = I_1^* \times I_2^*$ with $I_1^* \cong S_4$ and $I_2^* \cong S_3$. Also $J(T)$ is not normal in I, so by 39.1.3, $A_1 = J(O_2(I))$ char I. Let X be of order 3 in I with $X^* = O_3(I^*)$. Then $I_1^* = O^{2'}(C_{I^*}(X^*))$ and $I_2^* = C_{I^*}(I_1^*)$ are A-invariant, so as $S_k = Aut(S_k)$ for $k = 2, 3$, $A = IC_A(I^*)$. Then by a Frattini argument, $A = IB$, where $B = C_A(I^*) \cap N_A(X)$. As $N_I(X)$ is a complement to A_2 in I, $[B, N_I(X)] = 1$.

Next $U_1 \cap A_2 = C_{A_2}(O_2(I))$ is A-invariant and I_2 is absolutely irreducible on $U_1 \cap A_2$, so $B \leq C(U_1 \cap A_2)$. Further there is Y of order 3 in $N_{I_1}(X)$ with $C_{A_2}(Y) = U_1 \cap A_2$. Then B acts on $[Y, A_2]$ and as $N_I(Y) \cap N_I(X)$ is absolutely irreducible on $[Y, A_2]$, B centralizes $[Y, A_2]$. But now B centralizes $I = N_I(X)A_2$, so $B = 1$ and $A = I$.

Lemma 41.4: If $\bar{G}$ is of type M_{24} or $L_5(2)$ and the same type as G with uniqueness system $\bar{\mathcal{U}}$ then $\mathcal{U}$ is equivalent to $\bar{\mathcal{U}}$.

Proof: Let $I = G_{13}$. By 41.3, $Aut(I) = I$. Further if $x \neq y \in \Gamma_1(l)$ then $G_{x,y}$ is the pointwise stabilizer of $\Gamma_1(l)$, so from 40.5, $G_{x,y} = N_{G_1}(A_1)^\infty$ is the split extension of A_1 by $L_3(2)$.

We first use Theorem 37.11 to show $\mathcal{U}$ is similar to $\bar{\mathcal{U}}$. By 39.3, we have an isomorphism $\alpha : G_1 \to \bar{G}_1$ and by 39.1, $A_1 \alpha = \bar{A}_1$, so from the previous paragraph, $G_{x,y}\alpha = N_{G_1}(A_1)^\infty \alpha = N_{\bar{G}_1}(\bar{A}_1)^\infty = \bar{G}_{\bar{x},\bar{y}}$. Also $G_{1,3}$ is determined up to conjugation in G_1 by its isomorphism type, so we may choose α with $G_{1,3}\alpha = \bar{G}_{1,3}$.

Next either G is of type M_{24} and G_3 is the split extension of A_2 by $S_6/\mathbf{Z}_3$ or G is of type $L_5(2)$ and the split extension of A_2 by $L_3(2) \times S_3$. In particular in either case, G_3 is determined up to isomorphism, so we have an isomorphism $\zeta : G_3 \to \bar{G}_3$ and arguing as in the previous paragraph, we may pick ζ with $G_{1,3}\zeta = \bar{G}_{1,3}$ and $G_3(\{x,y\})\zeta = \bar{G}_3(\{\bar{x},\bar{y}\})$.

Thus to show $\mathcal{U}$ is similar to $\bar{\mathcal{U}}$, it suffices to verify Hypothesis V of Section 37. As $I = Aut(I)$, (V1)-(V3) are satisfied. Further $G_{\pi,x,y}$

and $G_{x,y}$ are the subgroups of G_3 and G_1 fixing $\Gamma_1(l)$ pointwise, so $N_{G_{1,3}}(G_{\pi,x,y}) = G_{1,2,3} \leq G_{1,2} = N_{G_1}(G_{x,y})$, so (V4) also holds.

Finally to complete the proof we appeal to Theorem 37.9. Namely $G_{x,y}$ is the split extension of A_1 by $L_3(2)$ and $G_{\pi,x,y}$ contains a Sylow 2-subgroup of $G_{x,y}$, so 38.5 says $Aut(G_{x,y}) \cap C(G_{\pi,x,y}) = 1$.

Lemma 41.5: *If G is of type $L_5(2)$ then $G \cong L_5(2)$, while if G is of type M_{24} then $G \cong M_{24}$.*

Proof: First if G is of type $L_5(2)$ then by 41.4, $\mathcal{U}$ is equivalent to the uniqueness system $\bar{\mathcal{U}}$ of $L_5(2)$, so the lemma follows from the discussion in Example 38.4.

So assume G is of type M_{24}; here we appeal to Exercise 13.2. Namely let $\bar{G} = M_{24}$ and $\bar{\mathcal{U}}$ the uniqueness system for $\bar{G}$. By 41.4, $\mathcal{U}$ is equivalent to $\bar{\mathcal{U}}$, so by Exercise 13.2, $G \cong \bar{G}$.

Notice that 41.5 shows:

Theorem 41.6: *Up to isomorphism $L_5(2)$, M_{24} are the unique groups of type $L_5(2)$, M_{24}, respectively.*

42. Groups of type *He*

In this section we continue the hypotheses and notation of Sections 39 and 40. In addition we assume G is of type *He*. Let $t \in Z(J(T)) \cap T$ and define $B = B(t) = C_{Z(J(T))}(O^2(C_G(t) \cap N_G(Z(J(T)))))$. Let $Y \in Syl_3(N_G(Z(J(T))))$, $L_i = N_G(A_i) \cap C_G(B)$, and $L = \langle L_1, L_2 \rangle$. By 40.4.2, $B \cong E_4$, $L_i/A_i \cong L_2(4)$, and $Y_i = C_Y(L_i/A_i) \cong \mathbb{Z}_3$ with $C_{L_i}(Y_i)$ a complement to A_i in L_i and Y acting on B. Let $l_i \in C_{L_i}(Y_i)$ be an involution acting on Y. As $Y_1 \neq Y_2$ but l_i inverts $Y_0 = C_Y(B)$, we have:

Lemma 42.1: *$\langle l_1, l_2 \rangle$ induces S_3 on Y, fixes Y_0, and is transitive on the remaining three subgroups Y_1, Y_2, Y_3 of Y of order 3.*

Lemma 42.2: *$C_{L_i}(Y_i) = C_G(Y_iB)$ for $i = 1$ and 2.*

Proof: Let $T_i = T \cap C_{L_i}(Y_i)$, so that $T_i \in Syl_2(C_{L_i}(Y_i))$. There is a conjugate v_i of z in $T_iB \leq A_{3-i}$ (cf. the discussion in paragraph one of the proof of 41.2) and as $C_G(\langle Y_i, v_i \rangle) \leq N_G(A_{3-i})$, we have $C_G(Y_iB\langle v_i \rangle) = T_i$. Now the lemma follows from Exercise 16.6 in [FGT].

Lemma 42.3: *$\langle l_1, l_2 \rangle \cong S_3$, so $l_1^{l_2} = l_2^{l_1}$.*

Proof: As l_1 and l_2 are involutions, $W = \langle l_1, l_2 \rangle \cong D_{2n}$ for some n. As W induces S_3 on Y, 3 divides n. Let $l = l_1 l_2$. It remains to show $l^3 = 1$.

But by 42.1, $l^3 \in C_G(YB) \leq C_{L_i}(Y_i) \cap C_G(Y) = Y_0$ by 42.2. So we may assume $Y_0 = \langle l^3 \rangle$.

Next by 40.1 there is an involution $v \in N_T(B)$ inverting Y. By 42.2, $C_G(YB) = Y_0$, so $[v, l] \in C_G(YB) = Y_0$. Thus there is an element k of order 3 in $lY_0 \cap C(v)$. Then k centralizes $C_B(v) = \langle t \rangle$, and hence also B. But $\langle l \rangle$ is a Sylow 3-group of $C_G(B) \cap N(Y)$, so $E_9 \cong \langle k, Y_0 \rangle = \langle l \rangle \cong \mathbf{Z}_9$, a contradiction.

Lemma 42.4: $L = L_1 \cup L_1 l_2 L_1 = L_2 \cup L_2 l_1 L_2$.

Proof: It suffices to show $L_1^{l_2} \subseteq L_1 \cup L_1 l_2 L_1$. Let $K = N_{L_1}(J(T))$. Then $K = B_0 B_1 B_2 Y_0$, where $B \leq B_i$, $B_0 = Z(J(T))$, $A_i = B_0 B_i$, and B_0 / B and B_i / B are the Y-invariant 4-subgroups of A_i / B. Thus l_i interchanges B_0 and B_i via conjugation.

Next $l_2 K l_1 l_2 = l_2 A_2 Y_0 B_1 l_1 l_2 = l_2 A_2 Y_0 l_1 B_0 l_2 = l_2 A_2 Y_0 l_1 l_2 B_2 = A_2 Y_0 l_1^{l_2} B_2 = A_2 Y_0 l_2^{l_1} B_2 \subseteq L_1 l_2 L_1$. Therefore $(K l_1 B_2)^{l_2} = l_2 K l_1 l_2 B_0 \subseteq L_1 l_2 L_1$.

Finally $L_i = K \cup K l_i B_{3-i}$. So $L_1^{l_2} = (K \cup K l_2 B_2)^{l_2} = K^{l_2} \cup (K l_1 B_2)^{l_2}$. We have seen $(K l_1 B_2)^{l_2} \subseteq L_1 l_2 L_1$. Also $K^{l_2} \subseteq L_2 = K \cup K l_2 B_1 \subseteq L_1 \cup L_1 l_2 L_1$. So the proof is complete.

Lemma 42.5: $L/B \cong L_3(4)$ with $B \leq L = L^\infty$.

Proof: First $B \leq L_i = L_i^\infty$, so as $L = \langle L_1, L_2 \rangle$, $B \leq L = L^\infty$. Let $\mathcal{F} = \{L_1, L_2\}$ and $\Gamma = \Gamma(L, \mathcal{F})$ the coset geometry of Example 4 in Section 4. The largest subgroup of $L_1 \cap L_2$ normal in L_1 and L_2 is B, so B is the kernel of the action of L on Γ. We show Γ is the projective plane P over $GF(4)$, and $|L/B| = |L_3(4)|$. But by 5.2, $Aut(P)^\infty \cong L_3(4)$. So $L = Aut(P)^\infty \cong L_3(4)$.

By 42.4, L is 2-transitive on the points and lines of Γ. Thus each pair of points is incident with a line and as $L_2 = \langle A_1, A_1^g \rangle$ for $g \in L_2 - N_L(L_1)$ while $A_1 \leq L_2^h$ for each $h \in L_1$, L_2 is the unique line incident with L_1 and $L_1 g$. Similarly each pair of lines is incident with a unique point. So Γ is a projective plane of order $|L_2 : L_1 \cap L_2| - 1 = 4$. Then as $L_1/B \leq Aut(\Gamma)$, $\Gamma = P$ by 18.6. Further there are 21 points of Γ as $\Gamma = P$, so $|L/B| = 21|L_1/B| = |L_3(4)|$.

Lemma 42.6: $L = C_G(B)$.

Proof: Let $M = C_G(B)$ and $M^* = M/B$. Assume $L \neq M$. As L^* is transitive on its involutions and $C_{M^*}(z^*) = C_M(z)^* \leq L^*$, L^* is strongly embedded in M^* by 7.3. But L^* has $9 \cdot 5 \cdot 7$ involutions but no subgroup of odd order divisible by $5 \cdot 9 \cdot 7$, contradicting 7.6.

Lemma 42.7: *(1)* $C_G(t) = L\langle u \rangle$, *where u is an involution in* $N_G(J(T)) - C_G(B)$.
(2) $C_G(t)$ *is transitive on involutions in* uL.
(3) $u \in T$.
(4) $N_G(B) = YL\langle u \rangle$ *and* $N_G(B)/B \cong P\Gamma_3(4)$.

Proof: Let $M = C_G(t)$, $M^* = M/\langle t \rangle$, $K = L\langle u \rangle$, and $B = \langle t, b \rangle$. Then $K = N_M(B)$ by 42.6, so $K^* = C_{M^*}(b^*)$. Next K^* has five classes of involutions with representatives b^*, z^*, u^*, b^*u^*, and b^*z^*. Notice bu is of order 4 while b, z, u, and bz are involutions. By 40.1, $\langle t, z \rangle$ and $\langle t, bz \rangle$ contain conjugates of z but B contains no such conjugate. Also $tu \in u^B$, so once we show $u \in T$, $\langle u, t \rangle$ contains no conjugate of z. Use the argument in paragraph three of the proof of 41.2 to see $u \notin z^G$.

We have shown z^* and b^*z^* are not conjugate to b^*, u^*, or b^*u^* in M^*. Also $C_M(z)$ and $C_M(bz)$ are contained in K. Finally z^* is not fused to b^*z^* in M^* by 7.7 as z^* and b^*z^* are in the center of the Sylow 2-subgroup $C_T(t)^*$ of M^*, but z^* is not fused to b^*z^* in $N_{M^*}(C_T(t)^*)$. Thus by 7.3, K^* is the unique point of M^*/K^* fixed by z^* and b^*z^*, and then $M^* = K^*$ by 7.5.

So (1) and (3) are established. Part (2) follows as $T \cap L$ is transitive on the involutions in $u(T \cap L)$ by Exercise 2.8. As Y is transitive on $B^\#$, $|N_G(B) : C_G(t)| = 3$ and $N_G(B) = YC_G(t)$. Also $N_G(B)$ acts on the projective plane Γ constructed during the proof of 42.5, with B the kernel of this action. So 5.2 completes the proof of (4).

It follows from 42.7 that B is a TI-set in G. Let $\Delta = B^G$ and form the graph Δ with vertex set Δ and A adjacent to C if $A \neq C$ and $[A, C] = 1$. We refer to the members of Δ as *root 4-subgroups* of G.

Lemma 42.8: *(1)* $\Delta(B)$ *is an orbit of L of length 105.*
(2) For $A \in \Delta(B)$, $\langle A, B \rangle = Z(J(S))$ for some $S \in Syl_2(G)$, AB/B is the root group of a transvection of L/B, and $\{A, B\} = AB \cap \Delta$.

Proof: Let $A \in \Delta(B)$. Then $A \leq C_G(B) = L$, so conjugating in L we may take $A \leq J(T) \in Syl_2(L)$. Then as A_1 and A_2 are the maximal elementary abelian subgroups of $J(T)$, $A \leq A_i$ for $i = 1$ or 2. Then as $N_G(A_i)$ controls fusion in A_i and is 2-transitive on $B^{N_G(A_i)}$, L is transitive on $\Delta(B)$ and we may take $AB = Z(J(T))$. Now by 40.1, $\{A, B\} = AB \cap \Delta$ and $|\Delta(B)| = |L : N_L(A)| = 105$. As $Z(J(T))/B$ is a root group in L/B, the proof is complete.

Lemma 42.9: *(1)* H *is transitive on $H \cap \Delta$ of order 42.*
(2) $T = J(T)Q$.

(3) $\langle t \rangle = B \cap Q$.

(4) H has two orbits on $T \cap H$ with representatives t and $b \in B - Q$.

Proof: By 42.7, $z^G \cap C_G(t) \subseteq L = C_G(B)$. Also $N_G(B)$ is transitive on $z^G \cap L$, while $C_G(t)$ has two orbits on $z^G \cap L$. Therefore H is transitive on $H \cap \Delta$ and has two orbits on $T \cap H$ with representatives $t, b \in B$. By 39.1.5, $t^H = \mathcal{T}_Q = T \cap Q$, so $b \in B - Q$ and (3) holds. Also $|\Delta \cap H| = |H : N_H(B)| = 2|H : N_H(Z(J(T)))| = 42$.

Next $J(T) = C_H(b)$ is of index 4 in T. But as $b \notin Q$, $|Q : C_Q(b)| \geq 4$, so (2) holds.

Let Ξ be the set of subgroups K of G with $K \cong L_2(4)$ and $Syl_2(K) \subseteq \Delta$.

Lemma 42.10: *Let A, C be distinct members of $\Delta(B)$. Then L is transitive on $\Xi \cap L$ of order $2^4 \cdot 3 \cdot 7$ and one of the following holds:*

(1) $C \in \Delta(A)$ *and ABC is L-conjugate to A_1 or A_2.*

(2) $\langle A, C \rangle \in J(T)^L$.

(3) $\langle A, C \rangle \in \Xi$ *is conjugate to a complement to A_i in $N_L(A_i)$.*

Proof: Let $L^* = L/B$. By 42.8.2, $\Delta(B)^*$ is the set of root groups of L^*. Hence by Exercise 14.1, $\langle A^*, C^* \rangle \in A_i^{*L}$ for $i = 1$ or 2, or $\langle A^*, C^* \rangle \in J(T)^{*L}$, or $\langle A^*, C^* \rangle \cong L_2(4)$ is conjugate to a complement to A_i^* in $N_{L^*}(A_i^*)$. Further there are $2^4 \cdot 3 \cdot 7$ subgroups of L^* of the last type, and L is transitive on such subgroups. As $\{A, B\} = AB \cap \Delta$ and $J(T) = \langle A, C \rangle$ for $A \in (\Delta \cap A_1) - A_2$ and $C \in (\Delta \cap A_2) - A_1$, the lemma follows.

We now calculate the order of G using the Thompson Order Formula. For $u = z$ or x, let $a(u)$ be the number of pairs (x, y) such that $x \in z^G$, $y \in \mathcal{T}$, and $u \in \langle xy \rangle$. By the Thompson Order Formula 7.2:

$$|G| = |C_G(z)|a(t) + |C_G(t)|a(z).$$

Thus we need to calculate $a(z)$ and $a(t)$. This is tedious but not difficult given the right approach; the details appear in the next two lemmas.

Lemma 42.11: $a(t) = 5 \cdot 7 \cdot 9 \cdot 481$.

Proof: Let $L\langle u \rangle^* = L\langle u \rangle / B$. Thus $L\langle u \rangle^*$ is $L_3(4)$ extended by the field automorphism u. Let $x \in z^G$ and $y \in \mathcal{T}$ with $t \in \langle xy \rangle$. By 7.1, $\langle x, y \rangle$ is dihedral and $t \in Z(\langle x, y \rangle)$, so $\langle x, y \rangle \leq C_G(t) = L\langle u \rangle$. We calculated during the proof of 42.7 that L has four orbits on $T \cap L$ with representatives $t, b \in B - \langle t \rangle$, $s \in A \in \Delta(B)$, and u, and L has two representatives on $z^G \cap L$ with representatives $z = ts$ and $z_1 = bs$. Observe that as $t \in \langle xy \rangle$, $t \neq y$.

If $y = b$ then as $b \in Z(L)$, xb is an involution, so $t = xb$. Then $x = tb \in B$, contradicting $x \in z^G$.

Suppose $y = s$. If $A = A^x$ then $(sx)^2 \in A$, so as $t \in \langle sx \rangle$, $sx = t$ and $x = st = z$. There are $|s^L| = 9 \cdot 5 \cdot 7$ choices for s and hence $9 \cdot 5 \cdot 7$ pairs (x, y) arise is this subcase.

If $A \neq A^x$ but $\langle A, A^x \rangle$ is abelian then sx is of order 4 but the involution in $\langle sx \rangle$ is $ss^x \neq t$.

If $\langle A, A^x \rangle \in J(T)^L$ then $\langle x, A \rangle$ is a 2-group and $\langle A, A^x \rangle \in Syl_2(L)$, so $x \in \langle A, A^x \rangle$. But then $\langle A, A^x \rangle$ is abelian.

By 42.10, when $y = s$ the only subcase remaining satisfies $K = \langle A, A^x \rangle \cong L_2(4)$. Then $x \in N_L(K) = BK$, and as $x \in z^G$, $x = cr$ for some $c \in B^\#$ and some involution $r \in K$. Notice $(xs)^2 = (rs)^2 \in K$, so as $t \in \langle xs \rangle$, $|rs|$ is odd and $c = t$. Now there are $16 \cdot 21$ choices for K by 42.10. Further there are fifteen choices for s in K and as $|rs|$ is odd, there are twelve choices for r given s. Thus there are $16 \cdot 21 \cdot 180$ pairs (x, y) in this subcase.

Finally we have the case $y = u$. Here $x = cr$, $c \in B^\#$, $r \in A \in \Delta(B)$. If $[x, u] = 1$ then $x = tu$. But $tu \in T$ by 42.7.2.

If $A = A^u$ then as $[x, u] \neq 1$, xu is of order 4 and $t = [x, u] = [c, u][r, u]$. Thus $[r, u] = 1$ and $c = b$ or bt. As u induces a field automorphism, $C_{L^*}(u^*) \cong L_3(2)$ has 21 involutions, so u acts on 21 root groups. That is, there are 21 choices for r and two choices b and bt for c, so there are 42 pairs (x, u). Also $|u^L| = 16 \cdot 15$, so there are $32 \cdot 9 \cdot 5 \cdot 7$ pairs (x, y) in this subcase.

If $A \neq A^u$ but $\langle A, A^u \rangle$ is abelian then xu is of order 4 but $t \neq [x, u]$.

Suppose $\langle A, A^u \rangle = J(T)$. Then $A \leq A_i$ and as $A_i \trianglelefteq T \geq \langle u, J(T) \rangle$, $\langle A, A^u \rangle \leq A_i$, a contradiction.

Finally suppose $\langle A, A^u \rangle = K \in \Xi$. Then rr^u is of order 3 or 5 and the involution t in $\langle xu \rangle$ is cc^u. Thus $c = b$ or bt and as $A \neq A^u$ there are six choices for r in K. So there are $12N$ pairs (x, y) in this subcase, where N is the number of pairs (v, M) with $M \in K^L$ and $v \in u^L \cap N_G(M)$. Then $N = 20|L \cap \Xi| = 20 \cdot 16 \cdot 21$.

We have shown that

$$a(t) = (9 \cdot 5 \cdot 7)(1 + 3 \cdot 64 + 32 + 2^8) = 9 \cdot 5 \cdot 7 \cdot 481.$$

Lemma 42.12: $a(z) = 4 \cdot 9 \cdot 7 \cdot 19$.

Proof: By 42.9, we may take $y \in B$. If $[x, y] = 1$ then $x = yz$. But if $y \in Q$ then $yz \in y^Q$, so $y \notin Q$. Thus $y = b$ or bt and $x = yz$. As there are 42 choices for B by 42.9.1, there are 84 pairs (x, y) in this subcase.

Suppose $x \in N_H(B)$ but $[x, y] \neq 1$. Then $\langle x, y \rangle \cong D_8$ but the involution in $\langle xy \rangle$ is in B and hence not z.

Suppose $x \in T$. Recall $Z(J(T)) = B \times B_1$ for some $B_1 \in \Delta$. By the previous paragraph, $B^x = B_1$. Thus $z = yy^x$, which forces $y \in Q$ and hence $y = t$. Now there are 64 involutions in $T - N_H(B)$ and 16 of these involutions are in T, so there are 48 choices for x. As there are 42 choices for B, we have $42 \cdot 48$ pairs (x, y) in this subcase.

Suppose $y = t$ but $x \notin T$. Then as $N_H(\langle t, z \rangle) = T$ and z is in $\langle xt \rangle$, $\langle t, t^x \rangle \cong D_8$. This is impossible as $[v, v^x] = 1$ for each involution $v \in Q$ and each involution $x \notin Q$.

So $y \notin Q$ and $x \notin T$. Let $H^* = H/Q$. Then $x^* y^*$ is of order 3 or 4. Suppose $|x^* y^*| = 4$. Then $y^x \in T$ and $z \in \langle y, y^x \rangle$. As $y \neq t$, we have a contradiction as in paragraphs two and three.

So $|x^* y^*| = 3$. Thus there is h of order 3 in $\langle xy \rangle$ and y inverts h. Now $C_H(h) = \langle h \rangle \times C_Q(h)$ with $C_Q(h) \cong D_8$ and $[y, C_Q(h)] = 1$. Further each involution in $yC_Q(h)$ is in z^G, so the three elements of $y\langle h \rangle$ are the three choices for elements of $H \cap T$ inverting h, and $x = zyh$ is determined by y and h. Finally there are $7 \cdot 2^7$ choices for h, so there are $2^7 \cdot 3 \cdot 7$ pairs (x, y) in this subcase.

We have shown that

$$a(z) = 84 + 42 \cdot 48 + 2^7 \cdot 3 \cdot 7 = (4 \cdot 21)(1 + 24 + 32) = 4 \cdot 21 \cdot 57 = 4 \cdot 9 \cdot 7 \cdot 19.$$

Lemma 42.13: $|G| = 2^{10} \cdot 3^3 \cdot 5^2 \cdot 7^3 \cdot 17$.

Proof: We apply the Thompson Order Formula 7.2:

$$|G| = |C_G(z)|a(t) + |C_G(t)|a(z).$$

We have $|C_G(z)| = 2^{10} \cdot 3 \cdot 7$ and $|C_G(t)| = 2^9 \cdot 3^2 \cdot 5 \cdot 7$. So

$$|G| = (2^{10} \cdot 3 \cdot 7)(5 \cdot 7 \cdot 9 \cdot 481) + (2^9 \cdot 3^2 \cdot 5 \cdot 7)(4 \cdot 9 \cdot 7 \cdot 19)$$

$$= 2^{10} \cdot 3^3 \cdot 5 \cdot 7^2 \cdot (481 + 6 \cdot 19)$$

$$= 2^{10} \cdot 3^3 \cdot 5 \cdot 7^2 \cdot 595,$$

as claimed.

Lemma 42.14: *Let X be of order 5 in L. Then*

(1) X is contained in a unique member K of Ξ.

(2) $C_G(X) = C_G(K) \in \Xi$.

(3) $N_G(KC_G(K))$ interchanges K and $C_G(K)$ and is of index 2 in $S_5 wr \mathbf{Z}_2$.

Proof: First $C_G(\langle t, X \rangle) = BX$, so by Exercise 16.6 in [FGT], either $C_G(X) \leq N_G(B)$ or $C_G(X)/X \cong L_2(4)$. But by 42.13, $X \leq P \in Syl_5(G)$ with $|P| = 25$, so the latter holds. Then $C_G(X) = X \times I$ with $B \leq I \cong L_2(4)$. Thus $I \in \Xi$.

Next $X \leq K \in L \cap \Xi$. Then there is an involution $k \in K$ inverting X. Notice k centralizes BR, where R is of order 3 in $N_G(B) \cap C_G(K)$. Then k acts on $I = C_G(X)^\infty$, so as $Aut(I) \cong S_5$ and $[k, BR] = 1$, $[k, I] = 1$. Then $I \leq C_G(k)^\infty \leq C_G(B(k))$, so I centralizes $\langle X, B(k) \rangle = K$. Now by symmetry, $K = C_G(X_I)$ for $X_I \in Syl_5(I)$. So as $N_G(X) = IN_G(X_I)$ by a Frattini argument, $N_G(X) \leq N_G(K)$. Hence as K is transitive on $Syl_5(K)$, K is the unique member of Ξ containing X.

Next by 42.8, $Z(J(T)) \cap \Delta = \{A, B\}$ and we may take $A \in Syl_2(K)$. So there is $g \in N_G(J(T))$ with $B^g = A$ and $g^2 \in N_G(B)$. Now $I^g \in L \cap \Xi$ and as $N_G(B) \cap N_G(B^g)$ is transitive on $\{D \in L \cap \Xi : B^g \leq D\}$, we may take $I^g = K$. Then as $K = C_G(I)$, $K^g = C_G(I^g) = C_G(K) = I$, so g interchanges I and K. We conclude (3) holds as $|N_G(BX)| = 240$.

Lemma 42.15: *(1) Up to conjugation in L there is a unique subgroup S of L with $S \cap T$ nonempty and $S \cong S_3$.*
(2) $C_G(S) \cong S_5$ with $C_G(S)^\infty \in \Xi$.

Proof: By Exercise 14.2, up to conjugation in $L^* = L/B$ there is a unique subgroup $S_0^* \cong S_3$. Further $C_{L^*}(S_0^*) = 1$. Then $S_0 = B \times S$ with $S \cong S_3$ and if s is an involution in S then sB contains a unique member of T which we take to be s. Hence (1) is established.

Let $K \in L \cap \Xi$. Then K contains an S_3-subgroup and the involutions in K are in T, so we may take $S \leq K$. By 42.14, $C_G(K) = I \in \Xi$. Let $M = C_G(S)$, so that $I \leq M$. Now $C_G(S\langle t \rangle) = B\langle u \rangle \cong D_8$, with $u \in T$ inducing a field automorphism on L^*. Thus $I\langle u \rangle \cong S_5$. Then $C_I(u) \cong S_3$, so as $C_G(S\langle t \rangle)$ is a 2-group, $u \notin t^M$. Hence by a transfer argument (cf. 37.4 in [FGT]) $u \notin O^2(M)$. Then $B = C_{O^2(M)}(t)$, so $O^2(M) = I$ by Exercise 16.6 in [FGT]. Thus (2) holds.

Lemma 42.16: *Let X be of order 3 in L. Then*
(1) $N_G(X)/X \cong S_7$ with $C_G(X) = C_G(X)^\infty$.
(2) In the representation of $N_G(X)$ on $\{1, \dots, 7\}$, transpositions and products of two distinct commuting transpositions are in T, while products of three distinct commuting transpositions are in z^G.

Proof: Let $M = N_G(X)$ and $M^* = M/X$. Then $C_{M^*}(t^*) \cong D_8 \times S_3$. Let $s \in T \cap L$ with $X\langle s \rangle \cong S_3$. By 42.15, $C_{M^*}(s^*) \cong Z_2 \times S_5$.

Next we may take $XD_1 \trianglelefteq N_G(D_1)$ for some conjugate D_1 of A_1 with

$\langle s, B \rangle \leq N_M(D_1) = M_1$. Then $M_1 \cong S_6/Z_3$ with $X \leq C_{M_1}(X)^\infty$ by 40.4. Let $v = st$ and $R \in Syl_2(C_M(\langle s, t \rangle))$. Then $v \in z^G$ and $C_{M_1}(v)^* \cong Z_2 \times S_4$ with $O_2(C_{M_1}(v)) \leq O_2(C_G(v))$. Hence $C_{M_1}(v)$ is maximal among subgroups K of $C_G(v)$ with $R \in Syl_2(K)$. But $\Phi(R) = \langle t \rangle$, so as $R \in Syl_2(C_M(t))$, we conclude $R \in Syl_2(M)$. Hence $C_{M_1}(v) = C_M(v)$.

Therefore t^*, s^*, v^* are representatives for the conjugacy classes of involutions of M^*. Further we are in a position to apply the Thompson Order Formula 7.2. Namely the parameters in the order formula for M^* are the same as those for S_7 since the involution centralizers and fusion pattern in M^* are the same as in S_7. Therefore by the Order Formula, $|M^*| = |S_7|$. Then as $|M^* : M_1^*| = 7$ and $M_1^* \cong S_6$, we conclude $M^* \cong S_7$. Further in the representation of M^* on $\{1, \ldots, 7\}$, t^* is the product of two transpositions as $C_{M^*}(t^*) \cong S_3 \times D_8$, s^* is a transposition as $C_{M^*}(s^*) \cong Z_2 \times S_5$, and v^* is the product of three transpositions as $C_{M^*}(v^*) \cong Z_2 \times S_4$. So (2) holds.

43. The root 4-group graph for He

In this section we continue the hypotheses and notation of Section 42. In particular Δ is the commuting graph on the set of root 4-subgroups of G. We will show that Δ is 4-generated in the sense of Section 34. Throughout this section let $M = N_G(B)$.

Lemma 43.1: M has seven orbits on Δ:

(1) $\{B\}$.

(2) $\Delta(B)$ of order 105.

(3) $\Delta_i^2(B) = \{A \in \Delta : \langle A, B \rangle = J(S), S \in Syl_2(G), B \leq D \leq S, D \in A_i^G\}$, $i = 1, 2$, of order $8 \cdot 105$. Further $\Delta(A, B) = \Delta \cap Z(J(S))$ is of order 2.

(4) $\Delta_3^2(B) = \{A \in \Delta : \langle A, B \rangle \in \Xi\}$ of order $2^6 \cdot 21$. Further $\Delta(A, B) = Syl_2(I)$ is of order 5, where $I = C_G(\langle A, B \rangle) \in \Xi$.

(5) $\Delta_1^3(B) = \{A \in \Delta : \langle A, B \rangle \cong S_4\}$ of order $2^4 \cdot 3^2 \cdot 5$. Further $N_M(A) = N_A(B) \times N_B(A) \times L_3(2)$.

(6) $\Delta_2^3(B)$ consisting of those $A \in \Delta$ such that $\langle A, B \rangle = C_G(X)$ for some X of order 3, and of order $2^7 \cdot 35$. Further $N_M(A) = N_M(R) \cong E_4/3^{1+2}$ for some $R \in Syl_3(G)$.

Proof: By 42.8, $\Delta(B)$ is an orbit of M of length 105.

Next by 42.8, $\Delta \cap (A_i - Z(J(T)))$ is of order 4 and by 42.10, if $A \in \Delta \cap (A_i - Z(J(T)))$ and $C \in \Delta \cap (A_{3-i} - Z(J(T)))$ then $\langle A, C \rangle = J(T)$

and A is regular on $\Delta \cap (A_{3-i} - Z(J(T)))$. So G is transitive on the set S of pairs (A, C) from $\Delta \times \Delta$ with $A^g \le A_i$ and $\langle A, C \rangle^g = J(T)$, with $16|G : N_G(T)| = |S| = |\Delta|N_i^2$, where $N_i^2 = |\Delta_i^2(B)|$. Thus

$$N_i^2 = 16(3 \cdot 5^2 \cdot 7^3 \cdot 17)/(2 \cdot 5 \cdot 7^2 \cdot 17) = 8 \cdot 105,$$

and M is transitive on $\Delta_i^2(B)$. As $C_G(J(T)) = Z(J(T))$, $\Delta(A, C) = \Delta \cap Z(J(T))$ is of order 2 by 42.8.

Next if $A \in \Delta_3^2(B)$ then by 42.14, $C_G(\langle A, B \rangle) = K \in \Xi \cap L$ and $\langle A, B \rangle = C_G(K)$. So as L is transitive on $\Xi \cap L$ by 42.10, and as B is regular on $\langle A, B \rangle \cap \Delta - \{B\}$, L is transitive on $\Delta_3^2(B)$. Further $|\Delta_3^2(B)| = 4|\Xi \cap L| = 2^6 \cdot 21$ by 42.10.

By 42.7 there is $u = t^g \in C_G(t) - L$. Then B acts on $I = \langle u, t \rangle$ and $t \in C_G(u) \le N_G(B^g)$, so $[B^g, t] = \langle u \rangle$ and hence B^g acts on I. Thus $\langle B, B^g \rangle$ acts on I and from that action we conclude $\langle B, B^g \rangle \cong S_4$. That is, $B^g \in \Delta_1^3(B)$. Conversely if $A \in \Delta_1^3(B)$ then $N_A(B) = \langle v \rangle$ is of order 2 and $C_B(v) = \langle c \rangle$ for some $c \in B^\#$. Conjugating in M we may take $c = t$. Then as $[v, B] \ne 1$, $v \notin L$, so by 42.7.2, $v \in u^L$. Hence $A = B(v)$ is L-conjugate to B^g, so $\Delta_1^3(B)$ is an orbit under M. Further $|\Delta_1^3(B)| = |u^M| = 2^4 \cdot 3^2 \cdot 5$ and $N_M(B^g) = C_M(u) = \langle u \rangle \times \langle t \rangle \times L_3(2)$ as u induces a field automorphism on L/B (cf. 42.7.4).

By Exercise 14.2, L is transitive on its subgroups X of order 3 and by 42.16, $N_G(X)/X \cong S_7$. Represent $I = N_G(X)$ on $\Sigma = \{1, \ldots, 7\}$ and let $I^* = I/X$. Then B^* is the 4-group in I^* moving some set $\Sigma(B)$ of order 4. Further $C_G(X) = \langle A, B \rangle$ for some $A \in \Delta \cap I$ if and only if $\Sigma(A) \cup \Sigma(B) = \Sigma$. Then $N_M(A) = N_{I \cap M}(A) \cong E_4/3^{1+2}$ and B is regular on $\Delta_2^3(B) \cap I$. Hence $|\Delta_2^3(B)| = 4|X^M| = 2^7 \cdot 35$ and M is transitive on $\Delta_2^3(B)$.

To complete the proof, observe that the sum of the orders of the seven orbits listed in 43.1 is $8,330 = 2 \cdot 5 \cdot 7^2 \cdot 17 = |G : M| = |\Delta|$, so these are all the orbits of M on Δ.

Lemma 43.2: *All triangles in Δ are fused under G into A_1 or A_2.*

Proof: Let ABC be a triangle. We may take $AB = Z(J(T))$ by 42.8. Then $C \le C_G(AB) = J(T)$, so as A_1 and A_2 are the maximal elementary abelian subgroups of $J(T)$, $ABC \le A_i$ for $i = 1$ or 2.

Lemma 43.3: $\Delta^2(B) = \bigcup_{i=1}^3 \Delta_i^2(B)$.

Proof: Let $A \in \Delta$. If $A \in \Delta_i^2(B)$ then $\Delta(A, B) \ne \varnothing$ by 43.1 and hence $\Delta_i^2(B) \subseteq \Delta^2(B)$. Conversely assume $d(A, B) = 2$ and let $C \in \Delta(A, B)$. Then CB/B is a root group of L/B. But by 43.1, $C_{L/B}(A)$ contains a root group of L/B only if $A \in \Delta_i^2(B)$ for $1 \le i \le 3$.

Lemma 43.4: *Each square in Δ is fused under G into KI, where K, I are a pair of commuting members of Ξ.*

Proof: If $p = B_0 \cdots B_4$ is a square then $B_2 \in \Delta^2(B_0)$ and $B_1, B_3 \in \Delta(B_0, B_2)$, with $B_1 \in \Delta^2(B_3)$, so the remark follows from 43.1 and 43.3. For example, if $B_2 \in \Delta_i^2(B_0)$ for $i = 1$ or 2, then by 43.1.3, $\Delta(B_0, B_2) = \{A, B\}$ with $A \in \Delta(B)$ rather than $A \in \Delta^2(B)$.

Lemma 43.5: *If $A \in \Delta_i^2(B)$ for $i = 1$ or 2, then $B \in \Delta_{3-i}^2(A)$.*

Proof: By 43.2, $\langle A, B \rangle = J(S)$ for some $S \in Syl_2(G)$ and $B \leq D \leq S$ with $D \in A_i^G$. But then $A \in D' \leq S$ with $D' \in A_{3-i}^G$.

Remark 43.6. We sometimes view M/B as acting as $P\Gamma L_3(4)$ on its natural 3-dimensional projective module W over $GF(4)$. Observe that in this representation, each $X \in \Xi \cap L$ is a characteristic subgroup of the stabilizer of some decomposition $W = c(X) \oplus a(X)$ of W as the sum of a point $c(X)$ and a line $a(X)$.

Similarly each $A \in \Delta(B)$ is the root group of transvections determined by an incident-point–line pair $(e_1(A), e_2(A)) = (c(A), a(A))$. Further we may choose notation so that A_1 is the group of transvections of V with a common center and A_2 is the group of transvections with a common axis. Then $C \in \Delta(A)$ if and only if $c(C) = c(A)$ or $a(C) = a(A)$ and in the first case $ABC \in A_1^G$, while in the second $ABC \in A_2^G$.

Notice $A \leq X$ if and only if $c(A) \in a(X)$ and $a(A) = c(X) + c(A)$.

Finally $C \in \Delta_i^2(A)$ if and only if $e_i(A) \in e_{3-i}(C)$ but $e_i(C) \notin e_{3-i}(A)$.

Lemma 43.7: *(1) $\Delta^3(B) = \Delta_1^3(B) \cup \Delta_2^3(B)$.*

(2) Δ is of diameter 3.

Proof: Observe that 43.1, 43.3, and (1) imply (2), so it remains to prove (1). Further $d(B, B') \geq 3$ for $B' \in \Delta_i^3(B)$, $i = 1, 2$, by 43.3, and by 43.1, $\Delta_i^3(B)$ is an orbit of $N_G(B)$ on Δ, so it suffices to exhibit $A \in \Delta_i^3(B)$ with $d(A, B) \leq 3$.

Now by 43.8, $N_L(A_1)$ is the split extension of A_1 by $I \in \Xi$. Next by 40.5,

$$O^2(N_G(A_1))^* = O^2(N_G(A_1))/O_{2,3}(N_G(A_1)) \cong A_6.$$

Then $O^2(N_G(A_1))^*$ is represented as A_6 on a set Ω of six points and if $C_1 \in \Delta \cap I$ and $C_2 \in C_1^{N(A_1)}$ such that $\langle C_1, C_2 \rangle$ fixes no point of Ω, then $\langle C_1, C_2 \rangle^* \cong S_4$. Hence by 43.1, $C_2 \in \Delta_1^3(C_1)$. Therefore there exist B-invariant $D \in A_1^G$ and $A \in N_G(D) \cap \Delta_1^3(B)$. Further AB_1B_2B is a path in Δ, where $B_1 \in \Delta \cap C_D(A)$ and $B_2 \in \Delta \cap C_D(B)$. This shows $d(A, B) \leq 3$ for $A \in \Delta_1^3(B)$.

Let $A \in \Delta_3^2(B)$ and adopt the convention of Remark 43.6. Thus $BN_M(A)$ is the stabilizer of a nonincident-point–line pair (p, l) of W. Then one orbit of $N_M(A)$ on $\Delta(B)$ consists of those C with $(c(C), a(C)) = (q, k)$, where $q \neq p$, $k \neq l$, $q \notin l$, and $p \notin k$. Then $Y = N_M(A) \cap N_M(C) \cong S_3$ with $O_3(Y) = X$ having eigenspaces W_1 and W_2 on W of dimension 1 and 2, respectively. Then $X \not\leq L$, so X is faithful on B and $q \leq W_2$ and $W_1 \leq k$, so X centralizes C. We claim $C \in \Delta_2^3(A)$; then as $d(A, C) \leq 3$, the lemma is established.

As $A \in \Delta_3^2(B)$ and X is faithful on B, X is also faithful on A. As $C \not\leq N_M(A)$, $C \notin \Delta(A)$. As $[X, C] = 1$ and $A = [A, X]$, $C \notin \Delta_1^3(A)$. Finally we check using the representation of $N_G(C)$ on $\Delta(C)$ supplied by Remark 43.6 that if $A \in \Delta^2(C)$, then there exists no $B \in \Delta(C)$ such that $Y = N_M(A) \cap N_G(C) \cong S_3$ with $O_3(Y) \leq C_G(C)$.

In the remainder of this section let $\sim \; = \; \sim_{\mathcal{C}_4(\Delta)}$ be the invariant equivalence relation on the paths of Δ generated by $\mathcal{C}_4(\Delta)$ (cf. Chapter 12).

Lemma 43.8: *Let $p = B_0 \cdots B_4$ be a path with $B_2 \in \Delta_3^2(B_0)$. Assume there exists no path q of length 3 with $p \sim q$. Then*

(1) $B_4 \in \Delta_j^2(B_2)$ *for $j = 1$ or 2.*

(2) *For $C \in \Delta(B_2, B_4)$ and $X = C_G(\langle B_0, B_2 \rangle)$, $c(C) \notin a(X)$ and $c(X) \notin a(C)$.*

(3) *There exist three $P \in \Delta_3^2(B_3) \cap X$ with $B_0 P B_2 B_3 B_4 \sim p$.*

Proof: Without loss of generality $B = B_2$; we adopt the notation of Remark 43.6. As p is equivalent to no path of length 3, 34.10.1 says $B_{i+2} \in \Delta^2(B_i)$ for $0 \leq i \leq 3$. Similarly by 34.10.2, $d(\Delta(B_0, B), \Delta(B_4, B)) > 1$.

For $i = 0, 4$, let $X_i = C_L(B)_i$, and if $B_i \in \Delta_3^2(B)$ let $c_i = c(X_i)$ and $a_i = a(X_i)$. For example, by hypothesis, $B_0 \in \Delta_3^2(B)$. Assume $B_4 \in \Delta_3^2(B)$ and let c be a point on $a_0 \cap a_4$ and $C_i \in X_i \cap \Delta$ with $c(C_i) = c$. Then by Remark 43.6, $C_4 \in C_0^\perp$, contradicting $d(\Delta(B, B_0), \Delta(B, B_4)) > 1$.

So $B_4 \in \Delta_j^2(B)$ for $j = 1, 2$, and without loss of generality $j = 1$ and $\Delta(B, B_4) = \{C, D\}$ with $BCD = A_1$. Notice $B_3 = C$ or D by 43.1. As $d(C, \Delta(B, B_0)) > 1$, $c(C) \notin a_0$ and $c_0 \notin a(C)$. Thus (1) and (2) are established.

Next let $P \in \Delta \cap X_0$. If $c(B_3) \notin a(P)$ and $a(P) \notin c(B_3)$ then $P \in \Delta_3^2(B_3)$ and $p \sim B_0 P B B_3 B_4$. But by (2) and as $a(P) = c_0 + c(P)$ with $c(P) \in a_0$, there exists a unique $P \in \Delta \cap X_0$ with $c(B_3) \in a(P)$; namely that P with $c(P)$ the projection of $c(B_3)$ on a_0. Similarly the unique $P' \in \Delta \cap X_0$ with $c(P) \in a(B_3)$ satisfies $c(P) = a(B_0) \cap a_0$. So (3) holds.

In the remainder of this section assume Δ is not 4-generated, in the sense of Section 34. Thus there exists a nontrivial cycle $p = B_0 \cdots B_n$. Choose p with n minimal subject to this constraint. Then by 34.4, p is an n-gon.

Lemma 43.9: $B_{i+2} \in \Delta^2_{j(i)}(B_i)$ *with* $j(i) = 1$ *or* 2 *for each* i.

Proof: Assume not. Then without loss of generality $B_2 \in \Delta^2_3(B_0)$. Now by minimality of n and 34.10.3, $B_0 \cdots B_4$ is not equivalent to any path of length 3, so by 43.8, $B_4 \in \Delta^2_j(B_2)$ for $j = 1$ or 2, and $|\Delta^2_3(B_3) \cap \Delta(B_0, B_2)| \geq 3$. By symmetry, $B_{n-2} \in \Delta^2_k(B_0)$ for $k = 1$ or 2, and $|\Delta^2_3(B_{n-1}) \cap \Delta(B_0, B_2)| \geq 3$. Hence as $|\Delta(B_0, B_2)| = 5$, there exists $P \in \Delta^2_3(B_{n-1}, B_3) \cap \Delta(B_0, B_2)$. Now by 34.10.4, $p \sim B_0 P B_1 \cdots B_n$, so without loss of generality $P = B_1$. But then as $P \in \Delta^2_3(B_{n-1}, B_3)$, 43.8.1 supplies a contradiction.

Lemma 43.10: *(1)* $j(i) = 3 - j(i+1)$ *for all* $0 \leq i \leq n$.
(2) n *is even.*
(3) $n > 7$.

Proof: Without loss of generality $B_2 = B$. By 43.9, we can assume $B \in \Delta^2_2(B_0)$; that is, $j(0) = 2$. Then by 43.5, $B_0 \in \Delta^2_1(B)$. Then $\langle B, B_0 \rangle = J(S)$ for some $S \in \mathrm{Syl}_2(G)$ and $Z(J(S)) = B_1 B'_1$ with $\{B_1, B'_1\} = \Delta(B, B_0)$ and $BB_1 B'_1 \in A^G_1$.

Similarly $\langle B, B_4 \rangle = J(R)$ and $Z(J(R)) = B_3 B'_3$ with $BB_3 B'_3 \in A^G_i$ for $i = 1$ or 2. Finally $\langle B_1, B_3 \rangle = J(T)$.

We claim $J(T) = B_1 B'_1 B B_3 B'_3$. Assume not; then without loss of generality $B'_1 \not\leq T$. Then as $BB_1 B'_1$ is the member of A^G_1 containing BB_1, $B_1 \leq A_2$, so $B_3 \in \Delta^2_2(B_1)$. Similarly $B_3 \in \Delta^2_2(B'_1)$. That is, $c(B_3) \in a(B_1)$ and $c(B_1) \notin a(B_3)$. But $p \sim B_0 B'_1 B_2 \cdots B_n$, so also $\langle B'_1, B_3 \rangle = J(T')$ and as $B_3 \in \Delta^2_2(B'_1)$, $B_1 \not\leq T'$, so $c(B_3) \in a(B'_1)$. Therefore $c(B_3) \in a(B_1) \cap a(B'_1) = c(B_1)$, so $c(B_3) = c(B_1)$, contradicting $[B_1, B_3] \neq 1$.

Hence the claim is established. In particular $Z(\langle B, B_0 \rangle)B = A_1 = Z(\langle B_1, B_3 \rangle)B_1$ and $\Delta(B_0, B) \subseteq Z(\langle B, B_0 \rangle) \subseteq J(T)$. By symmetry $\Delta(B, B_4) \subseteq J(T)$, so $\langle B_1, B_3 \rangle = J(T) = \langle \Delta(B_0, B_2), \Delta(B_2, B_4) \rangle$.

Now as $Z(\langle B_1, B_3 \rangle)B_1 = A_1$, $j(1) = 1$. That is, (1) holds. Also by (1), $j(2r) = j(0) = 3 - j(1) = 3 - j(2s+1)$ for all integers r, s, so (2) holds.

Next let $A'_2 = B_0 B_1 B'_1$ be the member of A^G_2 in S and $Y = \langle A'_2, A_2 \rangle$. Then $A_1 \trianglelefteq Y$ and $A_1 \cap A_2$ and $A_1 \cap A'_2$ are of index 4 in A_1, so $E = A_2 \cap A_1 \cap A'_2$ is of order at least 4 and contained in $Z(Y)$. Then from the structure of $N_G(A_1)$, $Y = C_G(E) \cap N_G(A_1)$ is the split extension of A_1 by $\langle B_0, B_3 \rangle \cong S_4$ and $E^\# \subseteq Z^G$.

As $n \geq 5$ is even, to prove (3) and complete the proof of the lemma, we may assume $n = 6$. Notice $Y = \langle B_i : 0 \leq i \leq 3 \rangle$. Also $K = C_G(\langle B_0, B_3 \rangle) \cong L_3(2)$ by 43.1.5 and E is a 4-subgroup of K. As $E^\# \subseteq Z^G$, we may assume $z \in E$. Then as 3 divides the order of $Y \leq C_G(E)$, $E \leq Q$. Let $H^* = H/Q$. By 42.9, $B_i^* \cong \mathbf{Z}_2$ for all i, $0 \leq i \leq 3$, so $Y^* \cong S_4$. Then from the structure of H, $|Y| = |C_H(E)|$, so $Y = C_H(E) = C_G(E)$. As $A_2 \leq Y$ and A_2 is weakly closed in $N_G(A_2)$, there is no Y-invariant member of A_2^G. But $\bar{Y} = \langle B_3, B_4, B_5, B_0 \rangle$ satisfies the hypotheses of Y except that there is a member of A_2^G normal in $\bar{Y}$, so $\bar{E} = Z(\bar{Y})$ must live in the second class of 4-groups of K. In particular if $r = B_3 C_4 C_5 B_0$ is a path such that $E' = Z(\langle r \rangle) \in \bar{E}^K$, then $p' = B_0 B_5 B_4 B_3 r \sim 1$. Thus to complete the proof it suffices to show we can choose r subject to this constraint such that $B_0 \cdots B_3 \cdot r \sim 1$. Indeed we will show that this holds when $Z(\langle r \rangle) \cap E = 1$. Toward that end we let $G_2 = N_G(A_2)$ and show:

(a) $N_G(J(S)) \cap N_G(B)$ is transitive on those $D \in A_1^G$ with $B_3 \leq D \leq N_G(A_2)$ and $\langle A_1, D \rangle / A_2 \cong S_4$.

(b)   There exists a hexagon $q = C_0 \cdots C_6$ with $C_{i+2} \in \Delta_{j(i)}^2(C_i)$, $C_{i+3} \in \Delta_1^3(C_i)$ for $0 \leq i \leq 6$ and $\langle q \rangle = H$.

Now if q satisfies (b) then conjugating in H we may take $C_i = B_i$ for $0 \leq i \leq 3$, and then the hypotheses of (a) are satisfied with $D_q = Z(J(T))C_4$. Similarly our typical nontrivial hexagon p satisfies the hypotheses of (a) with $D_p = Z(J(T))B_4$, so by (a), $\langle p \rangle = \langle D_p, J(S) \rangle \in \langle D_q, J(S) \rangle^G = \langle q \rangle^G = H^G$. Therefore we have shown that if p is nontrivial then $E \cap \bar{E} \neq 1$, to complete the proof.

So it remains to establish (a) and (b). For (b), let $H^* = H/Q$ and $q^* = C_0^* \cdots C_6^*$ a hexagon in the commuting graph of involutions of H^*. Let q be any lift of q^* in H to Δ with $C_0 = C_6$.

Let $G_2^* = G_2/O_{2,3}(G_2)$ and observe that G_2^* acts on $A_1^{G_2}$ as $Sp_4(2)$ on the fifteen points of its symplectic space with $\langle A_1^*, D^* \rangle \cong S_4$, A_5, for D orthogonal, not orthogonal to A_1 in this space, respectively. Further $N_{G_2}(B_3)^* \cong S_5$ has two orbits of length 5, 10 on $A_1^{G_2}$ with the set θ of members of $A_1^{G_2}$ containing B_3 of order 5. So as $B_3 \not\leq A_1$ and $N_{G_2}(A_1) = N_{G_2}(J(T))$, a subgroup X of $C_{G_2}(B_3) \cap N_G(J(T))$ of order 3 is transitive on the three members D of θ with $\langle D^*, A_1^* \rangle \cong S_4$. But as $[B_3, X] = 1$, $[X, N_G(A_1)^\infty] \leq A_1$, so $X \leq N_G(J(S))$.

Lemma 43.11: Δ *is 4-generated.*

Proof: By 43.7.2 and 34.5, $n \leq 7$. Then 43.10.3 supplies a contradiction.

44. The uniqueness of groups of type He

In this section we continue the hypotheses and notation of Section 42. Our aim is to show that, up to isomorphism, there is at most one group of type He. Let $G_i = N_G(A_i)$ and $G_3 = M = N_G(B)$. Let $B \leq K_1 \in \Xi$ and $K_2 = C_G(K_1)$. Choose K_1 so that $Z(J(T)) \in Syl_2(K_1 K_2)$. Let $G_4 = N_G(K_1 K_2)$. Finally let $I = \{1, \dots, 4\}$ and $\mathcal{F} = (G_i : i \in I)$ and $\Gamma = \Gamma(G, \mathcal{F})$.

Lemma 44.1: *Up to isomorphism there is at most one quasisimple group L with $Z(L) \cong E_4$ and $L/Z(L) \cong L_3(4)$.*

Proof: Let $\tilde{L}$ be the universal covering group of $L_3(4)$ and $\tilde{Z} = Z(\tilde{L})$ (cf. Section 33 in [FGT]). Let $\tilde{Z}_0 = \{x^2 : x \in \tilde{Z}\}$ and $\hat{L} = \tilde{L}/\tilde{Z}_0$. Then $\hat{L}$ is the largest perfect central extension of $L_3(4)$ whose center $\hat{Z}$ is an elementary abelian 2-group, so there is a surjection of $\hat{L}$ onto L. Thus it suffices to show $|\hat{Z}| \leq 4$.

Let $I/\hat{Z}$ be a maximal parabolic of $\hat{L}/\hat{Z}$. Then I contains a Sylow 2-subgroup of $\hat{L}$, so by 33.11 in [FGT], I is a perfect central extension of $I/\hat{Z}$. Let $E = O_2(I)$. Then $\bar{E} = E/\hat{Z}$ is the natural module for $I^* = I/E \cong L_2(4)$. As $\bar{E}$ is the natural module for I^*, I is transitive on $\bar{E}^\#$, so each element in $E^\#$ is an involution, and hence E is elementary abelian. Hence there are involutions in $I - E$, so as I is perfect, $E = [E, I]$.

Thus $|\hat{Z}| \leq |H^1(\bar{I}, \bar{E})|$ by 17.12 in [FGT]. But as a $GF(4)$-module, $1 = dim(H^1(\bar{I}, \bar{E}))$ by Exercise 14.6, so indeed $|\hat{Z}| \leq |H^1(\bar{I}, \bar{E})| = 4$.

See Example 36.1 for the definition of the amalgam $\mathcal{A}(\mathcal{F})$ of the family $\mathcal{F}$ of subgroups of G.

Lemma 44.2: *If $\bar{G}$ is a group of type He then the amalgams $\mathcal{A}(\mathcal{F})$ and $\mathcal{A}(\bar{\mathcal{F}})$ are isomorphic.*

Proof: By 44.1, L is determined up to isomorphism, so as M is the split extension of L by S_3 with the involution inducing a field automorphism on L/B, there is an isomorphism $\alpha_3 : M \to \bar{M}$. Similarly for $i = 1, 2$, G_i is the split extension of A_i by $S_6/\mathbf{Z}_3$ and in particular is determined up to isomorphism. So there exist isomorphisms $\alpha_i : G_i \to \bar{G}_i$ for $i = 1, 2$. Finally by 42.14, G_4 is a determined subgroup of index 2 of the wreath product $L_2(4)wr\mathbf{Z}_2$, so there is an isomorphism $\alpha_4 : G_4 \to \bar{G}_4$.

Let $I = N_G(J(T))$ and $E = Z(J(T))$; notice $I = G_{12}$. Now $I/E = I_1/E \times I_2/E$, where $I_i/E \cong S_4$. So as $S_4 = Aut(S_4)$, the subgroup D of $Aut(I)$ normalizing I_1 factors as $D = IA$, where $A = C_{Aut(I)}(I/E)$.

Next for $X \in Syl_3(I)$, $X = X_1 \times X_2$ with $X_i \in Syl_3(I_i)$ and $E = [E, X_i]$. Then $E = I \cap A$ and $A = EC_A(X)$ with $C_A(X) = C_A(N_I(X)) \le C_A(E)$ as $N_I(X)$ is absolutely irreducible on E. Similarly $[C_{J(T)}(X_i), C_A(X)] \le C_E(X_i) = 1$ as $N_I(X)$ is absolutely irreducible on $C_{J(T)}(X_i)$. Hence as $J(T) = EC_{J(T)}(X_1)C_{J(T)}(X_2)$, $[C_A(X), J(T)] = 1$. But then $C_A(X)$ centralizes $I = N_I(X)J(T)$, so $C_A(X) = 1$. Thus $D = IC_A(X) = I$.

That is, $|Aut(I) : I| = 2$ and I is the stabilizer in $Aut(I)$ of I_1 and I_2 and hence also the stabilizer of A_1 and A_2.

In particular as $A_j \alpha_i = \bar{A}_j$ for $j = 1, 2$ and $i = 1, 2, 3$, $\alpha_1 \alpha_2^{-1}$ induces an inner automorphism on $I = G_{12}$, so by Exercise 14.3 we may pick α_1 and α_2 so as to agree on I.

Similarly G_{123} is a well-determined subgroup of index 2 of I and the argument above shows I is the stabilizer in $Aut(G_{123})$ of A_1 and A_2, and hence by Exercise 14.3 we may choose α_3 so that α_3 agrees with α_1 and α_2 on G_{123}.

Finally $G_{34} = N_{G_4}(B) \cong \mathbf{Z}_2/(A_4 \times A_5)$ and $G_{14} = G_{24} = G_{124} = N_{G_4}(E)$ is a well-determined subgroup of index 2 in $S_4 wr \mathbf{Z}_2$. Then it is an exercise to show $Aut(G_{34}) = N_{Aut(G_4)}(G_{34}) \cong S_5 \times S_4$, $Aut(G_{124}) = N_{Aut(G_4)}(G_{124}) \cong S_4 wr \mathbf{Z}_3$, and $C_{Aut(G_4)}(G_{1234}) = 1$, so by Exercise 14.3 we can choose α_4 so that $\alpha = (\alpha_i : i \in I)$ defines an isomorphism of the amalgams $\mathcal{A}(\mathcal{F})$ and $\mathcal{A}(\bar{\mathcal{F}})$.

Lemma 44.3: *Up to isomorphism there is at most one group of type He.*

Proof: Let $A = \mathcal{A}(\mathcal{F})$. Applying the construction of Example 36.2, we obtain a faithful completion $\beta : A \to G$ and by 36.4, a surjection $\varphi : G(A) \to G$ of the universal completion $G(A)$ onto G. Observe that $\Delta = \Delta(\beta, 1)$ is the collinearity graph of Γ defined in Section 36.

We now appeal to 36.7 applied to the completion $\beta : A \to G$ and the collinearity graph Δ. Observe that hypotheses $(*)$ and $(**)$ of 36.7 are satisfied. For example, 43.2, 43.4, and 43.11 say $(**)$ holds. So by 36.7, $G(A) \cong G$.

Next by 44.2, if $\bar{G}$ is a group of type He then $\mathcal{A}(\mathcal{F}) \cong \mathcal{A}(\bar{\mathcal{F}})$. So also $G(A) \cong \bar{G}$.

Theorem 44.4: *Assume G is a finite group containing an involution z such that $C_G(z)$ is isomorphic to the centralizer of a transvection in*

$L_5(2)$ *and* z *is not weakly closed in* $O_2(C_G(z))$ *with respect to* G. *Then one of the following holds:*

(1) *G is isomorphic to the maximal parabolic of $L_5(2)$ stabilizing a point of the natural module for $L_5(2)$.*

(2) $G \cong L_5(2)$.

(3) $G \cong M_{24}$.

(4) $G \cong He$.

Proof: By 39.10 either (1) holds or $z^G \cap Q = U_1^\# \cup U_2^\#$, and we may assume the latter. Then by 40.5, G is of type $L_5(2)$, M_{24}, or He. In the first two cases (2) or (3) holds by 41.5. In the third, G is determined up to isomorphism by 44.3, and hence as the group He of 32.5 is of type He, (4) holds.

Remarks. Held [He] was the first to consider groups G possessing an involution z such that $C_G(z)$ is isomorphic to the centralizer of a transvection in $L_5(2)$. He proves that if G is simple then G is isomorphic to $L_5(2)$ or M_{24}, or G has order $2^{10} \cdot 3^3 \cdot 5^2 \cdot 7^3 \cdot 17$ and determines much of the local structure of G. G. Higman and J. MacKay then proved the existence and uniqueness of He using the machine.

Our treatment of the problem bears little resemblance to Held's. For one thing, more local group theoretic techniques are available today than in 1969, making possible a more conceptual, less computational proof. For another, Held felt free to quote extensively from the literature, whereas the treatment here is self-contained. And of course we actually establish the uniqueness of He.

Exercises

1. Let $L = L_3(4)$, Ω be the set of root groups of transvections in L, and $A \in \Omega$. Prove $N_L(A)$ has six orbits on Ω:

(1) $\{A\}$.

(2) $\{A \neq B \in \Omega : center(A) = center(B)\}$ of order 4. Further $\langle A, B \rangle = A \times B$.

(3) $\{A \neq B \in \Omega : axis(A) = axis(B)\}$ of order 4. Further $\langle A, B \rangle = A \times B$.

(4) $\{B \in \Omega : center(B) \neq center(A) \in axis(B) \neq axis(A)\}$ of order 16. Further $\langle A, B \rangle \in Syl_2(L)$.

(5) $\{B \in \Omega : center(A) \neq center(B) \in axis(A) \neq axis(B)\}$ of order 16. Further $\langle A, B \rangle \in Syl_2(L)$.

(6) $\{B \in \Omega : center(A) \notin axis(B)$ and $center(B) \notin axis(A)\}$ of order 64. Further $\langle A, B \rangle \cong L_2(4)$.

2. Let $L = L_3(4)$. Prove
   (1) L is transitive on it subgroups of order 3.
   (2) L is transitive on its subgroups isomorphic to S_3, and if S_0 is such a subgroup then $C_L(S_0) = 1$.

3. Let A, B be subgroups of a group G, $\bar{A}, \bar{B}$ be subgroups of a group $\bar{G}$, $\alpha : A \to \bar{A}$, $\beta : B \to \bar{B}$ be group isomorphisms, $I = A \cap B$, $\bar{I} = \bar{A} \cap \bar{B}$, $I\alpha = \bar{I} = I\beta$, and $\gamma = \alpha\beta^{-1} \in Aut(I)$ be the automorphism induced on I by $\alpha\beta^{-1}$. Prove
   (1) If $\gamma \in Aut_B(I)$ then for some $c \in B$, $\beta_0 = \xi_c\beta : B \to \bar{B}$ with $\alpha = \beta_0$ on I, where $\xi_c : b \mapsto b^c$ is the automorphism of B induced by c.
   (2) Let $\alpha_i : A_i \to \bar{A}_i$, $1 \le i \le n$, be isomorphisms for subgroups $A_i \le G$, $\bar{A}_i \le \bar{G}$, $I_i = A_i \cap B$, $\bar{I}_i = \bar{A}_i \cap \bar{B}$, $J = \bigcap_i A_i \cap B$, $\bar{J} = \bigcap_i \bar{A}_i \cap \bar{B}$, $I_i\alpha_i = \bar{I}_i = I_i\beta$, and $J\alpha_i = \bar{J} = J\beta$. Assume $\alpha_i = \alpha_j$ on J for all i, j, $Aut_B(J_i) = Aut(J_i)$ for each i, $Aut_B(J) = Aut(J)$, and $C_B(J) = 1$. Prove there exists $c \in B$ such that $\xi_c\beta = \beta_0 = \alpha_i$ on I_i for each i.

4. Let V be a 6-dimensional orthogonal space over the field of order 2 and $L_3(2) \cong L \le G = O(V)$. Prove
   (1) V has sign $+1$ and L stabilizes a maximal totally singular subspace U of V.
   (2) $P = N_G(U)$ is transitive on the set $\mathcal{W}$ of eight totally singular complements W to U in V. Further $N_P(W) = L_1 \cong L$.
   (3) If $L \in L_1^P$ then L has orbits $\mathcal{O}_i$, $i = 1, 2, 3$, of length $7, 7, 21$ on singular points of V.
   (4) If $L \notin L_1^P$ then L is transitive on $\mathcal{W}$, $L \cap L_1$ is transitive on $\mathcal{O}_i$ for each i, and L has two orbits of length $7, 28$ on singular points of V.

5. Let G be a finite group, z an involution in G, and $H = C_G(z)$, and assume $Q = F^*(H)$ is extraspecial of order 128 with $H/Q \cong L_3(2)$ and $|z^G \cap Q| = 29$. Let $G_0 = L_5(2)$, z_0 a transvection in G_0, and $H_0 = C_{G_0}(z_0)$. Prove
   (1) $H/\langle z \rangle \cong H_0/\langle z_0 \rangle$, so in particular 39.1.2 and 39.1.5 hold.
   (2) H is transitive on the set $\mathcal{W}$ of pairs (W_1, W_2) such that W_i is a hyperplane of U_i with $z \notin W_i$.
   (3) Let $(W_1, W_2) \in \mathcal{W}$ and $L = N_H(W_1, W_2)$. Then $L \cong SL_2(7)$ or $L_3(2)$.

(4) If $L \cong SL_2(7)$, s is an involution in $Q - \langle z \rangle$, and $S \in Syl_2(C_H(s))$, then $\Phi^2(S) = \langle z \rangle$.

(5) $H \cong H_0$.

(Hint: Use Exercise 14.4 to prove (1).)

6. Let $L \cong L_2(4)$ and V the natural $GF(4)L$-module. Prove

$$dim(H^1(L, V)) = 1.$$

Chapter 15

The Group $U_4(3)$

The group $U_4(3)$ is the image $PSU_4(3)$ of the special unitary group $SU_4(3)$ in $PSL_4(9)$. Thus this unitary group is a classical group (cf. Chapter 7 in [FGT] for a discussion of the classical groups and unitary groups in particular), but it exhibits sporadic behavior and is crucial in the study of various sporadic groups. For example, we saw in Exercise 9.4 that $U_4(3)$ is the stabilizer of a point in the rank 3 representation of the McLaughlin group. Further we find in Chapter 16 that $U_4(3)$ is a section in a 3-local of Suz.

In Section 45 we prove a uniqueness theorem for $U_4(3)$ characterizing the group in terms of the centralizer of an involution. In the end we construct a geometric complex for the group which we prove to be isomorphic to the complex of singular points and lines in the 4-dimensional unitary space over the field of order 9. This last part of the proof can also be used in Exercise 9.4 to identify the point stabilizer of Mc in terms of the information in Lemma 24.7.

In Chapter 16 we use the characterization of Section 45 to pin down the structure of the centralizer of a certain subgroup of order 3 in groups of type Suz. This result is used in turn in the proof of the uniqueness of Suz and in 26.6 to determine the structure of the centralizer of the corresponding subgroup of order 3 in Co_1, while 26.6 is used in 32.4 to establish the existence of a subgroup of the Monster of type F_{24}.

45. $U_4(3)$

In this section we assume the following hypothesis:

Hypothesis 45.1: *G is a finite group, z is an involution in G, $H = C_G(z)$, $O^2(H) = H_1 H_2$ with $H_i \cong SL_2(3)$, $H_i \trianglelefteq O^2(H)$, and $H_1 \cap H_2 = \langle z \rangle$, $Q = O_2(O^2(H)) = F^*(H)$, and z is not weakly closed in Q with respect to G.*

Let $\tilde{H} = H/\langle z \rangle$ and $H^* = H/Q$.

Lemma 45.2: *(1) $[H_1, H_2] = 1$ and H permutes $\{H_1, H_2\}$.*
(2) $Q \cong Q_8^2$.
(3) $\tilde{Q}$ is a 4-dimensional orthogonal space of sign -1 over $GF(2)$ and $H^* \le O(\tilde{V}) \cong O_4^-(2)$.
(4) $O^2(H)$ is transitive on the eighteen involutions in $Q - \langle z \rangle$.

Proof: As $H_i \trianglelefteq O^2(H)$, $[H_1, H_2] \le H_1 \cap H_2 = \langle z \rangle \le Z(H_i)$, so as $H_i = O^2(H_i)$, $[H_1, H_2] = 1$. Then $\{H_1, H_2\}$ is the set of normal subgroups of $O^2(H)$ isomorphic to $SL_2(3)$ and hence is permuted by H. That is, (1) holds. Further (1) implies (2) while (2), 8.3, and 8.4 imply (3). By (3) there are 18 involutions in $Q - \langle z \rangle$ and if t is such an involution then $|O^2(H) : C_{O^2(H)}(t)| = 18$, so (4) holds.

By Hypotheses 45.1 and 45.2, there exists $u = z^g \in Q - \langle z \rangle$ with $\langle u, z \rangle = E \trianglelefteq T \in Syl_2(H)$.

Lemma 45.3: *Assume $Q \cap Q^g = E$ and let $N = \langle Q, Q^g \rangle$ and $R = C_Q(u) C_{Q^g}(z)$. Then*

(1)   *N is the split extension of R by $Y \cong S_3$ with $C_R(Y) = 1$.*
(2)   *R is isomorphic to a Sylow 2-subgroup of $L_3(4)$ and both members of $\mathcal{A}(R)$ are normal in N.*
(3)   *$R^* O^2(H^*)$ is the subgroup of $O_4^-(2)$ of index 2 whose Sylow 2-subgroup R^* is a 4-group containing no transvections.*
(4)   *All involutions in $RO^2(H)$ are in z^G.*
(5)   *For $X \in Syl_3(H)$, $N_{RO^2(H)}(X) = X T_X$ with $T_X \cong D_8$ and $\langle z \rangle = C_{T_X}(X)$.*

Proof: By 8.15, $N/R \cong S_3$ and R is special with center E and $R/E = R_1/E \oplus R_2/E$ the sum of natural modules for N/R.

Let $s \in Q - R$ be an involution. By the Baer-Suzuki Theorem (cf. 39.6 in [FGT]), s inverts some element y of order 3 in N, so $Y = \langle s, y \rangle \cong S_3$ is a complement to R in N. Further $[R/E, s] = (Q \cap R)/E$ and $Q \cap R = C_Q(E) \cong \mathbf{Z}_2 \times D_8$.

Next $R^* \cong R/(R \cap Q) \cong E_4$ and $[R, Q \cap R] \leq E$, so $[\tilde{E}^\perp, R^*] \leq \tilde{E}$. Now if $r^* \in R^*$ induces a transvection on $\tilde{Q}$ then $[\tilde{Q}, r^*]$ is a nonsingular point of $\tilde{Q}$. However, $\tilde{E}$ is singular and $[\tilde{E}^\perp, r^*] \leq \tilde{E}$, so either $\tilde{E} = [\tilde{E}^\perp, r^*] \leq [\tilde{Q}, r^*]$ or $[\tilde{E}^\perp, r^*] = 0$ so $\tilde{E}^\perp$ is the axis of r^* and hence $\tilde{E} = (\tilde{E}^\perp)^\perp = [\tilde{Q}, r^*]$. Therefore R^* contains no transvection and (3) holds.

Next there is exactly one more irreducible R_3/E for Y on R/E and $R_i = \langle r_i^Y \rangle E$, where 1 and r_i, $1 \leq i \leq 3$, are coset representatives for E in $Q \cap R$. Thus we may pick r_1 and r_2 to be involutions and r_3 of order 4. Then as Y is transitive on $(R_i/E)^\#$, all elements in $R_i^\#$ are involutions for $i = 1, 2$, while $\Omega_1(R_3) = E$. In particular $R_1 \cong R_2 \cong E_{16}$. If $x \in C_{R_1}(r_2) - E$ then $x^* \neq 1$ and $[x, Q \cap R] = 1$, so x^* induces a transvection on $\tilde{Q}$, contrary to (3). Hence $C_{R_1}(r_2) = E$ and then as Y is transitive on $(R_2/E)^\#$, $C_{R_1}(r) = E$ for all $r \in R_2 - E$. It is now an easy exercise to verify that R is isomorphic to a Sylow 2-subgroup of $L_3(4)$ with $\{R_1, R_2\} = \mathcal{A}(R)$. By construction, $R_i \trianglelefteq N$. So (2) is established.

By (2), if $1 \neq r^* \in R^*$ then $[\tilde{Q}, r^*] = C_{\tilde{Q}}(r^*)$, so by Exercise 2.8, if $\tilde{r}$ is an involution then all involutions in $\tilde{r}\tilde{Q}$ are conjugate to $\tilde{r}$. Thus all involutions in rQ are contained in $r\langle z \rangle$. Therefore each involution in $RO^2(H) - Q$ is conjugate to s_i, $1 \leq i \leq 3$, where 1 and s_i are coset representatives for E in $Q^g \cap R$. Thus 45.2.4 implies (4). Recall also that two of the s_i are involutions and the third is of order 4.

Finally for $X \in Syl_3(H)$, $RO^2(H) = QN_{RO^2(H)}(X)$ by a Frattini argument with $N_Q(X) = \langle z \rangle$, so $N_{RO^2(H)}(X) = XT_X$, where $T_X \in Syl_2(N_{RO^2(H)}(X))$ with $T_X/\langle z \rangle \cong E_4$. We have seen that if 1 and t_i, $1 \leq i \leq 3$, are coset representatives for $\langle z \rangle$ in T_X then as $\tilde{t}_i$ is an involution, two of the t_i are involutions and the third is of order 4. Thus $T_X \cong D_8$, completing the proof of (5).

Lemma 45.4: *Either*

(1) $Q \cap Q^g = E$, *or*

(2) $Q \cap Q^g = V \cong E_8$, *there exists* $z^k \in V - E$, *and if we set* $N = \langle Q, Q^g, Q^k \rangle$, $N/V \cong L_3(2)$ *is faithful on* V.

Proof: As Q has width 2, 8.15.7 says $V = Q \cap Q^g$ is of rank at most 3. So as $E \leq V$, either $V = E$ or $V \cong E_8$, and we may assume the latter. By 45.2.4, $V^\# \subseteq z^G$, so there is $z^k \in V - E$.

Now $\tilde{V}$ is a totally singular line in $\tilde{Q}$, so as each such line admits the action of a subgroup of $O^2(H)$ of order 3, $N_H(V)$ is transitive on $\tilde{V}^\#$. Thus $V \leq Q^k$ for $z^k \in V - E$, so 8.16 completes the proof, since $Q \cap H^g \cap H^k = C_Q(V) = V$.

Define G to be of *type* $U_4(3)$ if Hypothesis 45.1 is satisfied with $|Q \cap Q^g| = 4$ for $z^g \in Q - \langle z \rangle$ and $|H| = 2^7 \cdot 3^2$. In the remainder of this section assume G is of type $U_4(3)$. We will show $G \cong U_4(3)$.

Define R, R_i, $i = 1, 2$, and N as in the proof of 45.3. Recall $T \in Syl_2(H)$ with $E \trianglelefteq T$.

Lemma 45.5: *(1) $H = RO^2(H)$.*

(2) $T \in Syl_2(G)$, $T \in Syl_2(N)$, and $|T| = 2^7$.

(3) $R = J(T)$, $\mathcal{A}(T) = \{R_1, R_2\}$, and R_i is weakly closed in T with respect to G.

(4) G has one conjugacy class of involutions.

Proof: As $|H| = 2^7 \cdot 3^2 = |RO^2(H)|$, (1) holds. As $\langle z \rangle = Z(T)$ and $T \in Syl_2(H)$, $T \in Syl_2(G)$. As $E \trianglelefteq T$ and $|N|_2 = |T|$, $T \in Syl_2(N)$. Thus (2) is established.

If s is an involution in $T - R$ then $|R/E : C_{R/E}(s)| = 4 = |C_{R/E}(s)|$ and $\langle s \rangle E \cong D_8$, so by Exercise 2.8, s^R is the set of involutions in sR. But we saw during the proof of 45.3 that if we choose $s \in Q$ then $C_R(s) = C_{Q \cap R}(s) \cong D_8$, so $m(C_T(s)) = 3$. Thus as $m(R_i) = 4$ and $\mathcal{A}(R) = \{R_1, R_2\}$ with $R_i \trianglelefteq T$, (3) holds. Finally 45.3.4 implies (4).

Lemma 45.6: *$N_G(R_i)$ is the split extension of R_i by A_6 acting as $Sp_4(2)'$ on R_i.*

Proof: Let $M_i = N_G(R_i)$. First $C_G(R_i) = C_H(R_i) = R_i$, so M_i/R_i is faithful on R_i. Next as R_i is weakly closed in T it follows from 7.7 and 45.5.4 that M_i is transitive on $R_i^{\#}$. In particular $|M_i| = 15 \cdot |H \cap M_i| = 16 \cdot |A_6|$, so $|M_i/R_i| = |A_6|$. Finally the map

$$z^k \mapsto Q^k \cap M_i$$

is a bijection of $R_i^{\#}$ with the hyperplanes of R_i, so by Exercise 15.1, $M_i/R_i \leq Sp_4(2) \cong S_6$. Hence M_i/R_i is the unique subgroup A_6 of S_6 of index 2. The extension splits by Gaschutz's Theorem (cf. 10.4 in [FGT]) as T splits over R_i.

Lemma 45.7: *Let $X \in Syl_3(H)$ and $T_X \in Syl_2(N_H(X))$. Then*

(1) *If $\mathbf{Z}_3 \cong U \leq X$ is T_X-invariant and $|C_H(U)|_2 > 2$ then $C_H(U)$ is the split extension of X by $C_{T_X}(U) \cong E_4$.*

(2) *$N_G(X) = VT_X$, where $V = O_3(N_G(X)) \cong E_{81}$ and $T_X \in Syl_2(N_H(X))$ is isomorphic to D_8.*

(3) *If $\mathbf{Z}_3 \cong U \leq X$ with $C_{T_X}(U) \cong E_4$ then $V \trianglelefteq N_G(U)$ and $N_G(U)/V \cong S_4$.*

Proof: By 45.3.5, $N_H(X) = XT_X$, where $T_X \cong D_8$ and $\langle z \rangle = C_{T_X}(X)$. Thus $\langle z \rangle \in Syl_2(C_G(X))$, so $C_G(X) = V\langle z \rangle$, where $V = O(C_G(Z))$, and $N_G(X) = VT_X$. So to prove (2) it remains to show $V \cong E_{81}$.

Next by 45.6, $U_i = N_X(R_i) \cong \mathbf{Z}_3$ and $C_{R_i}(U_i) \cong E_4$, so as $\langle z \rangle \in Syl_2(C_G(X))$, $X = U_1 \times U_2$. Then (1) is established.

As z inverts V/X, V/X is abelian. By (1), $\mathbf{Z}_3 \cong U = U_1$ centralizes an involution $t \in T_X - \langle z \rangle$. By Exercise 8.1 in [FGT], $V = C_V(z)C_V(t)C_V(tz) = XC_V(t)C_V(tz)$. But as $t \in z^G$ and $|V|$ is odd, $|C_V(t)|$ divides 9. Further as tz is conjugate to t in T_X, $|C_V(t)| = |C_V(tz)|$. Thus as $U = C_V(\langle t, z \rangle) \cong \mathbf{Z}_3$, $V = X$ or $|V| = 81$.

Suppose $V = X$. Then $X \in Syl_3(G)$. So as $|N_G(R_1)|_3 = 9$, a Sylow 3-subgroup X_1 of $M_1 = N_G(R_1)$ is Sylow in G. This is impossible as an element of order 4 in $N_{M_1}(X_1)$ is faithful on X_1, whereas $T_X \in Syl_2(N_G(X))$ and T_X contains no such element.

So $|V| = 81$ and $V/X = V_t/X \oplus V_{tz}/X$, where $V_s = C_V(s)X \cong E_{27}$. Hence if V is not E_{81} then $a = [v, w] \neq 1$ for $v \in V_t - X$ and $w \in V_{tz} - X$ inverted by z. Notice $\langle a \rangle = [V, V]$. Further there is an involution $s \in T_X$ with $t^s = tz$. Then we may choose $v^s = w$, so $a^s = [w, v] = a^{-1}$. As $\langle a \rangle = [V, V]$, by symmetry between t and s, s inverts a also. Hence $st \in C_{T_X}(a)$ is of order 4, contradicting (1). This establishes (2).

Let $K = N_G(U)$. Then $N_{M_1}(U) = T_X W$, where $E_9 \cong W \in Syl_3(M_1)$ and $C_{M_1}(U) \cong \mathbf{Z}_3 \times A_4$. As $\langle z \rangle = Z(T_X)$ and $T_X \in Syl_2(N_H(U))$, $T_X \in Syl_2(K)$. Also $S = C_{T_X}(U) \cong E_4$ and as W is transitive on $S^\#$ with $C_H(U) = SX$, for all $s \in S^\#$, $C_G(\langle s, U \rangle) = SC_V(s)$. So W acts on $\langle C_K(\langle s, U \rangle) : s \in S^\# \rangle = SV$, and $I = T_X WV = \langle C_K(s), N_K(S) : s \in S^\# \rangle$. Similarly if $r \in T_X - S$ is an involution then $\langle z, r \rangle \in Syl_2(C_K(r))$ and $C_V(r) \in Syl_3(C_G(r))$, so as $C_G(r)$ is a $\{2, 3\}$-group, $C_K(r) = \langle r, z \rangle C_V(r) \leq I$. Thus if $I \neq K$ then I is strongly embedded in K in the sense of Section 7. This contradicts 7.6, as $r \notin z^K$ since r inverts U while z centralizes u. Therefore $I = K$ and (3) is established.

Lemma 45.8: Let $W = X \cap H_1 \in Syl_3(H_1)$. Then

(1) $F^*(N_G(W)) \cong 3^{1+4}$ has a complement $\bar{H}_2$ in $N_G(W)$, where $|\bar{H}_2 : H_2| = 2$ and $\bar{H}_2$ has quaternion Sylow 2-subgroups of order 16.

(2) $P = VF^*(N_G(W)) \in Syl_3(G)$, $|P| = 3^6$, and $V = \mathcal{A}(P)$.

Proof: Let $M = N_G(W)$. Then $N_H(W) = H_2\langle \xi \rangle$, where $\xi \in T_X$ is of order 4 inverting X. In particular $C_Q(W)\langle \xi \rangle = S \in Syl_2(N_H(W))$, where $C_Q(W) = Q \cap H_2 \cong Q_8$. Thus S is quaternion of order 16 and $N_H(W) = \bar{H}_2 W$, where $\bar{H}_2 = \langle \xi \rangle H_2$. In particular as $\langle z \rangle = Z(S)$,

$S \in Syl_2(M)$. Then as S is quaternion, by a result of Brauer and Suzuki (cf. Theorem 12.1.1 in [Go]), $M = D(H \cap M)$, where $D = O(M)$. Notice $\bar{H}_2$ is a complement to D in M, so to prove (1) it remains only to show $D \cong 3^{1+4}$.

Next as $W = D \cap H$, z inverts D/W, so D/W is abelian. Further $W \leq V$ and V is abelian, so $V \leq M$, and then $V \cap D = [V, z]W$ is of index 3 in V. Pick $U = U_1$ and U_2 as in the proof of 45.7. Then from 45.7.3, there is a conjugate F of U_2 under $N_G(U)$ with $F = [F, z]$. Hence $F \leq V \cap D$ and by 45.7.3, $N_G(F) \leq N_G(V)$ with $|N_G(F)| = 2^3 \cdot 3^5$. In particular as D is of odd order, $|C_D(F) : V \cap D| = 1$ or 3. But as D/W is abelian, $[D, F] \leq W$, so $|D : C_D(F)| = 1$ or 3. Thus $|D : D \cap V|$ divides 9.

Now if $D \leq V$ then $H_2 = [V \cap H_2, H_2] \leq C(D)$, contradicting $D \not\leq C(z)$. So $|D| = 3^a$, $a = 4, 5$. Then as z inverts D/W and $GL_2(3)$ contains no quaternion subgroup of order 16, $\bar{H}_2$ is irreducible on D/W and a is even, so $|D| = 3^5$ and $\Phi(D) \leq W$. Indeed $D = \langle (V \cap D)^{H_2} \rangle$ is generated by elements of order 3 and D is of class at most 2, so D is of exponent 3. Finally as $F \not\leq Z(D)$ and $\bar{H}_2$ is irreducible on D/W, $W = Z(D)$, so $D \cong 3^{1+4}$ and (1) is established.

Let $P = VD$, so that $P \in Syl_3(M)$. As $W = Z(P)$, $P \in Syl_3(G)$. Next $C_{D/W}(U) = (V \cap D)/W$, so for $d \in D - V$, $C_V(d) \leq D$. Then as $D \cong 3^{1+4}$, $|V : C_V(D_0)| > |D_0/(V \cap D)|$ for all $D_0 \leq V$ with $V \cap D$ proper in D_0. This shows $\{V\} = \mathcal{A}(P)$, and completes the proof of (2).

Lemma 45.9: $N_G(V)$ *is the split extension of V by A_6 acting faithfully as $\Omega_4^-(3)$ on V, with $W^G \cap V$ the set of singular points of V and $U_i^G \cap V$, $i = 1, 2$, the two classes of nonsingular points of V.*

Proof: Let $M = N_G(V)$ and $M^* = M/V$. Then from 45.8, $V = C_G(V) \cap N_G(W) = C_G(V)$, so M^* is faithful on V. Again by 45.8, $N_M(W)^* = N_M(P)^*$ is the split extension of $P^* \cong E_9$ by $\mathbf{Z}_4$ and $P \in Syl_3(M)$. By 45.8.2, V is weakly closed in M with respect to G, so by 7.7, $\bar{V}^G \cap V = \bar{V}^M$ for each $\bar{V} \leq V$.

We claim P^* is a TI-set in M^*. For if $d \in P \cap P^m - V$ for some $m \in M - N_G(P)$, then as $dim(C_V(d)) = 2$, $C_V(d) = WW^m$ is invariant under $\langle P, P^m \rangle \langle z \rangle$. Then $\langle P, P^m \rangle \langle z \rangle$ induces $GL_2(3)$ on $C_V(d)$, so $N_M(P)$ contains a 4-group, a contradiction.

Next $D_8 \cong T_X \in Syl_2(H \cap M)$, so as $\langle z \rangle = Z(T_X)$, $T_X \in Syl_2(M)$. Then as $M^* \leq GL_4(3)$, $|M^*| = 2^3 \cdot 3^2 \cdot 5^a \cdot 13^b$, with $a, b \in \{0, 1\}$. In particular $|M^* : N_{M^*}(P^*)| = 2 \cdot 5^a \cdot 13^b$. But as P^* is a TI-set in M^*, $|M^* : N_{M^*}(P^*)| \equiv 1 \mod 9$, so we conclude $|M^*| = 2^3 \cdot 3^2 \cdot 5 = |A_6|$.

In particular $|M^* : N_{M^*}(P^*)| = 10 = |P^*| + 1$, so as P^* is a TI-set in M^*, M^* is 2-transitive on $Syl_3(M^*)$ and hence on W^M as $N_M(W) = N_M(P)$.

Similarly from 45.7.3, $|U_i^M| = 15$, so as V contains 40 points, these 40 points are $W^M \cup U_1^M \cup U_2^M$. Now by Exercises 15.2 and 15.3, M^* acts as $\Omega_4^-(3)$ on V. The extension splits by Gaschütz's Theorem as we can find a complement to V in P contained in $F^*(N_G(W))$.

Let Γ be the rank 2 geometry with point set W^G and line set V^G, where incidence is defined by inclusion. Let Δ be the collinearity graph of Γ. It turns out Γ is the geometry of singular points and lines in the 6-dimensional orthogonal space for $P\Omega_6^-(3)$ ($\cong U_4(3)$) over $GF(3)$. This is essentially proved in the next two lemmas. See also 24.7, the Remark preceding 24.7, and Exercise 9.5.

Lemma 45.10: *(1) Each element of order 3 in P is contained in V or $D = F^*(N_G(W))$.*

(2) W is weakly closed in D with respect to G.

(3) Δ is the commuting graph on W^G.

(4) Each line has ten points and each point is on four lines.

(5) $|\Delta(W)| = 36$.

(6) D is regular on $\Delta^2(W)$ of order 3^5.

(7) For $W' \in \Delta^2(W)$, $\langle W, W' \rangle \in H_2^D$, so $\langle W, W' \rangle \cong SL_2(3)$.

(8) Γ is a generalized quadrangle.

(9) Δ is of diameter 2.

(10) $|G| = 2^7 \cdot 3^6 \cdot 5 \cdot 7$.

Proof: Let $x \in P - V$ be of order 3. Then from 45.9, x has Jordon blocks of size 1 and 3 on V with $[V, x, x] = W$. Then all elements of order 3 in xV are contained in $x C_V(x)[V, x] = x W^\perp = x(V \cap D)$, so (1) is established.

Let $K = N_G(W)$. By 45.9 there are ten points on the line V, while there are $|K : K \cap M| = 4$ lines through W. So (4) holds. Observe also from 45.9 that $W^M \cap V \cap D = \{W\}$.

Also the sets $((V^k \cap D)/W)^\#$, $k \in K$, contain 32 vectors of $(D/W)^\#$, with the remaining 48 vectors forming an orbit under K. We claim that if $w \in W$ and $y \in G$, then $w^y W$ is not such a vector and hence $w^y \in V^k$ for some $k \in K$. Then as $W^M \cap V \cap D = \{W\}$, (2) holds. Namely $|D : C_D(w^y)| = 3$ with $W \leq \Phi(C_D(W^y)) \leq D^y$, and $D^y \cap D$ is abelian so $|D \cap D^y| \leq 27$. Hence $C_{D^v}(W) \not\leq D$, establishing the claim.

By (1) and (2), each member of $W^G \cap K - \{W\}$ is contained in a unique member of V^K, so (3) holds and (3) and (4) imply (5). Further we see all triangles of Δ are contained in lines.

For $W_1 \in \Delta \cap V$, W_1 is transitive on the three lines through W distinct from V and $V \cap D$ is transitive on the nine points on any line $V^k \neq V$ distinct from W, so G is transitive on paths WW_1W_2 with $d(W, W_2) = 2$. In particular K is transitive on $\Delta^2(W)$.

Next let $W \neq W_2 \in Syl_3(H_1)$. Then $W_2 \in \Delta$ and $W_1 = X \cap H_2 \in \Delta(W, W_2)$, so $W_2 \in \Delta^2(W)$ with $\bar{H}_2 = N_K(W_2)$ a complement to D in K. Thus D is regular on $W_2^K = \Delta^2(W)$ of order 3^5. Further for each $k \in K$, and $W_0 \in \Delta \cap V^k$, $V^k \cap \bar{H}_2 \in \Delta(W_2, W_0)$, so $d(W_2, W_0) \leq 2$. This proves (8) and (9). Then $|\Delta| = 1 + |\Delta(W)| + |\Delta^2(W)| = 112$, so $|G| = 112 \cdot |K|$, and (10) holds.

Lemma 45.11: *Up to isomorphism, $U_4(3)$ is the unique group of type* $U_4(3)$.

Proof: Let G' be a second group of type $U_4(3)$. The group $\bar{H}_2$ of 45.8 is determined up to conjugacy in $Out(D) \cong GSp_4(3)$, so the split extension $K = N_G(W)$ of D by $\bar{H}_2$ is determined up to isomorphism, so there exists an isomorphism $\alpha : K \to K'$ with $H_2\alpha = H_2'$ and $V\alpha = V'$. Let $I = H_2 \cap V$ and $W \neq J \in \Delta \cap H_1$. Set $I' = I\alpha$ and let $W' \neq J' \in H_1' \cap \Delta'$. Define $\beta : \Delta \to \Delta'$ by $W\beta = W'$, $(I^k)\beta = I'^{k\alpha}$ for $k \in K$, and $(J^d)\beta = J'^{d\alpha}$ for $d \in D$. As D is regular on $\Delta^2(W)$ and K is transitive on $\Delta(W)$ with $N_K(I)\alpha = (VN_{\bar{H}_2}(V))\alpha = V'N_{\bar{H}_2'}(V') = N_{K'}(I')$, the map β is a well-defined bijection of Δ with Δ'.

We claim β is an isomorphism. By construction $\Delta(W)\beta = \Delta(W')$. Next by 45.10.8, I is adjacent to I^k if and only if I^k is on the unique line incident with W and I; that is, when $I^k \leq V$. So I is adjacent to I^k if and only if $k \in N_K(V)$ if and only if $k\alpha \in N_{K'}(V')$ if and only if I' is adjacent to $(I'^k)\beta$. Similarly I is adjacent to J^d if and only if $d \in D \cap V$ if and only if $d\alpha \in D' \cap V'$ if and only if I' is adjacent to $(J^d)\beta$. Thus $\Delta(I)\beta = \Delta(I')$.

Thus to complete the proof of the claim it suffices to show J is adjacent to J^d if and only if J' is adjacent to $(J'^d)\beta$. But J is adjacent to J^d if and only if the line $J + J^d$ through J and J^d contains a unique point of $\Delta(W)$, which we may take to be I. Then $[V, z] = D \cap D^y$, where $I = W^y$, and thus $[V, z]$ is regular on $J + I - \{I\}$, so J is adjacent to J^d if and only if $d \in [V, z]$ if and only if $d\alpha \in [V', z\alpha]$ if and only if J' is adjacent to $(J^d)\beta$. So the claim is established.

Now the isomorphism $\beta : \Delta \to \Delta'$ induces an isomorphism $\beta^* : Aut(\Delta) \to Aut(\Delta')$ (cf. Section 1). Finally W is the pointwise stabilizer in $Aut(\Delta)$ of $W^{\perp}$, so $W\beta^* = W'$ and hence $G\beta^* = \langle W^{Aut(\Delta)} \rangle \beta^* = G'$, completing the proof.

Remarks. The group $U_4(3)$ was first characterized via the centralizer of an involution by Phan in [Ph].

Exercises

1. Assume V is an n-dimensional vector space over the field F of order 2 and $h : v \mapsto h(v)$ is a bijection of $V^{\#}$ with the set of hyperplanes of V such that
   (1) $v \in h(v)$ for all $v \in V^{\#}$.
   (2) If $u, v \in V^{\#}$ with $u \in h(v)$ then $v \in h(u)$. Then the map $f :$ $(u, v) \mapsto f(u, v)$ is a symplectic form on V, where $f(u, v) = 0$ if $u \in f(v)$ and $f(u, v) = 1$ otherwise.
2. Assume G is a finite group, $S_4 \cong H \leq G$ with $|G : H| = 15$, $P \in Syl_3(G)$ with $E_9 \cong P$ a TI-set in G, and $N_G(P)/P \cong \mathbf{Z}_4$ is faithful on P. Prove:
   (1) $H = N_G(O_2(H))$ with $O_2(H)$ weakly closed in H with respect to G.
   (2) G is rank 3 on G/H with parameters $k = 6$, $l = 8$, $\lambda = 1$, and $\mu = 3$.
   (3) The rank 3 graph of G on G/H is isomorphic to the graph Λ of all 2-subsets of a 6-set, with s adjacent to t in Λ if $s \cap t = \varnothing$.
   (4) $G \cong A_6$.
3. Let $G = A_6$, F the field of order 3, $P \in Syl_3(G)$, $H = N_G(P)$, V_0 the 1-dimensional FH-module with $H/C_H(V_0) \cong \mathbf{Z}_2$, and $V = V_0^G$ the induced FG-module. Let I be the 4-dimensional orthogonal space of sign -1 over F regarded as an FG-module, recalling that $G \cong L_2(9) \cong \Omega_4^-(3)$. Prove
   (1) A nontrivial FG-module U is a homomorphic image of V if and only if there is a 1-dimensional FH-submodule U_0 of U with U_0 FH-isomorphic to V_0 and $U = \langle U_0 G \rangle$.
   (2) V is FG-isomorphic to the dual space V^*.
   (3) $C_V(P)$ is FH-isomorphic to the sum of two copies of V_0 and $dim(C_{[V,P]}(P)) = 1$.
   (4) $Soc(V) = \langle C_{[V,P]}(P)G \rangle$ is irreducible.
   (5) V is indecomposable with $Soc(V) \cong V/J(V) \cong I$.
   (6) If U is an irreducible FG-module and U_0 an FH-submodule of U which is FH-isomorphic to V_0, then $U \cong I$.

Chapter 16

Groups of Conway, Suzuki, and Hall–Janko Type

In this chapter we prove the uniqueness of the Conway group Co_1, the Suzuki group Suz, and the Hall–Janko group J_2. More precisely we consider groups of type $\mathcal{H}(w, \Omega_{2w}^\epsilon(2))$ and define G to be of *type* Co_1 if $(w, \epsilon) = (4, +1)$, G to be of *type Suz* if $(w, \epsilon) = (3, -1)$, and finally G to be of *type* J_2 if $(w, \epsilon) = (2, -1)$ and G has more than one class of involutions. Then we prove Co_1 is the unique group of type Co_1, Suz is the unique group of type Suz, and J_2 is the unique group of type J_2.

In the process we generate much information about the three groups. For example, for each group we obtain the group order and determine most of the conjugacy classes and normalizers of subgroups of prime order. This determination is completed in Chapter 17.

46. Groups of type Co_1, Suz, J_2, and J_3

In this section we assume G satisfies Hypothesis $\mathcal{H}(w, \Omega_{2w}^\epsilon(2))$ with $(w, \epsilon) = (2, -1)$, $(3, -1)$, or $(4, +1)$. (See the Preface for the definition of groups of type $\mathcal{H}(w, L)$.) In particular z is an involution, $H = C_G(z)$, and $Q = F^*(H)$.

Let $H^* = H/Q$, $\tilde{H} = H/\langle z \rangle$, $z^g \in Q - \langle z \rangle$, $E = \langle z, z^g \rangle$, and $R = (Q \cap H^g)(Q^g \cap H)$.

Let Δ_H be the set of subgroups X of H of order 3 such that $[Q, X] \cong Q_8$.

Lemma 46.1: *(1)* $\tilde{Q}$ *is an orthogonal space of dimension $2w$ and sign ϵ over $GF(2)$ and $H^* \cong \Omega_{2w}^\epsilon(2)$ acts as the commutator group $\Omega(\tilde{Q})$ of the orthogonal group $O(\tilde{Q})$ of $\tilde{Q}$.*

(2) H is transitive on the involutions in $Q - \langle z \rangle$, and all such involutions are in z^G.

Proof: By definition of Hypothesis $\mathcal{H}(w, \Omega_{2w}^\epsilon(2))$, $H^* \cong \Omega_{2w}^\epsilon(2)$ and Q is extraspecial of order 2^{1+2w}. Then by 8.3, $\tilde{Q}$ is an orthogonal space over $GF(2)$ of dimension $2w$ and H^* preserves this structure. In particular $H^* \leq O(\tilde{Q})$. As $H^* \cong \Omega_{2w}^\epsilon(2)$, this forces $\tilde{Q}$ to have sign ϵ and $H^* = \Omega(\tilde{Q})$. That is, (1) holds. As $\Omega(\tilde{Q})$ is transitive on the singular points of $\tilde{Q}$, which by 8.3.4 correspond to the involutions in Q distinct from z, (2) holds.

Lemma 46.2: *(1) $E = Q \cap Q^g$.*

(2) $(Q^g \cap H)^* = O_2(C_{H^*}(\tilde{E})) \cong E_{2^{2(w-1)}}$ with $C_{H^*}(\tilde{E})/(Q^g \cap H)^* \cong \Omega_{2w-2}^\epsilon(2)$ and $(Q^g \cap H)^*$ and $(Q \cap H^g)/E$ the natural module for

$$C_{H^*}(\tilde{E})/(Q^g \cap H)^*.$$

Proof: As $H^* = \Omega(\tilde{Q})$ and $\tilde{E}$ is a singular point in the orthogonal space $\tilde{Q}$, $O_2(C_{H^*}(\tilde{E})) \cong E_{2^{2w-2}}$ and $C_Q(E)/E$ are the natural modules for

$$C_{H^*}(\tilde{E})/O_2(C_{H^*}(\tilde{E})) \cong \Omega_{2w-2}^\epsilon.$$

In particular $C_H(\tilde{E})$ is irreducible on $O_2(C_{H^*}(\tilde{E}))$. But by 8.15, $1 \neq (Q^g \cap H)^* \trianglelefteq C_{H^*}(\tilde{E})$, so $(Q^g \cap H)^* = O_2(C_{H^*}(\tilde{E}))$ and then (1) holds by 8.15.8. Finally (1) and earlier remarks imply (2).

Lemma 46.3: *(1) R is special with center E.*

(2) $N_G(R)/R \cong \Omega_{2w-2}^\epsilon(2) \times S_3$ with R/E the tensor product of the natural module for its factors.

(3) $N_G(R)$ splits over R.

Proof: Parts (1) and (2) follow from 46.2 and 8.15. Let X be of order 3 in $N_G(R)$ with $XR \trianglelefteq N_G(R)$. Then $C_R(X) = 1$ so $N_G(X) \cap N_G(R)$ is a complement to R in $N_G(R)$, establishing (3).

Lemma 46.4: $H^1(\Omega_{2w}^\epsilon(2), \tilde{Q}) = 0$.

Proof: If $w = 3$ or 4 let X be of order 3 in $L = \Omega_{2w}^\epsilon(2)$ with $C_{\tilde{Q}}(X) = 0$. Then the commuting graph on X^L is connected as $L = \langle N_L(X), N_L(X^g) \rangle$ for $X \neq X^g \leq C_L(X)$. Hence the lemma follows from an observation of Alperin and Gorenstein which appears as Exercise 6.4 in [FGT].

So take $w = 2$. Here if V is a 5-dimensional $GF(2)L$-module with $[V, L] \cong \tilde{Q}$ then $D_{10} \cong Y \leq L$ satisfies $[V, L] = [V, Y]$, so if $C_V(Y) \neq C_V(L)$ then V is an image of the permutation module U for L on L/Y. As the 4-dimensional composition factor of U is not isomorphic to $\tilde{Q}$, this is a contradiction. Thus V splits over $[V, L]$, so 17.11 in [FGT] completes the proof.

Lemma 46.5: *Let* $w = 3$ *and* $Y \in \Delta_H$. *Then* $C_G(Y)$ *is quasisimple with* $C_G(Y)/Y \cong U_4(3)$ *and* $|N_G(Y) : C_G(Y)| = 2$.

Proof: Let $N_G(Y)^* = N_G(Y)/Y$, $K = C_G(Y)$, and $P = C_Q(Y)$. Then $F^*(C_H(Y)^*) = P^*$ with $O^2(C_H(Y)^*) \cong SL_2(3) * SL_2(3)$ and $4 = |C_H(Y) : O^2(C_H(Y))|$. By 8.13 and 46.1.2, all involutions in P are in z^K. In particular we may take $E \leq P$ and $g \in K$; then by 46.2.1, $E = P \cap P^g$. Thus K^* is of type $U_4(3)$, as defined in Section 45. Therefore by 45.11, $K^* \cong U_4(3)$.

Next as $|N_H(Y) : C_H(Y)| = 2$, $|N_G(Y) : K| = 2$. So it remains to show $Y \leq K^\infty$. But there is $h \in H$ with $Y^h \leq H_1 \trianglelefteq O^2(C_H(Y))$, $H_1 \cong SL_2(3)$, and $Y^h = [Y^h, s]$ for some 2-element $s \in C_H(Y)$. Then $Y^h = [Y^h, s] \leq K^\infty$ and if $Y \not\leq K^\infty$ then by 45.8, $Y^h \leq \Phi(S_0)$ for some $S_0 \in Syl_3(N_K(Y^h))$. Thus $Y \leq \Phi(S)$ for $S \in Syl_3(K)$, contradicting $K = Y \times K^\infty$.

Lemma 46.6: *Let* $w = 4$, $X \in \Delta_H$, *and* $X \leq Y \leq H$ *with* $Y \cong E_9$ *and* $C_Q(Y) \cong Q_8^2$. *Then*

    (1) $C_G(X)$ is quasisimple with $C_G(X)/X \cong Suz$.

    (2) $C_G(Y)$ is quasisimple with $C_G(Y)/Y \cong U_4(3)$.

Proof: To prove $C_G(X)/X \cong Suz$, argue as in 25.9. Actually this argument only shows $C_G(X)/X$ is of type Suz but proceeding by induction on the order G and appealing to 48.17 in an inductive context, up to isomorphism there is a unique group of type Suz. Notice (2) now completes the proof of (1), so it remains to prove (2). Let $P = F^*(C_G(z))$. As in 25.9, $C_Q(X) \cong Q_8^3$. Then there is $h \in N_H(Y)$ with $Y = XX^h$ and X^h inverted in $C_H(X)$. Thus by 46.5, $X^h \leq C_G(Y)^\infty$. Finally as $h \in N_G(Y)$, $X \leq C_L(Y)^\infty$, establishing (2).

Lemma 46.7: *Let* $w = 3$ *and* Γ *the set of involutions* $i \in H$ *such that* i^* *is of type* c_2. *Then*

    (1) H is transitive on Γ.

    (2)   For $i \in \Gamma$, $|C_H(i)| = 2^9$.

Proof: We recall from Exercise 2.11 that i^* is of type c_2 if $[\tilde{Q}, i]$ is 2-dimensional and not totally singular and that if $i \in \Gamma$ then $iz \in i^Q$, each involution in iQ is in iQ_i^+, and $i^Q = iQ_i^-$, where $\tilde{Q}_i^+ = C_{\tilde{Q}}(i)$ and $\tilde{Q}_i^- = [\tilde{Q}, i]$.

Pick $Y \in \Delta_H$; by 46.5, $C_G(Y)/Y \cong U_4(3)$, so there is an element $t \in C_H(Y)$ with $t^2 = z$ and $t^* \in c_2$. Now t centralizes $Y[Q, Y] \cong SL_2(3)$ with $Q_t^- = C_Q(Y) \cap Q_t^+$, so by the previous paragraph each involution $i \in tQ$ is conjugate under YQ to tu for some fixed $u \in [Q, Y]$ of order 4. In particular there are 24 involutions in tQ, so as H^* is transitive on c_2 of order 270, H is transitive on Γ of order $24 \cdot 270$. Thus $C_H(i) = |H|/(24 \cdot 270) = 2^9$.

Lemma 46.8: *H is determined up to isomorphism independently of G.*

Proof: Let H_0 ' e the universal covering group of H and

$$\hat{H} = H_0/O(H_0)\Phi(O_2(Z(H_0))).$$

Let $Z = Z(\hat{H})$ and $P = [O_2(\hat{H}), \hat{H}]$. Then by 46.4 and 8.17, $P \cong Q$ and $H = \hat{H}/U$ for some complement U to $Z(P)$ in Z and $\hat{H}/P \cong L$, where $L = L_0/O(L_0)\Phi(O_2(Z(L_0)))$ and L_0 is the universal covering group of H^*. Let $\langle \pi \rangle = Z(P)$.

Suppose first $w = 2$. Then $L \cong SL_2(5)$ (cf. 33.15 in [FGT]). Thus $Z \cong E_4$. Also H^* has one class of involutions and $m([\tilde{Q}, t]) = 2$ for each such involution t, so $\tilde{H}$ is transitive on involutions in $\tilde{H} - \tilde{Q}$ by Exercise 2.8. Thus if xZ is an involution in $\hat{H}/Z - PZ/Z$ then $x^2 = \sigma$ is the unique element of Z such that $t^2 = \sigma$ for some $t \in \hat{H} - PZ$ with $t^2 \in Z$. Notice as $Z \cong E_4$, $Z = \langle \sigma, \pi \rangle$ and $|U| = 2$.

We claim $U = \langle \sigma \rangle$. For if not $U = \langle \sigma\pi \rangle$, so for each $t \in H - Q$ with $t^2 \in \langle z \rangle$, $t^2 = z$. That is, Q contains all involutions in H. This contradicts 46.3.3 which says H contains an S_3 subgroup.

Suppose next that $w = 3$. The argument is similar. This time by Exercise 16.6, $Z \cong E_4$ and each involution in H^* of type c_2 lifts to an element of $\hat{H}/P$ of order 4. Now by 46.7 and its proof, $\hat{H}$ has two orbits $t_i^{\hat{H}}$, $i = 1, 2$, on the set $\hat{\Gamma}$ of $t \in \hat{H}$ with $t^2 \in Z$ and tPZ of type c_2 in $\hat{H}/PZ \cong H^*$ with $t_2 = t_1x$ for some element x of order 4 in $C_P(t_2)$, $|t_1^P| = 8$, and $|t_2^P| = 24$. Then as involutions of H^* of type c_2 lift to elements of order 4 in $\hat{H}/P$, $\sigma = t_1^2 \neq 1$; then $t_2^2 = \sigma x^2 = \sigma\pi$. In this case we claim $U = \langle \sigma\pi \rangle$. If not $U = \langle \sigma \rangle$ and $t = t_1U \in \Gamma$ (where Γ is defined in 46.7), so 46.7 supplies a contradiction as $|t^Q| = |t_1^P| = 8$.

Finally take $w = 4$. We lift $X \in \Delta_H$ to a Sylow 3-group in its preimage in $\hat{H}$, and hence regard X also as a subgroup of $\hat{H}$. Now X centralizes an involution $i \in H$ with i^* of type c_2, and by the previous paragraph i lifts to $t \in \hat{H}$ of order 4 with $t^2 \in U$.

By Exercise 2.11, H^* has two classes of involutions of type a_4 with representatives j_1 and j_2, where $[\tilde{Q}, j_i] = A_i$ are the two classes of maximal totally singular subspaces of $\tilde{Q}$. Further if α is a transvection in $O(\tilde{Q})$ then α induces an outer automorphism on H^* with $A_1^\alpha = A_2$ and $j_1^\alpha = j_2$. By Exercise 16.7, $m(Z) = 3$ and j_i lifts to an element t_i of order 4 in $\hat{H}/P$ with $Z = \langle \sigma_1, \sigma_2, \pi \rangle$, where $t_i^2 = \sigma_i$.

By 8.17.4, α lifts to an automorphism of $\hat{H}$, which we also denote by α. Then $t_i^\alpha \in t_{3-i}Z$, so α interchanges σ_1 and σ_2. Hence up to conjugation under α, $U = \langle \sigma_1, \sigma_2 \rangle$, $\langle \sigma_1, \pi\sigma_2 \rangle$, or $\langle \pi\sigma_1, \pi\sigma_2 \rangle$.

Next Q induces the full group of transvections on A_i with center $\langle z \rangle$, so $C_H(\tilde{A}_i) = QC_G(A_i)$ and $C_G(A_i)/A_i \cong C_{H^*}(\tilde{A}_i) \cong E_{64}$. Further the involutions in $C_{H^*}(\tilde{A}_i)$ are of type c_2 or j_1, so as $t^2 \in U$, $\Phi(C_G(A_i)) = 1$ if and only if $\sigma_i \in U$.

Now if $G = Co_1$ then (cf. the proof of 46.12) $\Phi(C_G(A_i)) = 1$ for exactly one of $i = 1, 2$, say $i = 1$. Thus in that case $\sigma_1 \in U$ but $\sigma_2 \notin U$. Therefore $U = \langle \sigma_1, \pi\sigma_2 \rangle$. Further $t^2 \in U$ is fixed by α, so $t^2 = \pi\sigma_1\sigma_2$. Now in the general case we know $t^2 \in U$, but of the three possibilities for U, only the second contains $t^2 = \pi\sigma_1\sigma_2$, so indeed H is determined up to isomorphism and the proof is complete.

Notice in the process of proving 46.8 we have also proved several other facts. First

Lemma 46.9: *Let $L \cong {\cdot}0 = Aut(\Lambda)$ be the automorphism group of the Leech lattice, $K \leq L$ be quasisimple with $K/Z(K) \cong Suz$, and $Z(L) \leq J \leq K$ with $J/Z(L)$ a root J_2-subgroup of $L/Z(L)$. Then the root 4-involutions of $K/Z(K) \cong Suz$, $L/Z(L) \cong Co_1$, and $J/Z(L) \cong J_2$ lift to elements of K, L, and J of order 4, and J is quasisimple.*

See 49.1.1 for the definition of a root J_2-subgroup of Co_1. Now $Z(K) \cong \mathbf{Z}_6$ and $Z(L) \cong \mathbf{Z}_2$, so from the proof of 46.8, $C_K(z)/O_3(Z(K)) = \hat{H}_K$ is $H_{0,K}/O(H_{0,K})$, where $H_{0,K}$ is the universal covering group of $C_K(z)/Z(K)$. Similarly $\hat{H}_L = \hat{H}_{0,L}/\langle \sigma_1 \rangle O(H_{0,l})$ in the language of the proof of 46.8. Our analysis in Section 48 will show that each root 4-involution i of $K/Z(K)$ contained in H is of type c_2 in H^* and hence a lift $\hat{i}$ in $\hat{H}_K$ or $\hat{H}_L$ is of order 4 from the proof of 46.8. Also $\hat{i}^2$ generates $Z(L)$, so as we can pick $\hat{i} \in J$, J is quasisimple.

Lemma 46.10: *If $w = 4$ and j_1 and j_2 are representatives for the two*

classes of involutions of H^* of type a_4 then there exist involutions $i \in H$ with $i^* = j_1$, but no involution $j \in H$ with $j^* = j_2$.

For if $t_i \in \hat{H}$ with $t_i PZ = j_i$ then we showed that $t_1^2 \in U$ but $t_2^2 \notin U$.

Lemma 46.11: *If $w = 2$ then H is transitive on the involutions in $H - Q$.*

Namely we saw that H is transitive on involutions in $\tilde{H} - \tilde{Q}$, while since an involution $t \in H - Q$ is of type c_2 on $\tilde{Q}$, $tz \in t^Q$ by Exercise 2.11, completing the proof.

Lemma 46.12: *Assume $w = 4$ and let $T \in Syl_2(H)$. Then*

(1) $J(T) = A \cong E_{2^{11}}$.

(2) $N_G(A)$ *is the split extension of A by M_{24} with A the Golay code module for $N_G(A)/A$.*

(3) $N_G(A)$ *has two orbits on $A^\#$: the octad involutions with representative z and the dodecad involutions.*

(4) *The isomorphism type of the amalgam*

$$H \leftarrow N_H(A) \rightarrow N_G(A)$$

is determined independently of the group G of type Co_1.

Proof: By 46.8, there is an isomorphism $\alpha : H \to \bar{H}$ of H with the centralizer $\bar{H}$ of a 2-central involution $\bar{z}$ in $\bar{G} = Co_1$. By 22.5, $\bar{G}$ contains a subgroup $\bar{M}$ which is the split extension of $\bar{A} \cong E_{2^{11}}$ by M_{24} with $\bar{A}$ the 11-dimensional Golay code module for M_{24}. By 23.10, $C_{\bar{A}}(\bar{A})$ is a point and then by 23.1 and 23.2, that point is $\tilde{\Lambda}_4^8$ and $\bar{N} = N_{\bar{G}}(\tilde{\Lambda}_4^8) = N_{\bar{G}}(\bar{A})$. By Exercise 7.4, $\bar{A} = J(\bar{M})$.

Let $A = \bar{A}\alpha^{-1}$. The isomorphism $\alpha : A \to \bar{A}$ induces an isomorphism $\alpha^* : GL(A)A \to GL(\bar{A})\bar{A}$ of the semidirect product of A by $GL(A)$ with the semidirect product of $\bar{A}$ by $GL(\bar{A})$. Now $\bar{M} = N_{\bar{G}}(\bar{A})$ satisfies conclusions (2) and (3) of the lemma. In particular we may regard $\bar{M}$ as a subgroup of $GL(\bar{A})\bar{A}$; let $M = \bar{M}\alpha^{-1} \leq GL(A)A$. Then $N_H(A) = N_{\bar{H}}(\bar{A})\alpha^{-1} \leq M$ is isomorphic to $N_{\bar{H}}(\bar{A})$ via α, so as $\bar{A} = J(\bar{M})$, also $A = J(T)$ and hence (1) holds. Similarly $N_H(A)$ is the split extension of A by the split extension H_0 of E_{16} by $L_4(2)$. Further $N_H(A)$ is the stabilizer of the octad involution z in the Golay code module for a complement M_0 to A in M containing H_0. Let $h \in O_2(H_0)$; then by 8.10, $C_{H_0}(h) = C_{M_0}(h)$ and hence by 8.8 is isomorphic to the centralizer of a transvection in $L_5(2)$.

Let $N = N_G(A)$ and $N^* = N/A$. We claim $C_{N^*}(h^*) = C_{H_0}(h)^* \cong C_{H_0}(h)$. First $C_G(A) = C_H(A) = A$, so N^* is faithful on A and

$H_0 \cong H_0^* = C_{N^*}(z)$. Now $h \in Q - A$, so h induces a transvection on $Q \cap A \cong E_{32}$ with center z and axis $B = C_{Q \cap A}(h)$. Further $V = C_{H_0}(B) \cong E_{16}$ with H_0/V the stabilizer of z in $GL(B)$. So if we set $H_1 = C_N(h) \cap N(B)$, $C_{H_1^*}(B) = C_{H_1^*}(B) \cap C(z) = V^*$ and H_0^*/V^* is maximal in $GL(B)$, so either $C_{H_0}(B)^* = H_1^*$ or $H_1^*/V^* = GL(B)$. The latter is impossible as H_1^* fixes h^* and $Aut(V^*) \cap C(h^*)$ contains no $L_4(2)$-section, so $H_1^* \leq C(V^*)$, whereas $V^* = C_{H_0^*}(V^*)$.

Thus $H_1^* = C_{H_0}(h)^*$. We next show $B = [A, h]$, so $C_{H_0}(h)^* = C_N(h)^*$. Indeed as A is the Golay code module for M_0, $[O_2(H_0), A] \leq Q \cap A$, so $[A, h] \leq C_{Q \cap A}(h) = B$. On the other hand M contains no transvection, so $m([A, h]) > 1$ and then as $C_{H_0}(h)$ is irreducible on $B/\langle z \rangle$, $B = [A, h]$.

So $C_N(h)^* \cong C_{H_0}(h)$ is the centralizer of a transvection in $L_5(2)$. Further $O_2(H_0)^\# = h^{H_0} \subseteq O_2(C_{H_0}(h)) = Q_0$, so by 44.4, either $N = N_H(A)$ or $N^* \cong L_5(2)$, M_{24}, or He. The first case is impossible as $A = J(T)$, so A is weakly closed in T with respect to G, and hence by 7.7, N controls fusion in A. Thus z is fused to z^g in N.

As $H_0 \cong H_0^*$ is a 2-local subgroup of N^*, 40.5 says N^* is not He. Suppose $N^* \cong L_5(2)$. Then $|N : N_H(A)| = 31 = |Q \cap A^\#|$, so as $Q \cap A^\# \subseteq z^G \cap A = z^N$, $Q \cap A^\# = z^G \cap N$. This is a contradiction as $N_G(R) \cap N_G(A)$ does not act on $Q \cap A$ from 46.3.

So $N^* \cong M_{24}$. As H_0 is a complement to A in $N_H(A)$ and $N_H(A)$ contains the Sylow 2-subgroup T of M, N splits over A by Gaschütz's Theorem (cf. 10.4 in [FGT]).

Next $A = \langle z^N \rangle$ with $H_0^* = C_{N^*}(z)$ and $m(Q \cap A) = 5$ with H_0^* the stabilizer in $GL(Q \cap A)$ of z. Thus by Exercise 7.2, A is the Golay code module as an N^*-module. Thus (2) holds. By 19.8, (3) holds.

We have shown there is an isomorphism $\zeta : N \to \bar{N}$ with $(H \cap N)\zeta = \bar{H} \cap \bar{N} = (H \cap N)\alpha$. To prove (4) it remains to show we can pick α and ζ so that $\alpha = \zeta$ on $H \cap N = I$. For this we observe $I/A = Aut(I/A)$. For example, $I/O_2(I) = GL(O_2(I)/A)$ and $H^1(GL(O_2(I)/A), O_2(I)/A) = 0$ by the Alperin-Gorenstein argument (cf. Exercise 6.4 in [FGT]) applied to the class of subgroups of order 3 that are fixed point free on $O_2(I)/A$. Then the claim is easy.

Thus by Exercise 14.3.1 we may assume $\alpha = \zeta$ on I/A. Finally A is generated by the orbits of I on z^N of length 30 and 240 (cf. 19.2) and as $\alpha = \zeta$ on I/A, $\alpha = \zeta$ on these orbits, so $\alpha = \zeta$ on I.

When $w = 4$, 46.12 says there are two G-classes of involutions in A corresponding to the octads and dodecads via the identification of A with the Golay code module. Further z is an octad involution and we refer to the second class as the *dodecad involutions* of G. We will see eventually that G has one more class of involutions: the root 4-involutions.

Lemma 46.13: *Let $w = 4$, $T \in Syl_2(H)$, $A = J(T)$, and $d \in A$ a dodecad involution. Then*

   (1)   *d is contained in a unique G-conjugate of A; in particular $C_G(d) \leq N_G(A)$.*

   (2)   *$C_G(d)$ is the split extension of A by $Aut(M_{12})$ and is determined up to isomorphism.*

Proof: By 46.12, $N = N_G(A)$ is the split extension of A by M_{24} with A the Golay code module, so as d is a dodecad involution, $C_G(d) \cap N$ satisfies (2); for example, see 19.8.3. Further by 46.12.1 and 7.7, N controls fusion in A. Therefore if $C_G(d) \leq N$ then (1) holds by 7.3. Thus we may assume $C_G(d) \nleq N$ and it remains to derive a contradiction.

By 26.3 the result holds if $G = Co_1$ so by 46.12.4, $C_H(d) \leq N$ and hence $C_G(\langle u, d \rangle) \leq N$ for $u \in z^N$. So as $C_N(d)$ controls $C_G(d)$-fusion in A, 7.3 says each $u \in z^G \cap A$ fixes a unique point of $C_G(d)/C_N(d)$. This puts us in a position to appeal to Exercise 2.10.

Let $d \in B \leq A$ with B maximal subject to $C_G(B) \nleq N_G(A)$. Let $K = C_G(B)$, $M = N \cap K$, and $K^* = K/B$. Then by Exercise 2.10, M^* is transitive on $A^{*\#}$. Thus $2^{11-m} - 1$ divides $|M|$, where $m = m(B)$. However, we see in a moment that $m \leq 3$, whereas $2^{11-m} - 1$ does not divide $|M_{24}|$ for $1 \leq m \leq 3$.

So it remains to show $m \leq 3$. But by Exercise 2.10, $z \notin B$, so each member of $B^{\#}$ is a dodecad involution. Thus $m \leq 3$ by Exercise 7.3.

Lemma 46.14: *Let $w = 2$ and $R \leq T \in Syl_2(H)$. Then*

   (1)   *R is isomorphic to a Sylow 2-subgroup of $L_3(4)$ and $N_G(R)$ is the split extension of R by $\mathbf{Z}_3 \times S_3$.*

   (2)   *$R = J(T)$ and $A(R) = \{A_1, A_2\}$ with $A_i \cong E_{16}$ and $R = N_T(A_i)$.*

   (3)   *$N_G(R) \cap N_G(A_i)$ is transitive on $A_i - E$ and $E^{\#}$.*

   (4)   *$N_G(A_i)$ controls fusion in A_i.*

   (5)   *Either*

      (a)   *$N_G(A_i)$ is the split extension of A_i by $GL_2(4)$ acting naturally on A_i and G has one conjugacy class of involutions, or*

      (b)   *$N_G(A_i) \leq N_G(R)$ and G has two classes of involutions with representatives z and $a \in A_i - E$. Moreover Q is the weak closure of z in H.*

Proof: By 46.3, $|R| = 64$ with $N_G(R)$ the split extension of R by $X = X_0 \times X_3$ with $X_0 \cong S_3$ and $X_3 \cong \mathbf{Z}_3$, and R/E is the tensor product of the natural modules for X_0 and X_3. Further X_0 is faithful on E, so $C_X(E) = X_3$.

Now X has two more conjugate subgroups X_1 and X_2 of order 3 and $R = C_R(X_1)C_R(X_2)E$ with $C_R(X_i) \cong E_4$ for $i = 1, 2$. Therefore $A_i = EC_R(X_i) \cong E_{16}$ with $R = A_1A_2$ and $A_1 \cap A_2 = E$. Moreover X_2 induces a $GF(4)$-structure on A_1 preserved by $C_R(X_2)X_1X_2$, so $C_R(X_2)X_1X_2$ acts as a Borel group of $GL_2(4)$ on A_1, establishing (1)–(3).

If $b, b^h \in A_1$ then $b \in A_1 \cap A_1^{h^{-1}}$, and then by (2), $A_1^{h^{-1}} = A_1^c$ for some $c \in C_G(b)$, so $ch \in N_G(A_1)$ with $b^{ch} = b^h$. Hence (4) holds.

Next by 46.11, H is transitive on involutions in $H - Q$, while by 46.12 each involution in Q is fused to z in G, so either G has one class of involutions or two classes with representatives z and $a \in A_1 - E$ and Q is the weak closure of z in H. In the first case by (4), $N_G(A_1)$ is transitive on $A_1^{\#}$, so $|N_G(A_1)| = 15 \cdot |N_H(A_1)| = 2^4 \cdot |GL_2(4)|$. Let $K = N_G(A_1)$. Then $|K : N_K(R)| = 5$ so the permutation representation π of K on $K/N_K(R)$ maps K into S_5. We conclude $A_1X_2 = ker(\pi)$ and $K\pi \cong A_5$. Thus (a) holds in this case.

In the second case K acts on $z^G \cap A_1 = E^{\#}$, so $K \leq N_G(R)$ and (b) holds.

If $w = 2$ and G has one class of involutions we say G is of *type J_3*. If $w = 2$ and G has two classes of involutions we say G is of *type J_2*. In previous sections we have already defined G to be of *type Co_1* if $w = 4$ and of *type Suz* if $w = 3$. Thus the hypotheses of this section can be restated to say that G is of type J_2, J_3, Suz, or Co_1.

In the remaining sections in this chapter we prove that up to isomorphism there is a unique group of type J_2, Suz, and Co_1, respectively. There is much parallelism in the proofs, so we could prove a few more lemmas simultaneously for each type, but soon the proofs would have to diverge. We choose to begin the separation after the next lemma, and even that lemma makes appeals to facts established later in this chapter.

Lemma 46.15: *G is simple.*

Proof: If G is not of type J_2 we appeal to Exercise 2.4. To apply that result we need only show Q is not the weak closure of z in H. This is true if G is of type J_3 since G has one class of involutions and we have seen there are involutions in $H - Q$. If G is of type Suz or Co_1 this follows from 48.4, 49.4, and their proofs.

So assume G is of type J_2. Then by 46.13, Q is the weak closure of z in H, so we need to work harder. Let $M = \langle z \rangle^G$. Then by 8.11, $M = F^*(G)$ is simple. Then $Q = \langle z^G \cap H \rangle \leq M$ and as Q is weakly closed in H, $G = MN_G(Q) = MH$ by a Frattini argument. Thus it remains to show $H \leq M$, so assume not.

As $Q \leq M$ and H/Q is simple, $Q = H \cap M$. Let $X \in \Delta_H$. By 47.3.1, $K = C_G(X) = \langle z^G \cap K \rangle$, so $K \leq M$. This is a contradiction as X is not contained in $Q = H \cap M$.

47. Groups of type J_2

In this section we assume G is of type J_2 as defined at the end of the previous section. Moreover we continue the notation of that section. In addition let $R \leq T \in Syl_2(H)$; by 46.14, $\mathcal{A}(T) = \{A_1, A_2\}$ with $A_i \cong E_{16}$, $R = A_1 A_2$, and $A_1 \cap A_2 = E$.

Lemma 47.1: *(1) For* $t \in A_1 - E$, $C_G(t) \cong E_4 \times A_5$ *and* $U(t) = O_2(C_G(t))$ *is a TI-set in* G *with* $N_G(U(t)) \cong A_4 \times A_5$, $z \in E(C_G(t))$, *and* X_3 *a Sylow 3-subgroup of* $C_G(E(C_G(t)))$ *in* Δ_H.
(2) $|G| = 2^7 \cdot 3^3 \cdot 5^2 \cdot 7$.

Proof: By 46.14.5, G has two classes of involutions with representatives z and t. This allows us to use the Thompson Order Formula 7.2 to calculate the order of G. For $x = z$ or t define $a(x) = |P(x)|$, where

$$P(x) = \{(u, v) : u \in z^G,\ v \in t^G,\ x \in \langle uv \rangle\}.$$

Then by the Thompson Order Formula,

$$|G| = a(z)|C_G(t)| + a(t)|C_G(z)|.$$

Thus we must determine $a(x)$ and $|C_G(x)|$. We recall that if $(u, v) \in P(x)$ then as $\langle u, v \rangle$ is dihedral and $x \in \langle uv \rangle$, $x \in Z(\langle u, v \rangle)$ so $\langle u, v \rangle \leq C_G(x)$.

Observe that $|C_G(z)| = |H| = 2^7 \cdot 3 \cdot 5$. We claim $a(z) = 2^6 \cdot 3 \cdot 5$. For if $(u, v) \in P(z)$ then by 46.14.5, $u \in Q$, while by 46.1, $v \in H - Q$. Then $uv \notin Q$, so $uv \neq z$ and hence $[u, v] \neq 1$. But if $\tilde{u} \in C_{\tilde{Q}}(v)$ then $[u, v] = 1$, so $[\tilde{u}, v] \neq 1$ and $\langle u, v \rangle \cong D_{16}$. Conversely for each involution $u \in Q$ and each involution $v \in H - Q$ with $[\tilde{u}, v] \neq 1$, $z \in \langle u, v \rangle$, so $(u, v) \in P(z)$. Thus $P(z)$ consists of such pairs.

Now $t^G \cap H = t^H$ is of order $2^3 \cdot 3 \cdot 5$, while $C_{\tilde{Q}}(t)$ contains a unique singular point, so there are eight choices for u with $v = t$. Hence $a(z) = 2^6 \cdot 3 \cdot 5$, as claimed.

Suppose that $C_G(t) \leq N_G(R)$. In the notation of the proof of 46.14, we may take $t \in C_R(X_1)$, so $C_G(t) = X_1 A_1$. Now $(u, v) \in P(t)$ if and only if $u \in E^{\#}$ and $v = ut$, so $a(t) = 3$. Also $|C_G(t)| = 3 \cdot 2^4$, so

$$
\begin{aligned}
|G| &= a(z)|C_G(t)| + a(t)|H| \\
&= 2^6 \cdot 3 \cdot 5 \cdot 2^4 \cdot 3 + 3 \cdot 2^7 \cdot 3 \cdot 5 \\
&= 2^7 \cdot 3^2 \cdot 5 \cdot (1 + 8) = 2^7 \cdot 3^4 \cdot 5.
\end{aligned}
$$

Now Sylow's Theorem supplies a contradiction. For let $P \in Syl_5(H)$. Then $N_H(P) \cong D_{20}$ so $\langle z \rangle \in Syl_2(C_G(P))$ and $|N_G(P)|_2 = 4$. But $|G : P| \equiv -2 \mod 5$, so as $|G : N_G(P)| \equiv 1 \mod 5$, $|N_G(P) : P| \equiv -2 \mod 5$. Thus as $|N_G(P)|_2 = 4$, $V = O_3(N_G(P)) \in Syl_3(N_G(P))$ is of order 3^3. As $N_H(P)$ has order prime to 3, z inverts V. Now we may take t to invert P and $V = C_V(t) \times C_V(tz)$ with $|C_V(t)| = 3^2$ and $|C_V(tz)| = 3$, contradicting $|C_G(t)|_3 = 3$.

So $K = C_G(t) \not\leq N_G(R)$. Let $K^* = K/\langle t \rangle$. Now $A \in Syl_2(K)$ and E is strongly closed in A, so by Thompson Transfer (or use the more general 37.4 in [FGT]) applied to A^*, K^* has a subgroup K_0^* of index 2 with $E^* \in Syl_2(K_0^*)$. Then as $E^* = C_{K_0^*}(z^*)$, Exercise 16.6 in [FGT] says $K_0^* \cong A_5$. As above we may take $t \in C_R(X_1) = U(t)$. As $U(t)$ centralizes X_1E, $[U(t), K_0^*] = 1$, so $K = U(t) \times E(K)$ with $E(K) \cong A_5$. Notice $E = [A_1, X_1] \in Syl_2(E(K))$. Further X_3 acts on $C_G(U(t)) = K$ and centralizes X_1E, so $[E(K), X_3] = 1$ and $N_G(U(t)) = X_3U(t) \times E(K) \cong A_4 \times A_5$. Now $z \in E \leq E(K)$ so $X_3 \leq C_G(z) = H$ and hence $X_3 \in Syl_3(H) = \Delta_H$. That is, (1) is established.

Now we use the Thompson Order Formula to complete the proof of (2). We have shown $|C_G(t)| = 2^4 \cdot 3 \cdot 5$, $|C_G(z)| = 2^7 \cdot 3 \cdot 5$, and $a(z) = 2^6 \cdot 3 \cdot 5$, so it remains to calculate $a(t)$. If $(u, v) \in P(t)$ then $u \in z^G \cap K \subseteq E(K)$ and $v \in t^G \cap K \subseteq K - E(K)$, so $v = wx$, $w \in U(t)^{\#}$ and $x \in z^K \cup \{1\}$. Then $t \in \langle uv \rangle = \langle wux \rangle$ and $(wux)^2 = (ux)^2 \in E(K)$, so $|ux|$ is odd and $w = t$. Thus $|ux| = 1$, 3, or 5 and there are 15, 60, 120 pairs (u, v) of the respective type. Therefore $a(t) = 195 = 3 \cdot 5 \cdot 13$. So by the Thompson Order Formula,

$$
\begin{aligned}
|G| &= a(z)|C_G(t)| + a(t)|C_G(z)| \\
&= 2^6 \cdot 3 \cdot 5 \cdot 2^4 \cdot 3 \cdot 5 + 195 \cdot 2^7 \cdot 3 \cdot 5 \\
&= 2^7 \cdot 3^2 \cdot 5^2 \cdot (8 + 13) = 2^7 \cdot 3^3 \cdot 5^2 \cdot 7
\end{aligned}
$$

establishing (2).

We call the conjugates of $U(t)$ the *root 4-subgroups* of G and the conjugates of t the *root 4-involutions* of G. Denote by Δ the set of G-conjugates of members of Δ_H; that is, Δ is the set of subgroups of G of order 3 centralizing a conjugate of z.

Lemma 47.2: *Denote by $5_A, 5_B$ the conjugacy class of subgroups of G of order 5 centralizing a root 4-involution, 2-central involution of G, respectively, and let $P \in Syl_5(G)$. Then*

(1) 5_A *and* 5_B *are the conjugacy classes of subgroups of order 5 in G.*

(2)   *For $X \in 5_A$, $N_G(X) \cong D_{10} \times A_5$, $E(C_G(X)) = \langle \Delta \cap N_G(X) \rangle$, X is inverted by a conjugate z^y of z, and $E(C_G(X))$ is a complement to Q^y in H^y.*

(3)   *For $Y \in 5_B$, $N_G(Y) = D_{10} \times D_{10}$.*

(4) $P = C_G(P)$ *and $N_G(P)$ is the split extension of P by D_{12}.*

Proof: Let $N = N_G(P)$. Then by 47.1 and Sylow's Theorem, $|P| = 5^2$ and $|N : P| \equiv |G : P| \equiv 2 \mod 5$.

Let t be a root 4-involution, $X \in Syl_5(C_G(t))$, $M = C_G(X)$, and $M^* = M/X$. By 47.1, $U(t) \in Syl_2(M)$ and $U(t)^* = C_{M^*}(t^*)$. Also taking $X \leq P$, $P \leq M$, so $M \not\leq C_G(t)$. Thus by Exercise 16.6 in [FGT], $M^* \cong A_5$. Further we may take $z \in E(C_G(t))$ to invert X, so z centralizes $C_G(E(C_G(t))) \cap N_G(U(t)) \cong A_4$ and hence $[z, E(M)] = 1$. Finally $N_G(X) = E(M)C_M(t) = E(M) \times X\langle z \rangle \cong A_5 \times D_{10}$, so (2) holds.

Next let $Y = P \cap E(M)$, so that $P = X \times Y \cong E_{25}$. Without loss of generality t inverts Y. Now $\langle z \rangle \in Syl_2(C_H(Y))$ with $N_H(Y) = Y\langle t, z \rangle$, so $C_G(Y) = O(C_G(Y))\langle z \rangle$ with z inverting $O(C_G(Y))/Y$. Therefore $O(C_G(Y)) = Y \times [O(C_G(Y)), z]$ is abelian, so $C_G(Y) \leq C_G(P) = C_M(P) = P$ and thus $N_G(Y) = P\langle t, z \rangle \cong D_{10} \times D_{10}$, so (3) holds.

As $\langle t, z \rangle \in Syl_2(N_G(Y))$ with X, Y the fixed points of $\langle t, z \rangle$ on P and $X \notin Y^G$, $\langle t, z \rangle \in Syl_2(N)$ and then by the Burnside p-Complement Theorem (cf. 39.1 in [FGT]), $N = O(N)\langle t, z \rangle$. As $|N : P| \equiv 2 \mod 5$, $|O(N) : P| \equiv 3 \mod 5$. Then as $N/P \leq GL_2(5)$, $N/P \cong D_{12}$. Therefore (4) holds and N has two orbits on the six subgroups of P of order 5, so (1) holds.

Lemma 47.3: *Denote by $3_A, 3_B$ the conjugacy class of elements of G of order 3 centralizing a 2-central, root 4-involution of G, respectively, and let $P \in Syl_3(G)$. Then*

(1)   *For $x \in 3_A$, $\langle x \rangle^G = \Delta$, $C_G(x)$ is quasisimple with $C_G(x)/\langle x \rangle \cong A_6$, and $N_G(\langle x \rangle)/\langle x \rangle \cong PGL_2(9)$.*

(2)   *For $y \in 3_B$, $N_G(\langle y \rangle) \cong S_3 \times A_4$.*

(3) $P \cong 3^{1+2}$, *$N_G(P)$ is the split extension of P by $\mathbf{Z}_8$ faithful on $P/\Phi(P)$, and $Z(P) \in \Delta$ is weakly closed in P with respect to G.*

(4) 3_A *and 3_B are the conjugacy classes of elements of G of order 3.*

Proof: First let $D \in Syl_7(G)$. Then $|N_G(D) : D| \equiv |G : D| \equiv 6$ mod 7. Further by 47.1 and 47.2, no element of order 2 or 5 centralizes an element of order 7, so $|N_G(D) : D| = 2^a \cdot 3^b$ with $a = 0, 1, 0 \le b \le 3$. Thus $|N_G(D) : D| = 6$ or 27.

Next let $K = E(C_G(t))$ and $X \in Syl_3(C_G(K))$. Thus $X \in \Delta_H$ and $X = \langle x \rangle$ for some $x \in 3_A$. Now $N_H(X) = SX$, where S is semidihedral of order 16 and $C_S(X) = C_Q(X) \cong D_8$. Let $X \le I \in Syl_3(XK)$ and $I \le P$. Choose I to be z-invariant. Then $I = X \times Y$, where $Y = [I, z]$ is generated by $y \in 3_B$. Indeed setting $M = C_G(Y)$ and $M^* = M/Y$, we have $U(t)^* = C_{M^*}(t^*)$, so by Exercise 16.6 in [FGT], either $U(t) \trianglelefteq M$, so (2) holds, or $M \cong Y \times A_5$. In either case $P \not\le C_G(Y)$, so P is nonabelian and as X and Y are the only subgroups of I fixed by z, $X = Z(P)$.

Now let $L = C_G(X)$. Then $K \le L$ and $K_5 \in Syl_5(K) \subseteq Syl_5(L)$ as a Sylow 5-subgroup of G is self-centralizing. Also $P \in Syl_3(L)$ and $C_Q(X) \in Syl_2(L)$, so $|L| = 2^3 \cdot 3^3 \cdot 5 \cdot 7^c$, $c = 0, 1$. If $c = 1$ we may take $D \le L$ and $|N_L(D) : D| \equiv |L : D| \equiv 2$ mod 7, so $DX < N_L(D)$. But by a Frattini argument in $N_G(X)$, $|N_G(X) \cap N_G(D) : N_L(D)| = 2$, so as $|N_G(D) : D| = 6$ or 27, $|N_G(D) : D| = 6$, contradicting $N_L(D) \ne DX$. Hence $c = 0$ and $|L : KX| = 6$. So as $K \cong A_5$ it follows that $L/X \cong A_6$. Then as P is nonabelian, L is quasisimple and $P \cong 3^{1+2}$. Also as S is semidihedral, $N_G(X)/X \cong PGL_2(9)$, so (1) is established.

Further we have seen that if (2) fails then $C_G(Y) \cong Z_3 \times A_5$. But by 47.2, only B of type 5_A centralizes an element of order 3 and a Sylow 3-group of $C_G(B)$ is of order 3. Thus $3_A = 3_B$, whereas such elements have nonisomorphic centralizers. Hence (2) holds.

As $X = Z(P)$, $N_G(P) \le N_G(X)$, so by (1), P is the split extension of P by Z_8 transitive on $(P/X)^\#$. Thus $N_G(P)$ is transitive on elements of P of order 3 in $P - X$, so all are in 3_B. Thus X is weakly closed in P, and (3) and (4) are established.

Notice in the process we have also shown:

Lemma 47.4: *The normalizer of a Sylow 7-subgroup of G is a Frobenius group of order 21.*

For we showed during the proof of the previous lemma that if $D \in Syl_7(G)$ then $|N_G(D) : D| = 6$ or 27. But by 47.1 and 47.3, elements of order 2 and 3 do not centralize elements of order 7, so the lemma holds.

Notice we have now determined the order of each group G of type J_2, its conjugacy classes and normalizers of subgroups of prime order, and the isomorphism type of the normalizers of Sylow groups of G.

The remainder of this section is devoted to a proof that, up to isomorphism, there is a unique group of type J_2. In the process we generate a great deal of information about the set Δ of subgroups of order 3 centralized by a 2-central involution. Thus in the remainder of this section let $X \in \Delta$ and $M = N_G(X)$.

Lemma 47.5: *(1) M has four orbits on Δ: $\{X\}$, $\Delta(X)$, $\Delta_1^2(X)$, and $\Delta_2^2(X)$.*

(2) $\Delta(X) = \{Y \in \Delta : \langle X, Y \rangle \cong A_4\}$, $|\Delta(X)| = 36$, and $N_M(Y) \cong A_5$ for $Y \in \Delta(X)$.

(3) $\Delta_1^2(X) = \{Y \in \Delta : \langle X, Y \rangle \cong SL_2(3)\}$, $|\Delta_1^2(X)| = 135$, and $N_M(Y) \in Syl_2(M)$ for $Y \in \Delta_1^2(X)$.

(4) $\Delta_2^2(X) = \{Y \in \Delta : \langle X, Y \rangle \cong A_5\}$, $|\Delta_2^2(X)| = 108$, and $N_M(Y) \cong \mathbf{Z}_2 \times D_{10}$ for $Y \in \Delta_2^2(X)$.

Proof: By 47.2, $X \leq E(C_G(F)) = L$ for some $F \in 5_A$, $L \cong A_5$, and $C_G(L) = D \cong D_{10}$. Then there exist $Y, X_2 \in X^L$ with $\langle X, Y \rangle \cong A_4$ and $\langle X, X_2 \rangle = L$. In particular $N_M(X_2) = (M \cap L) \times D \cong \mathbf{Z}_2 \times D_{10}$, so $|X_2^M| = |M : N_M(X_2)| = 108$ and $X_2 \in \Delta_2^2(X)$.

Next $O_2(\langle X, Y \rangle) = U$ is a root 4-subgroup of G, so by 47.1,

$$N_G(\langle X, Y \rangle) = \langle X, Y \rangle \times K$$

with $K \cong A_5$. Then $K = N_M(Y)$, $|Y^M| = |M : K| = 36$, and $Y \in \Delta(X)$.

Finally we may take $z \in M$. Then $C_M(z) = XT$, $T \in Syl_2(M)$, and $\langle X^Q \rangle = \langle X, X_1 \rangle \cong SL_2(3)$ for suitable $X_1 \in \Delta_H$. Then $N_G(\langle X, X_1 \rangle) = N_H(\langle X, X_1 \rangle) = T \langle X, X_1 \rangle$, so $N_M(X_1) = T$, $|X_1^M| = |M : T| = 135$, and $X_1 \in \Delta_1^2(X)$. Now

$$|\Delta| = |G : M| = 280 = 1 + 36 + 135 + 108$$

so $\Delta = \{X\} \cup Y^M \cup X_1^M \cup X_2^M$, and the lemma holds.

We now regard Δ as a graph with $\Delta(X)$ the set of vertices adjacent to X in Δ. In order to prove the uniqueness of groups of type J_2, we prove Δ is simply connected. By 35.14 it suffices to show each cycle of Δ is in the closure $C_3(\Delta)$ of the set of triangles of Δ. This is accomplished in a series of lemmas.

Lemma 47.6: *Let $Y \in \Delta(X)$ and $K = N_M(Y)$. Then*

(1) $\Delta(X) \cap \Delta_1^2(Y) = \{Y^e : e \in E(M) - KX\}$ is an orbit of KY of order 15.

(2) $\{Y^m : m \in M - E(M)\}$ is an orbit of KY of order 18 such that for each $M \in M - E(M)$, Y^{mX} contains one member

of $\Delta(X,Y)$ and two members of $\Delta(X) \cap \Delta_2^2(Y)$. In particular $|\Delta(X) \cap \Delta_2^2(Y)| = 12$.

(3) $|\Delta(X,Y)| = 18$.

(4) For each $U \in \Delta$, $z^G \cap C_G(\langle X, Y, U \rangle) \neq \varnothing$.

(5) Each triangle of Δ is conjugate to (X,Y,Y_1) or (X,Y,Y_2), where $Y_1 \in \Delta \cap \langle X,Y \rangle - \{X,Y\}$ and $Y_2 \in \Delta(X,Y) - Y^{E(M)}$.

Proof: Let $L = E(M)$. Then L acts 4-transitively as A_6 on L/KX, so $\Gamma = \{Y^e : e \in L - KX\}$ is an orbit under KX of order 15 with K acting as A_5 on Γ/X of order 5. Further for each $Y^e \in \Gamma$, $K \cap K^e \cong A_4$ and we may take $E = O_2(K \cap K^e)$. Then $\langle X, Y, Y^e \rangle \leq C_G(E) = RX$. Therefore $\langle X,Y \rangle = X[A_1, X]$, $\langle X, Y^e \rangle = X[A_2, X]$, and $\langle Y, Y^{ex} \rangle \cong SL_2(3)$ for each $x \in X$. Thus $\Gamma \subseteq \Delta_1^2(Y)$.

Similarly for $t \in M - L$, K acts as $L_2(5)$ on L/XK^t of order 6, so $\theta = Y^{tXK}$ is of order 18 and K acts as $L_2(5)$ on θ/X. Further for $Y^m \in \theta$, $K \cap K^m \cong D_{10}$ and $\langle X, Y, Y^m \rangle \leq C_G(K \cap K^m) \cong A_5$ by 47.2. In particular $C_G(K \cap K^m)$ acts on a set Γ of order 5 with $Fix_\Gamma(X)$ of order 2 and $Fix(X) \cap Fix(Y)$ and $Fix(X) \cap Fix(Y^m)$ distinct points of Γ as $Y, Y^m \in \Delta(X)$ but $Y^m \notin Y^X$. Thus Y^{mX} contains a unique $Y_1 \in \Delta(Y)$ (i.e., the Y_1 with $Fix_\Gamma(\langle Y, Y_1 \rangle) \neq \varnothing$) and two members of $\Delta_2^2(Y)$.

We have established (1), (2), (3), and (5). Further we showed (4) holds in the case $U \in \Delta(X)$. If $U \in \Delta_1^2(X)$ then $C_L(U) = S \in Syl_2(L)$, so each point of L/XK is fixed by some involution of S. In particular if $s \in S$ fixes XK then $s \in K \cap S$, so (4) holds in this case. Finally if $U \in \Delta_2^2(X)$ then $C_L(U) \cong D_{10}$ and again some involution in $C_L(U)$ fixes the point KX of L/KX, so (4) holds here too.

Lemma 47.7: (1) Δ is of diameter 2 with $\Delta^2(X) = \Delta_1^2(X) \cup \Delta_2^2(X)$.

(2) $|\Delta(X,Y)| = 4$ for $Y \in \Delta^2(X)$.

(3) If $Y \in \Delta_2^2(X)$ then $\Delta(X,Y) \subseteq \Delta \cap \langle X,Y \rangle$ and $\Delta(X,Y)$ is connected.

Proof: Count the set Ω_i of pairs (U,Y) with $Y \in \Delta_i^2(X)$ and $U \in \Delta(X,Y)$ in two ways. Let $\lambda_i = |\Delta(X,Y)|$ and $\alpha_i = |\Delta(U) \cap \Delta_i^2(X)|$ for $U \in \Delta(X)$. Then $|\Delta_i^2(X)|\lambda_i = |\Omega_i| = |\Delta(X)|\alpha_i$, while 47.5 and 47.6 say $|\Delta(X)| = 36$ and $|\Delta_i^2(X)|/\alpha_i = 9$. Therefore (2) holds and of course (2) implies (1). Further if $Y \in \Delta_2^2(X)$ then $K = \langle X,Y \rangle \cong A_5$ and by inspection $\Delta(X,Y) \cap K$ is of order 4 and connected, so (3) holds. Namely, as in the proof of the previous lemma, if we view K as acting on a set Γ of order 5 then $U \in \Delta \cap K$ is in $\Delta(X)$ if $Fix_\Gamma(X) \cap Fix_\Gamma(U) \neq \varnothing$ and $U \in \Delta_2^2(X)$ otherwise, which makes the calculation easy.

We now set up the machinery to analyze Δ_H. First $\tilde{Q}$ has five singular points which we label $S = \{s_1, \ldots, s_5\}$; then $s_i = \tilde{z}_i = \tilde{z}_i \tilde{z}$ is the image of two involutions of Q. We choose $X \leq H$. Then X is contained in four complements to Q in H conjugate under $C_Q(X)$, and if A is such a complement then $A^* \cong A$ acts as A_5 on S. Notice each $a \in A_5$ is determined by its action on S. For example, we may choose notation so that $X = \langle x \rangle$ with $x = (s_1, s_2, s_3)$, which we usually write as $x = (1, 2, 3)$. Then $C_Q(X) = \langle z_4, z_5 \rangle \cong D_8$ and $[Q, x] = \langle z_1 z_2, z_1 z_3 \rangle \cong Q_8$.

Notice $U \in A \cap \Delta$ is in $\Delta(X)$ if U and X have a common fixed point on Δ and $U \in \Delta_2^2(X)$ otherwise.

Lemma 47.8: $\Delta(X, Y)$ *is connected for* $Y \in \Delta_1^2(X)$.

Proof: Without loss of generality $\langle z \rangle = Z(\langle X, Y \rangle)$, so $X^* = Y^*$. Then $Y = X^a$ for some $a \in Q$ and without loss of generality $a = z_1$. Let A be a complement to Q in H containing X as above and let $U_i = \langle u_i \rangle \leq A$ with $u_i = (i, 2, 3)$, for $i = 4, 5$. Then $U_i \in \Delta(X)$ and $a \in C_Q(U_i)$ so $Y = X^a \in \Delta(U_i)$. Also $U_4 \in \Delta(U_5)$. Now $z_i \in C_Q(U_{9-i}) - C_Q(U_i)$ for $i = 4, 5$, and $C_Q(\langle X, Y \rangle) = \langle z_4, z_5 \rangle$, so since $|\Delta(X, Y)| = 4$ by 47.7.2, we conclude $\Delta(X, Y) = \{U_4, U_4^{z_4}, U_5, U_5^{z_5}\}$. Finally $A_4 \cong \langle U_4, U_5 \rangle \cong \langle U_4, U_5 \rangle^{z_i} = \langle U_i^{z_i}, U_{9-i} \rangle \cong \langle U_4, U_5 \rangle^{z_4 z_5} = \langle U_4^{z_4}, U_5^{z_5} \rangle$, so $\Delta(X, Y)$ is connected.

Lemma 47.9: Δ *is simply connected.*

Proof: By 47.7, Δ is of diameter 2, so by 34.5, it suffices to show each r-gon is in $C_3(\Delta)$ for $r \leq 5$. By definition of $C_3(\Delta)$ (in Section 34) triangles are in $C_3(\Delta)$. Further if $Y \in \Delta^2(X)$ then by 47.7 and 47.8, $\Delta(X, Y)$ is connected, so squares are in $C_3(\Delta)$ by 34.6. Thus it remains to show that if $p = X_0 \cdots X_5$ is a pentagon, then $p \in C_3(\Delta)$. By 47.6.4 we may take $\langle X_0, X_2, X_3 \rangle \leq H$. Let $X_i = \langle x_i \rangle$.

Suppose $\langle X_0, X_2 \rangle = A \cong A_5$ and $X_3 \in \Delta_1^2(X_0)$. Without loss of generality $x_0 = (1, 2, 3)$ and $x_2 = (3, 4, 5)$. Now $I = \langle X_2, X_3 \rangle \cong A_4 \cong \langle X_0, X_3 \rangle^*$, so we may take $x_3^* = (1, 3, 4)$. As $\langle x_0, x_3 \rangle \cong SL_2(3)$, $x_3 \notin A$. But as $I \cong A_4$, $I \leq A^b = B$ for some $b \in C_Q(X_2) = \langle z_1, z_2 \rangle$. Further $C_Q(I) = \langle z_2 \rangle$, so $I \leq B \cap B^{z_2}$ and $B = z_1$ or $z_1 z_2$. Let $v = (2, 3, 4) \in A$; then $V = \langle v \rangle \in \Delta(X_0, X_2)$. If $b = z_1$ then $b \in C_Q(v)$, so $V = V^b \leq B$ and then $V \in \Delta(X_3)$. Similarly if $b = z_1 z_2$ then $V = V^{z_1} \leq A^{z_1} = B^{z_1 z_2 z_1} = B^{z_2}$, so again $V \in \Delta(X_3)$. Therefore in any case $V \in X_0^\perp \cap X_2^\perp \cap X_3^\perp$, and then 34.8 says $p \in C_3(\Delta)$.

Suppose $X_2, X_3 \in \Delta_2^2(X)$. Without loss of generality, $x_0^* = (1, 2, 3)$, $x_2^* = (3, 4, 5)$, and $x_3^* = (1, 4, 5)$. Then there is $U = \langle u \rangle \in \Delta(X_2, X_3)$

with $u^* = (1,3,4)$ and by the previous paragraph, if $V \in \Delta(U, X_0)$ then $X_0 X_1 X_2 U V X_0$, $X_0 V U X_3 X_4 X_0$, and $U X_2 X_3 U$ are in $C_3(\Delta)$, so p is too by 34.3.

This leaves the case $X_2, X_3 \in \Delta_1^2(X_0)$. Now if $U \in \Delta(X_2, X_3)$ with $U \in \Delta_2^2(X_0)$, then as in the previous paragraph, $p \in C_3(\Delta)$, so assume otherwise. Then $X_0^* = X_2^*$ or X_3^*, say the former. But then up to conjugation in H, $\langle X_0, X_2, X_3 \rangle \leq C_H(E) = R X_0$, with $z \in \langle X_0, X_2 \rangle$ and $Z(\langle X_0, X_3 \rangle) = e = z^g$ or zz^g. Then replacing z by ze, we reduce to a previous case.

Lemma 47.10: *Up to isomorphism there is a unique group of type J_2.*

Proof: By 25.10, a root J_2-subgroup J of Co_1 is of type J_2 or J_3. (Cf. 49.1.1 for the definition of a root J_2-subgroup of Co_1.) Further J contains root 4-involutions and 2-central involutions of Co_1, so J has more than one class of involutions, and hence J is of type J_2. That is, there exists a group of type J_2. Thus it remains to show that if $\bar{G}$ is a group of type J_2 then $\bar{G} \cong G$. For this we use Corollary 37.8. Thus we must first construct a uniqueness system for G.

Let $Y \in \Delta(X)$; by 47.5, $G_{XY} \cong A_5$. Let $D_{10} \cong D \leq G_{XY}$ and $K = C_G(D)$; by 47.2, $K \cong A_5$. We let $\Delta_K = \Delta \cap K$ regarded as a subgraph of Δ. Finally let $\mathcal{U} = (G, K, \Delta, \Delta_K)$ and observe that $\mathcal{U}$ is a uniqueness system for G in the sense of Section 37. For example, $\langle M, K \rangle$ is of type J_2, so by 47.1.2, $G = \langle M, K \rangle$. Also $E(M)$ is the unique maximal subgroup of M containing G_{XY}, so as $K_X \not\leq E(M)$, $M = \langle G_{XY}, K_X \rangle$.

Next $E(M)$ is the unique covering of A_6 over $\mathbf{Z}_3$ (cf. 33.15 in [FGT]) and M is the split extension of $E(M)$ such that $M/X \cong PGL_2(9)$, so there exists an isomorphism $\alpha : M \to \bar{M}$, where of course $\bar{M} = N_{\bar{G}}(\bar{X})$ for some $\bar{X} \in \bar{\Delta}$. Further as M is transitive on its A_5-subgroups (cf. Exercise 5.1 in [FGT]), $G_{XY}\alpha = \bar{G}_{\bar{X}\bar{Y}}$ for some $\bar{Y} \in \Delta(\bar{X})$. Let $\bar{K} = C_{\bar{G}}(D\alpha)$ and form the uniqueness system $\bar{\mathcal{U}} = (\bar{G}, \bar{K}, \bar{\Delta}, \bar{\Delta}_{\bar{K}})$ just as $\mathcal{U}$ was constructed. Then $K \cong A_5 \cong \bar{K}$ and $K_X \cong S_3 = Aut(K_X)$, so by Exercise 14.1.1, $\alpha : K_X \to \bar{K}_{\bar{X}}$ extends to an isomorphism $\zeta : K \to \bar{K}$.

Now $C_G(G_{XY}) \cong A_4$ is a Borel group of K, so $C_G(G_{XY})\zeta$ is one of the two Borel groups of $\bar{K}$ containing $\bar{X}$. These two groups are interchanged by $\tau \in C_{Aut(\bar{K})}(\bar{K}_{\bar{X}})$, so replacing ζ by $\zeta\tau$ if necessary, we may assume $C_G(G_{XY})\zeta = C_{\bar{G}}(\bar{G}_{\bar{X}\bar{Y}})$.

Next for some $\bar{x} \in \bar{X}$, $Y\zeta c_{\bar{x}} = \bar{Y}$, where $c_{\bar{x}}$ is conjugation by $\bar{x}$. Thus replacing α, ζ by $\alpha c_{\bar{x}}, \zeta c_{\bar{x}}$, we may assume $Y\zeta = \bar{Y}$. In particular

$K(\{X,Y\}) = \langle t \rangle$, where $t \in C_G(G_{XY})$ with $X^t = Y$. Therefore $\bar{Y} = Y\zeta = (X^t)\zeta = \bar{X}^{t\zeta}$. Hence the pair α, ζ forms a similarity of $\mathcal{U}$ with $\bar{\mathcal{U}}$, in the sense of Section 37.

Indeed as $t \in C_G(G_{XY})$ and $t\zeta \in C_{\bar{G}}(\bar{G}_{\bar{X}\bar{Y}})$, $(b^t)\alpha = b\alpha = (b\alpha)^{t\zeta}$ for each $b \in G_{XY}$, so the pair forms an equivalence of $\mathcal{U}$ and $\bar{\mathcal{U}}$. By 47.6.5, each triangle of Δ is G-conjugate to a triangle of Δ_K. By 47.9, Δ is simply connected. Thus we have achieved the hypotheses of Corollary 37.8, and that result completes the proof.

48. Groups of type *Suz*

In this section we assume G is a group of type *Suz* as defined at the end of Section 46. Further we continue the notation of Section 46. As usual let Δ be the set of G-conjugates of members of Δ_H and regard Δ as a graph with vertices adjacent if they commute.

Lemma 48.1: *Let* $X \in Syl_5(H)$. *Then*

(1) *X is contained in a unique subgroup L of H with $L \cong A_5$ and $C_Q(L) = C_Q(X)$.*

(2) *$C_G(X) = XC_G(L)$ with $C_G(L) \cong A_6$ and $C_G(C_G(L)) = L$. A Sylow 3-subgroup Y of L is in Δ.*

(3) *$N_G(L) = (L{\times}C_G(L))\langle \alpha \rangle$, where α is of order 4 with $\langle \alpha \rangle C_G(L) = C_G(N_L(Y)) \cong M_{10}$ and $\langle \alpha \rangle L/\langle \alpha^2 \rangle \cong S_5$.*

Proof: First $C_Q(X) = D \cong D_8$ and we may take $E \leq D$. Now by 46.3, X is contained in a complement L_0 to R in $N_G(E)$ and $L_0 = L \times L_1$ with $L \cong A_5$ and $L_1R = \langle Q, Q^g \rangle$. Then $X \leq L \leq H$ and $N_G(XE) = N_L(X) \times L_1E$ with $L_1E \cong S_4$ and $D \leq L_1E$.

Now $C_H(D) = C_Q(D)L$ and as $N_H(X) = N_L(X) \times D \leq N_H(L)$, L is the unique complement to $C_Q(D)$ in $C_H(D)$ containing X, so (1) holds.

Let $I = C_G(X)$. Then $L_1E \leq I$ and $D \in Syl_2(I)$ as $D \in Syl_2(I \cap H)$. Let E' be the second 4-group in D. By the uniqueness property in (1), L also centralizes $S_4 \cong L_1'E' \leq N_G(E')$, so $K = \langle L_1E, L_1'E' \rangle \leq C_I(L)$. Now $Y \in Syl_3(L) \subseteq \Delta_H$. Let $S = N_L(Y)$ and $M = C_G(Y)$. Observe that $D_8 \cong D = C_Q(S)$ as $C_Q(DY) \cong D_8$ and for $s \in S$ an involution, $\langle s \rangle C_Q(DY)$ is semidihedral. Further as s inverts Y, s induces an outer automorphism on $M/Y \cong U_4(3)$. Therefore by Exercise 16.1.5, $C_M(S) \cong M_{10}$. Therefore as $K \leq C_M(S)$, $K \cong A_6$. Further $N_I(D_0) \leq KX$ for each $1 \neq D_0 \leq D$, so KX is strongly embedded in I. However, KX has no subgroup of odd order transitive on $|z^K|$, so by 7.6, $I = KX$. Therefore $K = C_G(L)$. Now $N_G(K) = KN_H(D) = (K \times L)\langle \alpha \rangle$, where

$E^\alpha = E'$ and α induces an outer automorphism on L. We may choose $\alpha \in C_G(S)$, so as $C_G(S) \cong M_{10}$, (3) holds.

Lemma 48.2: *Let* $Y \in \Delta_H$. *Then*

(1) $C_G(Y)$ *has one class of involutions* $z^{C_G(Y)}$.

(2) $C_G(Y)$ *is transitive on involutions* s *inverting* Y *and each such* $s \in H$ *is of type* c_2 *in* H^*. *Further* $C_G(\langle s \rangle Y) \cong M_{10}$, $N_G(Y) = \langle s \rangle C_G(Y)$, *and* $N_G(Y)/Y \cong \mathbf{Z}_2/U_4(3)$ *is determined up to isomorphism.*

(3) $Y^G \cap H = Y^H = \Delta_H$.

Proof: Part (1) follows from 46.5 and the fact that $U_4(3)$ has one class of involutions. By (1), $z^G \cap C_G(Y) = z^{C_G(Y)}$, so (3) holds. We saw during the proof of 48.1 that there exists an involution $s \in H$ inverting Y with $C_G(\langle s \rangle Y) \cong M_{10}$; indeed by Exercise 16.1.5, $\langle s \rangle C_G(Y)/Y$ is a uniquely determined subgroup of $\mathrm{Aut}(U_4(3))$ transitive on the involutions not in $U_4(3)$, so $N_G(Y)/Y$ is determined up to isomorphism and $C_G(Y)$ is transitive on involutions inverting Y. Finally each involution in H^* is of type a_2 or c_2 (cf. Exercise 2.11) and if $i \in H$ inverts Y then $[\tilde{Q}, Y, i]$ is nonsingular, so i^* is not of type a_2. Thus (2) is established.

Lemma 48.3: *Let* $X \in Syl_5(H)$, $A_4 \cong I \le C_G(E(C_G(X)))$, *and* $U = O_2(I)$. *Then*

(1) $C_G(I) \cong L_3(4)$.

(2) *For* $u \in U^\#$, $C_G(u) \le N_G(I)$, U *is a TI-set in* G, *and* u^* *is of type* c_2 *in* H^*.

(3) $N_G(I) = IC_G(I)\langle s \rangle$, *where* s *is an involution inducing a graph-field automorphism on* $C_G(I)$ *and* $I\langle s \rangle \cong S_4$.

(4) *Each involution in* $N_G(I) - C_G(I)$ *is fused into* U *under* G, *while* $z^{C_G(I)}$ *is the set of involutions in* $C_G(I)$.

Proof: Let $L_0 = C_G(C_G(X))$ be the the A_5-subgroup of H containing X supplied by 48.1; thus $I \le L_0$. From 48.1, a Sylow 3-group Y of I is in Δ_H. Also L_0 has one class of involutions, so $u \in U$ is conjugate to an element inverting Y and hence u^* is of type c_2 in H^* by 48.2. From the action of L_0 on $\tilde{Q}$ there is a unique singular point in $[\tilde{Q}, U]$, which we choose to be $\tilde{E}$. Thus $U \le R = C_H(\tilde{E}^\perp/\tilde{E}) \cap C_G(E)$ and Y centralizes E.

Now $R = [R, Y]C_R(Y)$ and using the structure of $N_G(R)$ described in 46.3, the argument of 46.14 shows the product is central with factors isomorphic to Sylow 2-subgroups of $L_3(4)$ and with $C_R(u) = C_R(U) = U \times S$, where $S = C_R(Y)$. In particular $\mathcal{A}(S) = \{A_1, A_2\}$ with $S = A_1 A_2$

and $A_1 \cap A_2 = E$. Notice $N_G(E) \cap N_G(U) = IS\langle Y_1, s \rangle$, where $\langle Y_1, s \rangle \cong S_3$ is faithful on E, $\mathbf{Z}_3 \cong Y_1$ centralizes I, and $I\langle s \rangle \cong S_3$. We may pick s to centralize u. Then $|US\langle s \rangle| = 2^9$, so by 46.7.2, $US\langle s \rangle = C_H(u)$.

Let $M = C_G(I)$; then $S \in Syl_2(M)$ and $S = M \cap H$. Let $K_0 = C_G(L_0)$; then $K_0 \le M$ and by 48.1, $K_0 \cong A_6$. Without loss of generality $z \in S \cap K_0 \in Syl_2(K_0)$, so $A_1 \cap A_2 \cap K_0 = \langle z \rangle$ and $N_{K_0}(A_i \cap K_0) \cong S_4$ for $i = 1, 2$. Then arguing as in 46.14, we conclude $N_M(A_i)$ is the split extension of A_i by $L_2(4)$ acting naturally on A_i.

Let $L = C_G(Y)$, so that $L/Y \cong U_4(3)$. Let $S \le S_0 \in Syl_2(L)$. By 45.5 and 45.6, $S = J(S_0)$ and $N_L(A_i)/A_i Y \cong A_6$ acts as $Sp_4(2)'$ on A_i. In particular there exists a unique subgroup N_i of $N_L(A_i)$ with $N_i/A_i \cong L_2(4)$ and A_i the natural module for N_i/A_i. Therefore $K = \langle N_1, N_2 \rangle$ is a uniquely determined subgroup of $U_4(3)$. We claim $K \cong L_3(4)$. For from 24.4.0, $L_3(4) \cong G_5 \le G_7$, in the notation of Section 24, while by Exercise 9.4, $G_7 \cong U_4(3)$. Further as $S = J(S_0)$ is isomorphic to a Sylow 2-subgroup of $L_3(4)$ we may take $S \le G_5$, and then by uniqueness of N_i, $K = \langle N_1, N_2 \rangle = G_5 \cong L_3(4)$.

So $K \cong L_3(4)$. In particular K has one class of involutions and $S = C_M(z) \le K$, so K is strongly embedded in M. Hence as K has no subgroup of odd order transitive on its involutions, $M = K$ by 7.6. So (1) is established.

Also as $I\langle s \rangle \cong S_4 \cong Aut(I)$, $IM\langle s \rangle = N_G(I)$. By 48.1 there is $t \in N_G(I)$ acting on $K_0 L_0$ with $K_0 \langle t \rangle \cong M_{10}$ and t inducing an outer automorphism on I. It follows that s induces a graph-field automorphism on K. Thus (3) is established. Each involution in $N_G(I) - IC_G(I)$ inverts a conjugate of Y and hence is fused into U under H by 48.2. Each involution in $UE - E$ is fused into U under $[R, Y]$, and z^K is the set of involutions in K, so (4) holds.

Thus it remains to show $C_G(u) \le N_G(I)$, so assume not. Let $D = C_G(u)$ and $B = N_D(I)$. Then $S = C_D(z) \le B$ and $z^G \cap B = z^K = z^B$, so by 7.3, z fixes a unique point of D/B. Then by Exercise 2.10.1, D has a normal subgroup D_0 with $D = BD_0$ and $S \in Syl_2(D_0)$. But now $K = B \cap D_0$ is strongly embedded in D_0, while K has no subgroup of odd order transitive on its involutions, so 7.6 supplies a contradiction.

We term the conjugates of the subgroup U of 48.3 *root 4-subgroups* of G and involutions fused into U under G as *root 4-involutions*. We call conjugates of the groups I and L of 48.3 and 48.1, *root A_4-subgroups* and *root A_5-subgroups*, respectively.

Lemma 48.4: *(1) G has two conjugacy classes of involutions: the class*

z^G of 2-central involutions and the root 4-involutions.

(2) Let $t_1 = z$ and t_2 be a root 4-involution. Let $\bar{G}$ be a group of type Suz with corresponding involutions $\bar{t}_i$, $i = 1, 2$. Then there exist isomorphisms $\alpha_i : C_G(t_i) \to C_{\bar{G}}(\bar{t}_i)$ such that $(t_j^G \cap C_G(t_i))\alpha_i = \bar{t}_j^{\bar{G}} \cap C_{\bar{G}}(\bar{t}_i)$ for all $i, j \in \{1, 2\}$.

(3) $|G| = |\bar{G}|$.

(4) Each involution inverting a member of Δ is a root 4-involution.

Proof: Part (3) follows from (1) and (2) and the Thompson Order Formula 7.2. Thus it remains to prove (1), (2), and (4).

Let $u, \bar{u}$ be root 4-involutions of $G, \bar{G}$, respectively; take $u \in H$. By 48.3 there is an isomorphism $\beta = \alpha_2 : C_G(u) \to C_{\bar{G}}(\bar{u})$ and setting $\bar{z} = z\beta$, we have $z^G \cap C_G(u) = z^{C_G(u)}$, so $(z^G \cap C_G(u))\beta = \bar{z}^{\bar{G}} \cap C_{\bar{G}}(\bar{u})$ is the set of 2-central involutions of $\bar{G}$ centralizing $\bar{u}$.

Next by 46.8 there is an isomorphism $\alpha = \alpha_1 : H \to \bar{H} = C_{\bar{G}}(\bar{z})$. By 46.1 the involutions in $Q, \bar{Q}$ are 2-central in $G, \bar{G}$, respectively. Thus it remains to consider involutions in $H - Q$.

Now $H^* \cong \Omega_6^-(2)$ has two classes of involutions: those of type a_2 and c_2 (cf. Exercise 2.11). By 48.3.2 and 46.7, each involution of H of type c_2 is a root 4-involution, so it remains to deal with the involutions of type a_2. Also this observation together with 48.2.2 implies (4).

Let $Y \in \Delta_H$. Then $C_H(Y)$ contains an involution i of type a_2, and in the notation of Exercise 2.11, $Q_i^+ = Q_i^- * [Q, Y]$. But $[Q, Y] \cong Q_8$, so z is its unique involution and hence by Exercise 2.11, the set of involutions $I(iQ)$ in iQ satisfies $I(iQ) \subseteq Q_i^- \subseteq C_G(Y)$. Therefore each involution of type a_2 is 2-central in G by 48.2.1, completing the proof.

Lemma 48.5: *Let $Y \in \Delta$ and $P \in Syl_3(N_G(Y))$. Then*

(1) $J(P) = A \cong E_{3^5}$ and $C_G(A) = A$.

(2) $N_G(A)$ is the split extension of A by M_{11} and is determined up to isomorphism.

(3) $P \in Syl_3(G)$.

(4) $N_G(A)$ is 4-transitive on $\Delta \cap A$ of order 11.

(5) $N_G(A)$ is transitive on the remaining 110 points of A.

Proof: Let $M = N_G(Y)$ and $M^* = M/Y$. Thus by 48.2.2, M^* is $E(M)^* \cong U_4(3)$ extended by an involution u^* inverting Y, and $C_{M^*}(u^*) \cong M_{10}$. In particular if A is the preimage of $J(P^*)$ then $E_{81} \cong A^*$ with $N_{M^*}(A^*)$ the split extension of A^* by M_{10} and A^* has the structure of a 4-dimensional orthogonal space over $GF(3)$ with

$N_M(A)$ transitive on the ten singular points and 30 nonsingular points of A^* (cf. 45.9).

Without loss of generality $z \in N_M(A)$. Then $C_A(z) = B \cong E_{27}$ with B^* a nondegenerate subspace of A^* containing two singular points B_1^* and B_2^*. Now from the structure of H, B is abelian and by 48.2.3, $B \cap \Delta = \{Y, Y_1, Y_2\}$ with $Y_i^* = B_i^*$. Thus $N_M(A)$ is 3-transitive on $\Delta \cap A - \{Y\}$ of order 10 and $A \cong E_{3^5}$. Now for $1 \neq P_0^* \leq P^*$, $|A : C_A(P_0)| > |P_0^*|$, so $\{A\} = \mathcal{A}(P)$, establishing (1). Further as $|\Delta \cap A|$ is prime to 3, $P \in Syl_3(G)$, so (3) holds.

By 7.7, $N = N_G(A)$ controls fusion in A, so $|N : N_M(A)| = |\Delta \cap A| = 11$ and N/A is a 4-transitive extension of M_{10}. Therefore by Exercise 6.6, $N/A \cong M_{11}$. As $N_M(A)$ splits over A, N splits over A by Gaschütz's Theorem (cf. 10.4 in [FGT]). So to show $N_G(M)$ is determined up to isomorphism it remains to show the representation of $L = M_{11}$ on A is determined. But $A = V/W$ for some L-submodule W of the induced module $V = \zeta_K^L$, where $\zeta : K \to GF(3)$ is the sign representation of $K \cong M_{10}$. In particular V is monomial with basis X and L preserves the quadratic form on V making X orthonormal. As $N_M(A)$ is irreducible on A/Y but A does not split over Y, A is not self-dual as an N/A-module, and hence W is not a nondegenerate subspace of V. Thus W has a totally singular L-subspace U-isomorphic to the dual of A and V is uniserial as a $GF(3)L$-module. In particular A is the unique 5-dimensional image of V and hence is determined as a $GF(3)L$-module. So (2) is established.

As N is 2-transitive on $Y^G \cap A$ of order 11 there are $\binom{11}{2} = 55$ 2-dimensional subspaces $Y_1 + Y_2$ with $Y_i \in \Delta \cap A$ and each such subspace contains two conjugate points not in $\Delta \cap A$. Hence N is transitive on the remaining 110 points of A.

Lemma 48.6: *Let I be a root A_4-subgroup of G and $B \in Syl_7(C_G(I))$. Then $N_G(B) \leq N_G(I)$, so $C_G(B) = BI$ and $N_G(B)$ is the split extension of BI by α of order 6 with $\alpha^2 \in C_G(I)$ and $\langle \alpha^3, I \rangle \cong S_4$.*

Proof: Let $U = O_2(I)$; by 48.3, $C_G(\langle u \rangle B) = BU$ for each $u \in U^\#$, so by Exercise 16.6 in [FGT], $C_G(B) = BI$ or $B \times K$ with $K \cong A_5$. Now in the latter case $Y \leq S \leq K$ with $S \cong S_3$. But by 48.2.2, G is transitive on such subgroups and indeed $C_G(S) \cong M_{10}$, contradicting $B \leq C_G(S)$.

Lemma 48.7: $|G| = 2^{13} \cdot 3^7 \cdot 5^2 \cdot 7 \cdot 11 \cdot 13$.

Proof: Let $M = Co_1$ act on $\tilde{\Lambda}$ the Leech lattice modulo 2. By 25.9 there is a quasisimple subgroup $\hat{G}$ of M with $Z(\hat{G}) = D$ of order 3 and $\hat{G}/D$ of type Suz. So by 48.4.3 we may take $G = \hat{G}/D$.

We observe first that $C_{\tilde{\Lambda}}(D) = 0$ by Exercise 9.6. Therefore D centralizes no element m of M of order 23. For $C_{\tilde{\Lambda}}(m)$ is a 2-dimensional nondegenerate subspace of the orthogonal space $\tilde{\Lambda}$ containing two singular points and hence admits no automorphism of order 3.

On the other hand $|G|$ divides $|M|$, so $|G| = 2^a \cdot 3^b \cdot 5^c \cdot 7^d \cdot 11^e \cdot 13^f$. As z is 2-central and $|H|_2 = 2^{13}$, $a = 13$. By 48.5, $b = 7$ and $e > 0$. So as $|M|_{11} = 11$, $e = 1$. By 48.6, $d = 1$. We interject a lemma:

Lemma 48.8: *(1) G has two classes 5_A and 5_B of subgroups of order 5.*

(2) G has a subgroup $K = (L_A \times L_B)\langle \alpha \rangle$ with $L_A \cong A_5$ a root A_5-subgroup of G, $L_B \cong A_6$, $\langle \alpha \rangle L_A / \langle \alpha^2 \rangle \cong S_5$, and $\langle \alpha \rangle L_B \cong M_{10}$, $X_C \in Syl_5(L_C)$ with $N_G(X_C) \leq K$, and $X_C \in 5_C$ for $C = A, B$.

(3) For $P \in Syl_5(G)$, $N_G(P)$ is the split extension of P by $\mathbf{Z}_4 \times S_3$ faithful on P.

Proof: In the notation of 48.1, let $X = X_A$, $L = L_A$, $L_B = C_G(L)$, $X_B \in Syl_5(L_B)$, and $P = X_A X_B$. Then by 48.1, $K = N_G(L)$ has the structure described in (2) and $N_G(X) \leq K$. Further L contains a root 4-subgroup U and $C_G(X_B) \cap C_G(u) = U X_B$ for $u \in U^{\#}$ by 48.3, so by Exercise 16.6 in [FGT], $X_B L = C_G(X_B)$ and then $N_G(X_B) \leq K$.

Therefore $P = C_G(P) \cap C_G(X) = C_G(P)$ and $N_K(P)$ is the split extension of P by $\mathbf{Z}_2 \times \mathbf{Z}_4$. In particular if $S \in Syl_2(N_K(P))$ then $S \in Syl_2(N_G(P))$ as X_A and X_C are the eigenspaces of S on P and $X_A \notin X_B^G$.

Similarly if $P \notin Syl_5(G)$ then P is of index 5 in $P_1 \leq G$ and as a Sylow 2-subgroup S of $N_G(P)$ is abelian, we conclude from the structure of $GL_2(5)$ that $P_1 \trianglelefteq N_G(P)$. Thus S acts on $Z(P_1)$, so $Z(P_1) = X_C$ for $C = A$ or B, contradicting $N_G(X_C) \leq K$. So $P \in Syl_5(G)$. Thus either the lemma holds or $N_G(P) \leq K$. In the latter event $1 \equiv |G : N_G(P)| \equiv 2^{13} \cdot 3^7 \cdot 7 \cdot 11 \cdot 13^f / 2^3 \equiv 13^f \equiv 3^f \mod 5$, so $f = 0$. But by 48.6, if $B \in Syl_7(G)$ then $|N_G(B) : B| = 72$, so if $f = 0$ then $1 \equiv |G : N_G(B)| = 2^{13} \cdot 3^7 \cdot 5^2 \cdot 11/72 \equiv -1 \mod 7$, a contradiction. Thus $f = 1$ and the lemma holds.

Notice by 48.8 that $c = 2$ and we also showed $f = 1$ during the proof of 48.8, so the proof of 48.7 is complete.

In the remainder of this section let $Y \in \Delta$ and $M = N_G(Y)$. As usual $\Delta(Y)$ denotes the vertices in Δ adjacent to Y but distinct from Y; that is, those $X \in \Delta$ with $\langle X, Y \rangle \cong E_9$.

Lemma 48.9: *(1) M has five orbits $\Delta^i(Y)$, $0 \leq i \leq 4$, on Δ, with $\Delta^0(Y) = \{Y\}$ and $\Delta^1(Y) = \Delta(Y)$.*

(2) $|\Delta(Y)| = 2^3 \cdot 5 \cdot 7 = 280$ and $N_M(Y_1) \cong \mathbf{Z}_2/(\mathbf{Z}_4 * SL_2(3))/3^{2+4}$ for $Y_1 \in \Delta(Y)$.

(3) $\Delta^2(Y) = \{Y_2 \in \Delta : \langle Y, Y_2 \rangle \cong SL_2(3)\}$ is of order $3^5 \cdot 5 \cdot 7 = 8{,}505$ and $N_M(Y_2)$ is a complement to $C_M(z)$, when z is the involution in $\langle Y, Y_2 \rangle$.

(4) $\Delta^3(Y) = \{Y_3 \in \Delta : \langle Y, Y_3 \rangle \cong A_5\}$ is of order $2^3 \cdot 3^5 \cdot 7 = 13{,}608$ and $N_M(Y_3) \cong \mathbf{Z}_2 \times M_{10}$.

(5) $\Delta^4(Y) = \{Y_4 \in \Delta : \langle Y, Y_4 \rangle \cong A_4\}$ is of order $2 \cdot 3^5 = 486$ and $N_M(Y_4) \cong L_3(4)$ extended by a graph-field automorphism.

Proof: The argument is much the same as 47.5. Let $M^* = M/Y$, so that M^* is $U_4(3)$ extended by an involution s with $C_{M^*}(s^*) \cong M_{10}$. We find members Y_i of $\Delta^i(Y)$ such that $N_M(Y_i)$, and hence also $|Y_i^M| = |M : N_M(Y_i)|$, is as claimed.

For example, by 48.2 and 48.5, the map $Y_1 \mapsto Y_1^*$ is a bijection of $\Delta(Y)$ with the set of root subgroups of transvections of $U_4(3)$, so (2) holds (cf. 45.8 and 45.10). We choose $Y \in \Delta_H$ and let $Y_2 \in Y^Q$. We pick Y_3 and Y_4 so that $\langle Y, Y_i \rangle$ is a root A_5-, root A_4-subgroup of G, for $i = 3, 4$, respectively. Finally we observe that

$$|\Delta| = |G : M| = 2^5 \cdot 5 \cdot 11 \cdot 13 = 22{,}880$$

while also

$$22{,}880 = 1 + 280 + 8{,}505 + 13{,}608 + 486$$

so $\Delta^i(Y) = Y_i^M$ and the lemma holds.

Lemma 48.10: (1) $\Delta^i(Y) = \{X \in \Delta : d_\Delta(Y, X) = i\}$.

(2) G is transitive on triples (Y_0, Y_2, Y_3) with $d(Y_0, Y_i) = 2$ for $i = 2, 3$ and $Y_3 \in \Delta(Y_2)$. Moreover $Y_0^\perp \cap Y_2^\perp \cap Y_3^\perp \neq \varnothing$ for each such triple.

(3) For $Y_3 \in \Delta^3(Y)$, M is transitive on $\Delta^2(Y) \cap \Delta(Y_3)$ of order 90.

Proof: By construction (1) holds for $i = 0, 1$. Further if we set $M^* = M/Y$, $\Delta(Y)^*$ is the set of root groups of transvections in $E(M)^* \cong U_4(3)$ with the map $Y_1 \mapsto Y_1^*$ a bijection of $\Delta(Y)$ with the set of such root groups, so for each $A, B \in \Delta(Y)$, $\langle A, B \rangle^* \cong E_9$ or $SL_2(3)$. Hence $\Delta^2(Y)$ is the set of members of Δ at distance 2 from Y by 48.9. In particular the members of $\Delta^3(Y) \cup \Delta^4(Y)$ are at distance at least 3 from Y.

Let $Y_0 \in \Delta^2(Y)$; without loss of generality $z \in \langle Y, Y_0 \rangle = S$. Then by 48.9, $N_M(Y_0)$ is a complement to Y in $H \cap M$. So by Exercise 16.2, $N_M(Y_0)$ has three orbits Γ_i, $1 \leq i \leq 3$, on $\Delta(Y)$ with $\Gamma_1 = \Delta(Y) \cap H = \Delta(Y, Y_0)$ of order 8, Γ_2 of order 128 and consisting of those $Y_2 \in \Delta(Y)$ with $Y_2^z \in \Delta(Y_2)$, and Γ_3 of order 144 consisting of those $Y_3 \in \Delta(Y)$

with $Y_3^z \in \Delta^2(Y_3)$. As $N_G(Y_0)$ is transitive on vertices at distance 2 from Y_0 and there exist vertices at distance 3, $d(Y_0, Y_i) = 3$ for $i = 2$ or 3.

Let $P \in Syl_3(M)$ and $A = J(P)$. Then from 48.5, $A \cong E_{3^5}$ and $N_{M^*}(A^*)$ and $N_{M^*}(Z(P^*))$ are the two maximal parabolics of M^* with $Z(P^*) = X^*$ for some $X \in \Delta(Y)$. Then picking $m \in C_M(Z(P)) - N_M(A)$ and letting $K = \langle A, A^m \rangle$, we have from 48.9.2 that $K = O^2(C_G(XY)) \cong SL_2(3)/3^{2+4}$. Choose $X_0 \in A \cap \Delta - \{X, Y\}$ and set $X_2 = X_0^m$. Then $X_2 \in \Delta^2(X_0)$ and we choose notation so that $z \in \langle X_0, X_2 \rangle$. Indeed for $X_3 \in A^m \cap \Delta - \{Y, X, X_2\}$, $X_3 \neq X_3^z \leq A^m$, so $X_3 \in \Delta^2(X_0)$ and $X_3^z \in \Delta(X_3)$. Thus (X_0, X_2, X_3) is a triple with $d(X_0, X_2) = 2$, $X_3 \in \Delta(X_2)$, and $X_3^z \in \Delta(X_2, X_3)$. But we showed in the previous paragraph that G is transitive on such triples, so $(Y_0, Y, Y_2) \in (X_0, X_2, X_3)^G$. Then as $d(X_0, X_3) = 2$, also $d(Y_0, Y_2) = 2$. Therefore by the previous paragraph, $d(Y_0, Y_3) = 3$ and M is transitive on the members of Δ at distance 3 from Y. Finally $Y \in X_0^\perp \cap X_2^\perp \cap X_3^\perp$, so (2) is established.

To complete the proof of (3) we show $Y_3 \in \Delta^3(Y_0)$. Then we count the set Ω of pairs (Y, Y_3) with $Y \in \Delta^2(Y_0)$ and $Y_3 \in \Delta(Y) \cap \Delta^3(Y_0)$ in two ways: $(3^5 \cdot 5 \cdot 7) \cdot (2^4 \cdot 3^2) = |\Delta^2(Y_0)|\alpha = |\Omega| = |\Delta^3(Y_0)|\beta = 2^3 \cdot 3^5 \cdot 7 \cdot \beta$, where $\alpha = |\Delta(Y) \cap \Delta^3(Y_0)| = |\Gamma_2| = 144$ and $\beta = |\Delta(Y_3) \cap \Delta^2(Y_0)|$. We conclude $\beta = 90$, so that (3) holds.

It remains to show $Y_3 \in \Delta^3(Y_0)$. We consider the action of $H^* \cong \Omega_6^-(2)$ on the orthogonal space $\tilde{Q}$. Now the stabilizer L^* of a nonsingular point $\tilde{n}$ of $\tilde{Q}$ is isomorphic to S_6. Represent L^* on $\{1, \ldots, 6\}$ and let $X_i = \langle x_i \rangle \in L \cap \Delta$ for $i = 0, 2, 3$, with $x_3 = (1, 2, 3)$, $x_0 = (3, 4, 5)$, and $x_2 = (4, 5, 6)$. Then $\langle X_0^*, X_3^* \rangle \cong A_5$, so by 48.9, $X_3 \in \Delta^3(X_0)$. Similarly $X_3 \in \Delta(X_2)$. Finally $\langle X_0^*, X_2^* \rangle \cong A_4$, so replacing X_2 by a suitable conjugate under Q, we may take $X_2 \in \Delta^2(X_0)$. Thus (X_0, X_2, X_3) is conjugate to (Y_0, Y, Y_3), completing the proof.

Lemma 48.11: *Let* $Y_3 \in \Delta^3(Y)$. *Then*

(1) $N_M(Y_3)$ *is transitive on* $\Delta(Y) \cap \Delta^i(Y_3)$ *of order 90, 180, 10, for* $i = 2, 3, 4$, *respectively.*

(2) $\Delta(Y) \cap \Delta^2(Y_3)$ *is connected.*

(3) *For* $X \in \Delta(Y) \cap \Delta^3(Y_3)$, $\Delta(Y, X) \cap \Delta^2(Y_3) \neq \emptyset$.

Proof: We represent $M^* = M/Y$ on the set $\mathcal{L}$ of singular lines of a 6-dimensional orthogonal space V over $GF(3)$, with $\Delta(Y)$ corresponding to $\mathcal{L}$ via the equivariant map $X \mapsto [V, X^*]$. Notice $X_1 \in \Delta(X)$ if and only if $[V, X_1] \cap [V, V_2] \neq 0$.

Let $N = N_M(Y_3)$. By 48.9.4, $\mathbf{Z}_2 \times M_{11} \cong N^* = C_{M^*}(\beta)$ for some involution $\beta \in M^* - E(M^*)$. Then by Exercise 16.1.8, N^* has three orbits $\mathcal{L}_j$, $1 \le j \le 3$, on $\mathcal{L}$ of order 10, 180, 90, which therefore correspond to the orbits of N on $\Delta(Y)$. By 48.10.3, the orbit of length 90 is $\Delta(Y) \cap \Delta^2(Y_3)$. Then (2) holds by Exercise 16.1.9.

Further if $X \in \Delta(Y)$ with $[V, X^*] \in \mathcal{L}_2$ then by Exercise 16.1.8, $X \in \Delta(X_1)$ for some $X_1 \in \Delta(Y) \cap \Delta^2(Y_3)$, so $d_\Delta(X, Y_3) = 3$. On the other hand by 48.10, $\Delta^4(Y) \ne \varnothing$, so as M is transitive on $\Delta^3(Y)$, $\Delta^4(Y_3) \cap \Delta(Y) \ne \varnothing$. We conclude (1) holds. Moreover $X_1 \in \Delta(Y, X) \cap \Delta^2(Y_3)$, so (3) holds.

Lemma 48.12: *Let $Y_4 \in \Delta^4(Y)$. Then $N_M(Y_4)$ is transitive on $\Delta(Y_4)$ and $\Delta(Y_4) \subseteq \Delta^3(Y)$.*

Proof: Let $\lambda = |\Delta(Y_4) \cap \Delta^3(Y)|$ and $\alpha = |\Delta^4(Y) \cap \Delta^3(Y)|$ for $Y_3 \in \Delta^3(Y)$. Then counting pairs (Y_3, Y_4) with $Y_i \in \Delta^i(Y)$ and $Y_3 \in \Delta(Y_4)$ in two ways, we get

$$2 \cdot 3^5 \cdot \lambda = |\Delta^4(Y)|\lambda = |\Delta^3(Y)|\alpha = 2^3 \cdot 3^5 \cdot 7 \cdot 10$$

from 48.9 and 48.11. We conclude $\lambda = 280 = |\Delta(Y_4)|$, so $\Delta(Y_4) \subseteq \Delta^3(Y)$. Further by 48.11, G is transitive on triples (Y, Y_3, Y_4) with $Y_i \in \Delta^i(Y)$ and $Y_4 \in \Delta(Y_3)$, so $N_M(Y_4)$ is transitive on $\Delta(Y_4) \cap \Delta^3(Y) = \Delta(Y_4)$.

Lemma 48.13: Δ *is simply connected.*

Proof: By 48.10, Δ has diameter 4, so by 35.14 and 34.5 it suffices to show each r-gon $p = X_0 \cdots X_r$ is in $\mathcal{C}_3(\Delta)$ for $4 \le r \le 9$. If $r = 4$ this holds by 34.6, once we show $\Delta(Y, Y_2)$ is connected for $Y_2 \in \Delta^2(Y)$. But taking $z \in \langle Y, Y_2 \rangle$, $\langle Y, Y_2 \rangle = Y[Q, Y] = S$ and $\Delta(Y, Y_2) = \Delta_1 \cup \Delta_2$, where $\Delta_i = \Delta \cap S_i$ and $O^2(C_H(S)) = S_1 * S_2$ with $S_i \cong SL_2(3)$. So as $\Delta_i \subset \Delta(X)$ for $X \in \Delta_{3-i}$, $\Delta(Y, Y_2)$ is connected.

If $r = 5$ then $p \in \mathcal{C}_3(\Delta)$ by 34.8 and 48.10.2. If $r = 6$ then $p \in \mathcal{C}(\Delta)$ by 48.11.2 and Exercise 12.2.1, while if $r = 7$ we appeal to 48.11.3 and Exercise 12.2.2. Finally by 48.12 there are no 9-gons in Δ and 48.12 plus Exercise 12.2.1 handle the case $r = 8$.

Lemma 48.14: *Let $L \cong A_6$, V a 4-dimensional orthogonal space of sign -1 over $GF(3)$ regarded as a faithful $GF(3)L$-module of dimension 4, $A = N_{GL(V)}(L)$, and $U = U(L, V)$ the largest $GF(3)L$-module such that $V \le U$, $C_U(L) = 0$, and $[U, L] \le V$. Then*

(1) L *is absolutely irreducible on* V, *and* $A/L \cong D_8$.
(2) $\dim(H^1(L, V)) = 2$.

(3)   *A has two orbits on hyperplanes of U containing V with representatives U_1 and U_2 such that $N_A(U_1) \cong \mathbf{Z}_2 \times S_6$ and $N_A(U_2) \cong \mathbf{Z}_2 \times M_{10}$.*

Proof: The first statement in part (1) follows as a Sylow 3-subgroup of L fixes a unique point of V. For the second observe that from Section 45, the semidirect product $S = LV$ is a local subgroup of $X = U_4(3)$ and if $B = Aut(X)$ then $B/X \cong D_8$, $B = XN_B(S)$, and V is self-centralizing in $N_B(S) = D$. Thus $D/V \leq A$ with $D/LV \cong D_8$. Further as L is absolutely irreducible on V, $|C_A(L)| = 2$, while $|Out(L)| = 4$, so indeed $D/V = A$.

Next L has two conjugacy classes of subgroups K^G and K^{aG} with $K \cong A_5$, for some $a \in A$ (cf. Exercise 5.1 in [FGT]). By Exercise 15.3, $V \cong [V_I, L]/C_{V_I}(L)$, where V_I is the permutation module for L over F on the cosets of $I = K$ or K^a. Notice, however, that $U_K = V_K/C_{V_K}(L)$ is not isomorphic to $U_{K^a} = V_{K^a}/C_{V_{K^a}}(L)$ as K fixes a point in the former but not the latter. Thus $dim(H^1(L, V)) \geq 2$.

Observe next that $dim(H^1(K, V)) = 1$. This is because $W/C_W(K) = U(K, V)$, where W is the 6-dimensional permutation module for K, since a D_{10}-subgroup of K fixes a point in $V' - V$ for any extension V' of V by the trivial K-module.

Thus if Z is an $GF(3)L$-module with $C_Z(L) = 0$ and $[Z, L] = V$, and $dim(Z/V) > 2$ then there exist points X, Y of Z fixed by K, K^a, respectively, with $V + X = V + Y$. Thus $V + X \cong U_K \cong U_{K^a}$, a contradiction. So (2) is established.

Finally by the universal property of U (cf. 17.11 in [FGT]), the action of $A = N_{GL(V)}(L)$ extends to U. Let $e \in A$ induce scalar action on V. Then $\langle e \rangle = C_A(L)$ and as $C_U(L) = 0$, $C_U(e) = 0$, so $A/L \cong D_8$ is faithful on U/V and hence has two orbits on the points of U/V with representatives U_i/V, $i = 1, 2$. We may pick $U_1 \cong U_K$, so $N_A(U_1) = LN_A(K) \cong \mathbf{Z}_2 \times S_6$. As U_2 is the restriction to L of the 5-dimensional irreducible for M_{11} discussed in the next lemma, $N_A(U_2) \cong \mathbf{Z}_2 \times M_{10}$.

Lemma 48.15: *Let $L \cong M_{11}$, V a faithful 5-dimensional $GF(3)L$-module induced from the sign representation of M_{10}, and $K = LV$ the semidirect product. Then*

(1)   *V is determined up to isomorphism as a $GF(3)L$-module, so K is determined up to isomorphism.*

(2)   *L has two orbits on the points of V of length 11 and 110.*

(3)   *If X is a point in the orbit of length 11 then*

$$Aut(K_X) = N_{Aut(K)}(K_X).$$

Proof: Parts (1) and (2) were established during the proof of 48.5. Moreover $L_X \cong M_{10}$ and $E = E(L_X) \cong A_6$ and from the proof of 48.5, V is isomorphic to the dual of the module U_2 of 48.14.3 as an E-module. Thus by 48.14.3, $N_{GL(V)}(E) = L_X\langle t \rangle$, where t is an involutory automorphism of K centralizing L and inverting V. So $Aut(K_X) = L_X\langle t \rangle U$, where $U = U(L_X, V)$ in the language of Section 17 of [FGT]. But by 17.11 in [FGT] there is an injection $U/X \to U(L_X, V/X)$ and $U(L_X, V/X)/(V/X) \cong (U(E, V/X)/(V/X)) \cap C(L_X) \cong \mathbf{Z}_3$ by 48.14.3. Thus $|U : V| \le 3$.

Now take S to be the quasisimple subgroup of Co_1 with $Z(S) \cong \mathbf{Z}_3$ and $S^* = S/Z(S) \cong Suz$, supplied by 46.6.1. Then there is $K_0^* \le S^*$ with $K_0^* \cong K$, and as K_0 contains a Sylow 3-subgroup of S, K_0 does not split over $Z(S)$. Thus $H^1(L, V) \ne 0$, so $V < U(L, V)$. But as L_X contains a Sylow 3-subgroup of L, $U(L, V) \le U$, so as $|U : V| \le 3$, we conclude $U = U(L, V)$. Thus $Aut(K_X) = L_X\langle t \rangle U = N_{Aut(K)}(K_X)$, completing the proof of (3).

Lemma 48.16: *Let* $P \in Syl_3(G)$ *and* $A = J(P)$. *Then*

(1) $J = \langle \Delta \cap P \rangle$.

(2) G *is transitive on triangles of* Δ *and each such triangle is fused into* $A \cap \Delta$.

Proof: By 48.9 if $X, Y \in \Delta$ with $\langle X, Y \rangle$ a 3-group then $[X, Y] = 1$, so $B = \langle \Delta \cap P \rangle$ is abelian. Then as $C_G(A) = A \le B$, (1) holds. Then (2) follows from (1) and 48.5.

We now establish the main theorem of this section:

Lemma 48.17: *Up to isomorphism there is a unique group of type* Suz.

Proof: By 25.9 there exists a group of type Suz. Thus it remains to prove that if $\bar{G}$ is a group of type Suz then $G \cong \bar{G}$. As usual we construct a uniqueness system $\mathcal{U}$ for G and appeal to Corollary 37.8.

Let $P \in Syl_3(N_G(Y))$, $A = J(P)$, and $K = N_G(A)$. Let $\Delta_K = A \cap \Delta$; by 48.5, K is 4-transitive on Δ_K of order 11 and Δ_K is a complete subgraph of Δ . Thus Δ_K contains a second member X in $Z(P)$. Let $\mathcal{U} = (G, K, \Delta, \Delta_K)$ and observe that $\mathcal{U}$ is a uniqueness system in the sense of Section 37.

Pick $\bar{Y} \in \bar{\Delta}$; by 48.2.2 and Exercise 16.3 there exists an isomorphism $\alpha : G_Y \to \bar{G}_{\bar{Y}}$. Let $\bar{P} = P\alpha$, $\bar{A} = J(\bar{P})$, $\bar{K} = N_{\bar{G}}(\bar{A})$, $\bar{\Delta}_{\bar{K}} = \bar{\Delta} \cap \bar{A}$, and $\bar{X} = X\alpha$ the second member of $\bar{\Delta}$ in $Z(\bar{P})$. Then $\bar{\mathcal{U}} = (\bar{G}, \bar{K}, \bar{\Delta}, \bar{\Delta}_{\bar{K}})$ is also a uniqueness system.

By 48.5.2 there is an isomorphism $\zeta : K \to \bar{K}$ with $X\zeta = \bar{X}$ and $Y\zeta = \bar{Y}$. Thus $K_Y \zeta = K\zeta_{Y\zeta} = \bar{K}_{\bar{Y}} = K_Y \alpha$. Now we appeal to Theorem 37.12 with $Z(Y) = Y$. By construction, hypotheses (1) and (2) of Theorem 37.12 hold. By 48.15.3, $Aut(K_Y) = Aut_{Aut(K)}(K_Y)$. This verifies hypothesis (3) of Theorem 37.12. As K is 2-transitive on K/K_X, hypothesis (4) of Theorem 37.12 holds. Thus Theorem 37.12 says our uniqueness systems are similar.

Next by 48.9.2, $F^*(G_{XY}) = O_3(G_{XY})$ and $Z(K_{XY}) = 1$. So as $P \leq K_{XY}$, 38.10 says $C_{Aut(G_{XY})}(K_{XY}) = 1$. Therefore Theorem 37.9 says our uniqueness systems are equivalent. Now by 48.16, each triangle in Δ is fused into Δ_K under G. Also by 48.13, Δ is simply connected. So Corollary 37.8 completes the proof of the main theorem of this section.

We close this section with several results which establish the existence of certain subgroups of G and various properties of the graph Δ. These results will be used in the next section in our proof of the uniqueness of groups of type Co_1.

Lemma 48.18: *Let U be a root 4-subgroup of G and $V \in Syl_3(C_G(U))$. Then*

    (1) $C_G(V) = V \times E(C_G(V))$ *with* $E(C_G(V)) \cong A_6$ *and* $N_G(V) = T_0 C_G(V)$, *where* $Q_8 \cong T_0 \in Syl_2(E(C_G(U)) \cap N_G(V))$ *and* $T_0 E(C_G(V))/C_{T_0}(E(C_G(V))) \cong S_6$.

    (2) $N_G(V_0) \leq N_G(V)$ *for each* $1 \neq V_0 \leq V$.

Proof: Let $L = E(C_G(U))$ and $I = C_G(L)$, so that $L \cong L_3(4)$, I is a root A_4-subgroup of G, and $N_L(V) = VT_0$ with $Q_8 \cong T_0$ faithful on V. Further $N_G(V) \cap N_G(I) = I\langle t \rangle \times VT_0$, where $t = u^h$ induces a graph-field automorphism on L and $u \in U^\#$.

Without loss of generality $\langle z \rangle = Z(T_0)$. Then as $z^G \cap N_G(I) \subseteq L$, $V\langle z \rangle \leq \langle z^G \cap C_G(t) \rangle = L^h$, so $\langle I, I^h \rangle = K \leq C_G(V\langle z \rangle)$. Next as U is weakly closed in $T \in Syl_2(I\langle t \rangle)$, $T \in Syl_2(C_G(V))$. By 46.1, all involutions in Q are in z^G, so as no involution in T is in z^G, $T \cap Q = 1$. So $K \cap Q = 1$ and $K \cong K^* \leq H^*$. Let K_0 be a root A_5-subgroup of G containing I, as described in 48.1. By 48.1, $N_G(K_0)$ contains a conjugate of VT_0, which we take to be VT_0. Notice that by 48.1, $C_{T_0}(K_0) = \langle t_0 \rangle$ is of order 4. Then $t_0 \in C_H(K_0) \leq Q$, so $[t_0, K] \leq Q \cap K = 1$, and therefore $K^* \leq C_{H^*}(\bar{t}_0) \cong S_6$, so $K \cong A_6$.

Now KV is strongly embedded in $C_G(V_0)$ for each $1 \neq V_0 \leq V$ and no subgroup of KV of odd order is transitive on the involutions of K_0, so by 7.6, $KV = C_G(V_0)$. Thus the lemma is established. Notice $T_0 K/\langle t_0 \rangle \cong S_6$ as $T_0 K_0/\langle t_0 \rangle \cong S_5$.

We refer to conjugates of the subgroup $E(C_G(V))$ of 48.18 as *root A_6-subgroups* of G.

Lemma 48.19: *(1) G is transitive on triples (X_0, X_1, X_2) from Δ with $d(X_0, X_1) = 4$ and $X_2 \in \Delta(X_1)$.*

(2) For each such triple, $\langle X_0, X_1, X_2 \rangle$ is a root A_6-subgroup of G, $X_2 \in \Delta^3(X_0)$, and $C_G(\langle X_0, X_1, X_2 \rangle) \cong \mathbf{Z}_4/E_9$.

Proof: Part (1) is 48.12. Then (1) and 48.18 imply $K = \langle X_0, X_2, X_3 \rangle$ is a root A_6-subgroup of G and $\mathbf{Z}_4/E_9 \cong V\langle t_0 \rangle \leq C_G(K)$, with the notation chosen as in the proof of 48.18. As $\langle t_0 \rangle = C_H(K)$, $C_G(K) = O(C_G(K))\langle t_0 \rangle$ with z inverting $O(C_G(K))$. Therefore $O(C_G(K))$ is abelian so $O(C_G(K)) = O(C_G(K)) \cap N_G(V) = V$.

Lemma 48.20: *For each $X \in \Delta$:*

(1) $\Delta(X) \cap H \neq \varnothing$, *and*
(2) $X^z \in \Delta^{\leq 2}(X)$.

Proof: If $X \in \Delta^i(Y)$ for $i > 2$ then $z^G \cap N_G(\langle X, Y \rangle) \subseteq C_G(\langle X, Y \rangle)$ by 48.3.4 and 48.1. Thus $X^z \in \Delta^{\leq 2}(X)$ and it remains to establish (1).

If $z \in N_G(X)$ then by 48.4.4, $X \leq H$, so (1) holds. Suppose $X^z \in \Delta(X)$. Then by 48.9.2, $O^2(N_G(XX^z)) \cong SL_2(3)/3^{2+4}$, so z centralizes some conjugate z_1 in $O^2(N_G(XX^z))$ and lies in $C_G(z_1) \cap N_G(XX^z) \cong D_8/(E_9 \times SL_2(3))$, where we check that z centralizes some member of $\Delta(X, X^z) \cap C(z_1)$.

Finally suppose $X^z \in \Delta^2(X)$ and let $\langle z_2 \rangle = Z(\langle X, X^z \rangle)$. Then $z \in Q_0 = O_2(C_G(z_2)) = Q_1 * Q_2 * Q_3$, where $Q_1 = [Q_0, X]$ and Q_2 and Q_3 are the two quaternion subgroups of $C_{Q_0}(X)$. Then $z = xy$, where $x \in Q_1 - \langle z_2 \rangle$ and $y \in C_{Q_0}(X)$. Then x is of order 4 so as $z = xy$ is an involution, y is also of order 4. Therefore $y \in Q_i$ for $i = 2$ or 3, and hence z centralizes $X_i \in \Delta(X)$ with $[X_i, Q_0] = Q_{5-i}$.

49. Groups of type Co_1

In this section we assume G is of type Co_1 as defined as the end of Section 46, and we continue the notation of Section 46. Again Δ denotes the set of G-conjugates of members of Δ_H, regarded as a graph whose edges are pairs of commuting members of Δ.

Recall that for $X \in \Delta$, $C_G(X)$ is quasisimple with $C_G(X)/X \cong Suz$ by 46.6. Define the root 4-involutions, root 4-subgroups, root A_4-subgroups, etc. of G to be the G-conjugates of the corresponding elements or subgroups of $C_G(X)$.

Lemma 49.1: *Let B be of order 5 in H with $C_Q(B) \cong Q_8 D_8$. Then*

(1) $E(C_G(B)) = L \cong J_2$.

(2) $C_G(L) \cong A_5$ *is a root A_5-subgroup of G.*

(3) $N_G(L) = N_G(C_G(L)) = (C_G(L) \times L)\langle \beta \rangle$, *where β is an involution inducing an outer automorphism on L and $C_G(L)$.*

Proof: As in 25.10, $C_G(B)/B$ is of type J_2, so by 47.10, $C_G(B)/B \cong J_2$. We claim $C_G(B)$ splits over B, so $L = E(C_G(B)) \cong J_2$. For by 47.2 there is $L_0 \le C_G(B)$ with $L_0/B = L_1/B \times L_2/B \cong D_{10} \times A_5$. Now both L_1 and L_2 split over B; for example, L_2 splits over B as L_2/B is perfect of 5-rank 1 (cf. 33.14 in [FGT]) while L_1 splits as $O(L_1) = C_{O(L_1)}(d) \times [O(L_1), d]$ for d an involution in L_1. Thus as L_0 contains a Sylow 5-subgroup of L, L splits over B by Gaschutz's Theorem.

Next L contains some $X \in \Delta$; let $M = C_G(X)$, so that $M/X \cong Suz$. Then by 48.1, $C_M(B)/X \cong \mathbf{Z}_5 \times A_6$ and $C_M(E(C_M(B))) = K \times X$ with $K \cong A_5$ a root A_5-subgroup.

Now $C_L(X)$ contains $E(C_M(B))$ and hence an A_6-section, so by 47.3, $X \le E(C_M(B))$. Then also $X \le E(C_M(B)) \le E(C_M(I))$ for a root A_4-subgroup I of K, so we have shown:

Lemma 49.2: *If $X \in \Delta$ and I is a root A_4-subgroup of $C_G(X)$ then $X \le C_G(IX) \cong SL_3(4)$.*

Returning to the proof of 49.1, we have $[Q, B] = [Q \cap M, B]$ is K-invariant so K acts on $C_Q([Q, B]) = C_Q(B)$ and then $K = [K, B]$ centralizes $C_Q(B)$. Therefore K centralizes $\langle C_Q(B), E(C_M(B)) \rangle = L$. Then by a Frattini argument and 48.1, $N_G(L) = LN_M(L) = LK\langle \alpha \rangle$, where α is of order 4 in M and induces an outer automorphism on K, and $\langle \alpha \rangle E(C_M(B))/X \cong M_{10}$. Now by 47.3, $N_L(X) = E(C_M(B))\langle \gamma \rangle$ with $N_L(X)/X \cong PGL_2(9)$ so there is an involution $\beta \in \gamma \alpha E(C_M(B))$ inducing an outer automorphism on K and L.

We call the conjugates of the group L of 49.1.1 *root J_2-subgroups* of G.

Recall the definition of octad involutions and dodecad involutions from Section 46; in particular z^G is the set of octad involutions of G.

Lemma 49.3: *(1) The classes of octad involutions, dodecad involutions, and root 4-involutions are distinct.*

(2) $H \cap \Delta = \Delta_H = X^H$ *for $X \in \Delta_H$.*

Proof: With notation as in 46.12, as A is weakly closed in T, $N_G(A)$ controls fusion in A by 7.7, so octad and dodecad involutions form different classes.

Let $X \in \Delta$, U a root 4-subgroup of $C_G(X)$, $u \in U^{\#}$ a root 4-involution, and $K = E(C_G(UX))$. Then $K \cong SL_3(4)$ by 49.2 and 48.3. Now if $u = z^j$ for some $j \in G$ then setting $J = H^j$ and $J^* = J/Q^j$, we have either $U \le C_{Q^j}(K)$ or $K^* \le C_{J^*}(U^*) \le P^*$ for some maximal parabolic P^* of $J^* \cong \Omega_8^+(2)$. In the former case $K^* \le C_{J^*}(\bar{U}) = P^*$, a maximal parabolic. So in any event $K^* \cong SL_3(4)$ is contained in a maximal parabolic of $\Omega_8^+(2)$, which is not the case as a Levi factor of such a parabolic is solvable or A_8.

So octad involutions are not root 4-involutions. Similarly by 46.12, the centralizer of a dodecad involution contains no $SL_3(4)$-section, so dodecad involutions are not root 4-involutions. Therefore (1) holds. Then (1) and 48.4 imply $z^{C_G(X)} = z^G \cap C_G(X)$ and therefore (2) holds.

Lemma 49.4: *(1) G has three classes t_j^G, $1 \le j \le 3$, of involutions: the octad involutions, the dodecad involutions, and the root 4-involutions.*

(2) If $\bar{G} = Co_1$ and $\alpha_i : H_i \to \bar{H}_i$ is the isomorphism of amalgams supplied by 46.12.4, then $\alpha_i(t_j^G \cap H_i) = \bar{t}_j^{\bar{G}} \cap \bar{H}_i$ for $i = 1, 2$ and $j = 1, 2, 3$.

(3) $X \in \Delta$ is inverted by dodecad involutions and root 4-involutions, but not by octad involutions.

(4) H is transitive on its root 4-involutions and if u is a root 4-involution in H then $|C_H(u)| = 2^{15} \cdot 3 \cdot 5$.

Proof: First by Exercise 2.11, H^* has five classes of involutions of type a_2, c_2, c_4, a_4, and a_4'. We choose $T \in Syl_2(H)$, let $A = J(T)$, and choose notation so that $(A \cap Q)/\langle z \rangle = [a, \bar{Q}]$ for some a of type a_4. (This is possible from the proof of 46.12.) Then the members of A^* are of type a_2 and a_4. By 46.10, if $j^* \in a_4'$ then j is not an involution. Thus

(a) If i is an involution in H then $i \in Q$ or $i^* \in a_2$, c_2, a_4, or c_4.

Next by Exercise 2.11:

(b) H has (at most) two orbits on involutions of type a_4 with representatives a and az, while H is transitive on involutions of type c_4.

(c) Each involution of type a_2 and a_4 is fused into A under H.

For a_4-involutions this follows from (b). Similarly there is an involution of type a_2 in A such that each involution in iQ^+ (we use the notation of Exercise 2.11) is fused into A under $C_H(\tilde{i})$, so (c) follows for a_2-involutions from Exercise 2.11.

(d) H is transitive on its involutions v of type c_2, each such involution is a root 4-involution, and $|C_H(v)| = 2^{15} \cdot 3 \cdot 5$.

Namely we may choose B as in 49.1, F the root A_5-subgroup centralizing $E(C_G(B))$, and $u \in H \cap F$ a root 4-involution inverting B. Then $u^* \in c_2$ and u centralizes $K = C_H(B)^\infty$ with $Q_u^+ = C_Q(u) = O_2(K) * Q_u^-$. Next $C_H(B\langle u \rangle)$ contains $X \in \Delta$ and each involution in uQ_u^+ is fused to u or uj under KQ, where $j \in C_Q(BX) - \langle z \rangle$. Finally from 46.7, u and uj are fused in $C_H(X)$. Thus each c_2-involution of H is conjugate to the root 4-involution u. Further $|u^H| = |c_2|\lambda$, where $\lambda = 48$ is the number of involutions in uQ. So $|C_H(u)| = |C_{H^*}(u^*)||Q|/48 = 2^{15} \cdot 3 \cdot 5$.

We now adopt the notation of (2); then $H = H_1$ and $H_2 = N_G(A) = M$. Let $\alpha = \alpha_2$, $\zeta = \alpha_1$, and $M^* = M/A$. Now $\alpha(z^M) = \alpha(z)^{\alpha(M)} = \bar{z}^{\bar{M}}$ is the orbit of octad involutions in $\bar{A}$ under $\bar{M}$ and similarly if $d \in A$ is a dodecad involution then $\alpha(d^M) = \bar{d}^{\bar{M}}$ is the orbit of dodecad involutions in $\bar{A}$ under $\bar{M}$. Therefore by (c), all involutions of type a_2 and a_4 are octad or dodecad involutions. Indeed $H \cap M$ has two orbits on those $i \in A - Q$ of type a_k for each k, one orbit octad and the other dodecad, so

(e) H has two orbits on involutions of type a_k for $k = 2, 4$, and in each case one orbit consists of octad involutions and the other of dodecad involutions.

Further as the pair α, ζ is an isomorphism of amalgams, a of type a_2 or a_4 is octad in G if and only if $\zeta(a)$ is octad in $\bar{G}$. Similarly as α, ζ is an isomorphism of amalgams and each involution of M is fused into $H \cap M$ under M, once we show $\zeta(t_j^G \cap H) = \bar{t}_j^{\bar{G}} \cap \bar{H}$ for each j, the same holds for M. Thus it remains to prove all involutions of H of type c_4 are dodecad, since by (d) the c_2 involutions are root 4-involutions.

Next each involution i inverting X centralizes a member of $z^G \cap C_G(X)$, so by 49.3, we may take $i \in H$. But only involutions of type c_2 and c_4 invert X^* in H^*, so to prove (3) it also suffices to show c_4-involutions are dodecad.

Finally observe that $C_G(d) = C_M(d)$ has more than two orbits on $C_G(d) \cap z^G$, so H has at least three orbits on $d^G \cap H$, forcing involutions of type c_4 to lie in d^G. Namely $H \cap M$ has two orbits on dodecad involutions in A, so $C_G(d)$ has two orbits on octad involutions in A. Further there are octad involutions in $C_G(d) - A$ fused into $Q - A$ under M.

Lemma 49.5: *Let I be a root A_4-subgroup of H and $U = O_2(I)$ a root 4-subgroup of G. Then*

(1) $N_G(U_0) \le N_G(I)$ *for each* $1 \neq U_0 \le U$.

(2) $N_G(I) = (I \times E(N_G(I)))\langle \gamma \rangle$, *where* $E(N_G(I)) \cong G_2(4)$, γ *induces a field automorphism on* $E(N_G(I))$, *and* $I\langle \gamma \rangle \cong S_4$.

(3)   *The octad involutions in* $N_G(I)$ *are the long root involutions*

of $E(N_G(I))$, the dodecad involutions are of the form uv for $u \in U^\#$ and v a short root involution of $E(C_G(I))$, and all other involutions are root 4-involutions of G.

Proof: The argument is much like that of 48.3. First from the proof of 48.3 we may choose $X \in \Delta_H$ to centralize I and with $U \leq C_R(X)$. Let $Y \in Syl_3(I)$, so also $Y \in \Delta_H$. Arguing as in the proof of 48.3, we have $R = C_R(Y) * [R, Y]$ with $[Y, R]$ a Sylow 2-subgroup of $L_3(4)$ and $C_R(Y)$ the "R" for $C_G(Y)$. Moreover $U = C_R(Y_1)$ for some Y_1 of order 3 with $N_G(E) \cap C_G(U) = U \times (C_R(Y)(Y_1 \times C_G(YY_1E)))$ and $C_G(YY_1E) \cong L_2(4)$. Further for $u \in U^\#$, $C_R(U)C_G(YY_1E)\langle\tau\rangle \leq C_H(u)$, where $\tau \in N_G(E)$ with $E\langle\tau\rangle \cong D_8$ and $I\langle\tau\rangle \cong S_4$. Then by 49.4.4, this containment is an equality.

First $C_G(I) \leq C_G(Y)$. Let $K_1 = C_G(I) \cap N_G(X)$. Then by 48.3 and 49.2, $K_1 = E(K_1)\langle\alpha\rangle$, where $E(K_1) \cong SL_3(4)$ and α induces an outer automorphism on $E(K_1)$ inverting X.

Now pick B and L as in 49.1 with $X \leq L$ and I contained in the root A_5-subgroup $C_G(L)$. Then by 47.3, as we saw during the proof of 49.1, $N_L(X) = E(N_L(X))\langle\alpha_0\rangle$ with $N_L(X)/X \cong PGL_2(9)$. Notice $C_{E(N_L(X))}(\alpha) = D \cong D_{10}$ and by 47.2, $K_2 = C_L(D) \cong A_5$. Let $K_0 = \langle K_1, K_2 \rangle$, so that $K_0 \leq C_G(I)$. We will show $C_G(I) \cong G_2(4)$.

As α induces an outer automorphism on $E(K_1)$ inverting X, α induces a graph or field automorphism. As α centralizes an element of order 5 in $D \leq E(K_1)$, α induces a graph automorphism.

Now $K_1 \leq C_G(XY)$ and by 46.6.2, $C_G(XY)$ is quasisimple with $C_G(XY)/XY \cong U_4(3)$. We saw during the proof of 48.3 that $E(K_1)$, and hence also K_1, is determined up to conjugation in $C_G(XY)$ and hence also in $C_G(Y)$. Further by 48.8, $K_2 = E(C_G(YD))$, so $K_0 = \langle K_1, K_2 \rangle$ is determined up to conjugacy in $C_G(Y)$. Thus as $C_G(Y)$ is determined up to isomorphism independent of G, without loss of generality $G = Co_1$. Let G act on $\tilde{\Lambda}$, the Leech lattice modulo 2, and set $V = C_{\tilde{\Lambda}}(U)$. Now by Exercise 9.6, $C_{\tilde{\Lambda}}(Y) = 0$ and each $u \in U^\#$ inverts a conjugate of Y, so $dim(C_{\tilde{\Lambda}}(u)) = 12$. Thus $0 \neq dim(V) \leq 12$. Further letting $F = GF(4)$, we have that as $C_V(Y) = 0$, Y induces an F-space structure on V preserved by K_0.

Similarly $C_{\tilde{\Lambda}}(X) = 0$, so K_1 is faithful on V. Then as $dim_F(V) \leq 6$ and α induces a graph automorphism on $E(K_1) \cong SL_3(4)$, $V = V_1 \oplus V_2$, where V_1 is the natural $FE(K_1)$-module and $V_2 = V_1\alpha$ is dual to V_1. In particular $dim_F(V) = 6$ and $V = C_{\tilde{\Lambda}}(u)$ for each $u \in U^\#$. Moreover K_1 is determined up to conjugacy in $GL(V)$.

Next $V = [V, D] \oplus C_V(D)$ with $dim_F(C_V(D)) = 2$. Thus

$$C_{GL(V)}(D) = L_+ \times L_-,$$

where $L_\epsilon \cong GL_2(4)$ with $[V, D, L_+] = 0$ and $[C_V(D), L_-] = 0$. Now $L_2(4) \cong K_2 \leq L_+L_-$, so as $\langle\alpha\rangle X \leq K_2$ with $C_V(X) = 0$, K_2 is a full diagonal subgroup of L_+L_-. Then as $\langle\alpha\rangle X$ is contained in a unique such subgroup, K_2 is uniquely determined. Thus $K_0 = \langle K_1, K_2 \rangle$ is determined up to conjugation in $GL(V)$.

Finally $K_* = G_2(4) \leq GL(V)$ is generated by subgroups K_1 and K_2 as above (cf. [A3]) so $K_0 \cong K_*$, completing the proof that $K_0 \cong G_2(4)$.

Next by 48.3 there is an involution γ fused into U in $C_G(X)$ with $\langle\gamma\rangle I \cong S_4$ and γ inverting Y and inducing a graph-field automorphism on $E(K_1)$. We pick $\gamma \in N_H(E)$ and let $\langle u \rangle = C_U(\gamma)$. By the first paragraph of this proof, $\langle\gamma\rangle U C_{K_0}(z) = C_H(u)$. In particular γ acts on $\langle K_1, C_H(U)\rangle = U \times K_0$ and then on K_0, and setting $M = C_G(u)$ and $K = U K_0 \langle\gamma\rangle$, we have $C_M(z) \leq K$.

As γ induces a graph-field automorphism on K_1, γ induces a field automorphism on $K_0 \cong G_2(4)$. Then as $G_2(4)$ is transitive on its field automorphisms, K is transitive on involutions in $K - UK_0$, and each is a root 4-involution.

Next $K_0 \cong G_2(4)$ has two classes of involutions: the long root elements in z^{K_0} and the short root elements v fused into K_2 in K_0, which are therefore root 4-involutions of G. Further, from 48.3 each member of Uz is a root 4-involution of $C_G(X)$. Thus all involutions in K are root 4-involutions except those in z^K and possibly involutions fused into $vU^\#$. But $C_K(z) = C_H(u)$ contains dodecad involutions, so these involutions are dodecad.

It remains to show $K = C_G(u)$. Assume not. We have shown $z^G \cap K = z^K$ and $C_M(z) \leq K$, so K is the unique point of M/K fixed by z by 7.3. Thus if $M \neq K$ we can apply Exercise 2.10 to obtain a contradiction. Namely by Exercise 2.10, M has a normal subgroup M_0 with $M = M_0 K$ and $M_0 \cap K = K_0$. Then if V is maximal in the set $\mathcal{U} = \mathcal{U}(M_0)$ of Exercise 2.10, then as $z^K \cap V = \varnothing$, V is either $\langle v \rangle$ or the root group of v in K_0 for some short root element v of K_0. But then $C_{K_0}(V)/V$ does not have one class of involutions, contradicting Exercise 2.10.2.

Lemma 49.6: $|G| = 2^{21} \cdot 3^9 \cdot 5^4 \cdot 7^2 \cdot 11 \cdot 13 \cdot 23$.

Proof: Let $\bar{G} = Co_1$. By 46.12.4 there is an isomorphism $\alpha_i : H_i \rightarrow \bar{H}_i$, $i = 1, 2$, of amalgams, where $H_1 = H$, $H_2 = M = N_G(A)$, and $A = J(T)$ for $T \in Syl_2(H)$. By 46.13, $C_G(d) \leq M$ for $d \in A$ a dodecad involution.

Then by 49.4, G has three classes of involutions t_j^G, $1 \leq j \leq 3$, and α_1 and α_2 induce isomorphisms $\alpha_i : C_G(t_i) \to C_{\bar{G}}(\bar{t}_i)$, $i = 1, 2$, such that $\alpha_i(t_j^G \cap C_G(t_i)) = \bar{t}_j^{\bar{G}} \cap C_{\bar{G}}(\bar{t}_i)$ for each i, j. Finally by 49.5, there is an isomorphism $\alpha_3 : C_G(t_3) \to C_{\bar{G}}(\bar{t}_3)$ with $\alpha_3(t_j^G \cap C_G(t_3)) = \bar{t}_j^{\bar{G}} \cap C_{\bar{G}}(\bar{t}_3)$. Therefore by the Thompson Order Formula 7.2, $|G| = |\bar{G}|$. Then 22.12 completes the proof.

Lemma 49.7: *Let $X \in \Delta$, $P_X \in Syl_3(N_G(X))$, and $J = J(P_X)$. Then*

(1) $J \cong E_{3^6}$ *and* $J = C_G(J)$.

(2) $N_G(J)$ *is the split extension of a group K with $Z(K)$ of order 2 inverting J and $K/Z(K) \cong M_{12}$.*

(3) K *is 5-transitive on $J \cap \Delta$ of order 12.*

Proof: The proof is much like that of 48.5. Let $M = N_G(X)$ and $M^* = M/X$. By 48.5, $J(P_X^*) \cong E_{3^5}$ and if J_0 is the preimage of $J(P_X^*)$ then $(C_G(X) \cap N_G(J_0))/J_0 \cong M_{11}$ with $C_G(X) \cap N_G(J_0)$ 4-transitive on $Y^{*M} \cap J_0^*$ of order 11, where $Y \in \Delta_H$ centralizes X. Now by a Frattini argument $|N_M(J_0) : C_G(X) \cap N_G(J_0)| = 2$, so as $M_{11} = Aut(M_{11})$, $N_M(J_0)/J_0 \cong \mathbf{Z}_2 \times M_{11}$. Let t be an involution in $N_M(J_0)$ with $[t, N_M(J_0)] \leq J_0$. Then t inverts X and as $N_M(J_0)$ is irreducible on J_0^*, t either inverts or centralizes J_0^*. As no involution has an $M_{11}/3^5$ subgroup in its centralizer, it is the former. Then $N_M(J_0) \cap C_G(t) = K_0$ is a complement to J_0 in $N_M(J_0)$ and as t inverts J_0, $J_0 \cong E_{3^6}$. Further for each $1 \neq P_0 \leq P_X$, $|J_0 : C_{J_0}(P_0)| > |P_0|$, so $J = J_0$. As $C_G(J) \leq C_M(J) = J$, $J = C_G(J)$, so (1) is established. Also $K = N_G(J) \cap C_G(t)$ is a complement to J in $N_G(J)$.

Next we may take $z \in K_0$ and $C_J(z) \cong E_{81}$ with $\Delta_H \cap C_J(z) = Y^{N_{H \cap M}(J)} \cup \{X\}$ of order 4 and $\Delta_H \cap XY = \{X, Y\}$ for some $Y \in \Delta_H \cap C_J(z)$. So by 49.3, $\Delta \cap C_J(z) = Y^{N_{H \cap M}(J)} \cup \{X\}$. But from 48.5, each subgroup of J of order p is fused into $C_J(z)$ under K_0, so $\Delta \cap J = Y^{K_0} \cup \{X\}$ and $Y^k \mapsto Y^{*k}$ is a bijection of Y^{K_0} with Y^{*K_0}. Therefore as $|Y^{*K_0}| = 11$, $|\Delta \cap J| = 12$. Now K_0 is 4-transitive on $\Delta \cap J - \{X\}$ and J is weakly closed in $N_M(J)$, so $N_G(J)$ is transitive on $J \cap \Delta$. Hence K is 5-transitive on $\Delta \cap J$. From 48.5, $\langle t \rangle$ is the kernel of the action of K_0 on $\Delta \cap J$ and hence also of the action of K. Then by Exercise 6.6.7, $K/\langle t \rangle \cong M_{12}$.

In the remainder of this section let $X \in \Delta$ and $M = N_G(X)$.

Lemma 49.8: *M has five orbits $\Delta_i(X)$, $0 \leq i \leq 4$, on Δ, where*

$\Delta_0(X) = \{X\}$ *and*

(1) $\Delta_1(X) = \Delta(X)$ *is of order* $2^5 \cdot 5 \cdot 11 \cdot 13 = 22,880$ *with* $N_M(X_1) = E(N_M(XX_1))$ *extended by* E_4 *and* $E(N_M(XX_1))/XX_1 \cong U_4(3)$.

(2) $\Delta_2(X) = \{X_2 \in \Delta : \langle X, X_2 \rangle \cong SL_2(3)\}$ *is of order* $3^4 \cdot 5 \cdot 7 \cdot 11 \cdot 13 = 405,405$ *with* $N_M(X_2)$ *a complement to* X *in* $H \cap M$, *when* $z \in \langle X, X_2 \rangle$.

(3) $\Delta_3(X) = \{X_3 \in \Delta : \langle X, X_3 \rangle \cong A_5\}$ *is of order* $2^5 \cdot 3^5 \cdot 11 \cdot 13 = 1,111,968$ *with* $N_M(X_3) \cong \mathbf{Z}_2 \times J_2$.

(4) $\Delta_4(X) = \{X_4 \in \Delta : \langle X, X_4 \rangle \cong A_4\}$ *is of order* $2 \cdot 3^5 \cdot 11 = 5,346$ *with* $N_M(X_4) \cong Aut(G_2(4))$.

Proof: The proof is entirely analogous to that of 47.5 and 48.9, and is left as an exercise.

Lemma 49.9: *Define J as in 49.7 and let $J \leq P \in Syl_3(G)$. Then*

(1) $J = \langle P \cap \Delta \rangle$.

(2) G *is transitive on triangles in* Δ.

Proof: As $N_G(J)$ is 5-transitive on $J \cap \Delta$, (1) implies (2). By 49.8, if $Y \in \Delta$ with $\langle X, Y \rangle$ a 3-group then $[X, Y] = 1$, so $B = \langle P \cap \Delta \rangle$ is abelian. Then as $C_G(J) = J \leq B$, (1) holds.

Lemma 49.10: Δ *is of diameter 2.*

Proof: This follows from 49.8, which says $\Delta(X, Y) \neq \varnothing$ for each $y \in \Delta$.

Lemma 49.11: *(1) G is transitive on triples (X_0, X_2, X_3) from Δ with $X_2 \in \Delta_4(X_0)$ and $X_3 \in \Delta(X_2) \cap \Delta^2(X_0)$.*

(2) $\langle X_0, X_2, X_3 \rangle = K \cong A_6$ is a root A_6-subgroup of G, $X_3 \in \Delta_3(X_0)$, and $N_G(K) = (K \times C_G(K))\langle \tau \rangle$, where $C_G(K) \cong U_3(3)$, $K\langle \tau \rangle \cong S_6$, and $C_G(K)\langle \tau \rangle \cong G_2(2)$.

(3) $X_0^{\perp} \cap X_2^{\perp} \cap X_3^{\perp} \neq \varnothing$.

Proof: Let $X_4 \in \Delta_4(X)$, so that $I = \langle X_4, X \rangle \cong A_4$ is a root A_4-subgroup and $L = C_G(I) \cong G_2(4)$. Then $\Delta(X, X_4)$ consists of the centers Y of Sylow 3-groups of L and from the proof of 49.5, $N_L(Y) \cong \mathbf{Z}_2 \times SL_3(4)$, so L is transitive on $\Delta(X, X_4)$ of order $|L : N_L(Y)| = 2^5 \cdot 5 \cdot 13$. Thus $\Gamma = \Delta(X) - \Delta(X, X_4)$ is of order $2^5 \cdot 5 \cdot 10 \cdot 13$. Pick $X_3 \in \Gamma$ such that there is $X_2 \in \Delta(X, X_4, X_3)$. By 48.19, $K = \langle X, X_3, X_4 \rangle$ is a root A_6-subgroup of G, $X_3 \in \Delta_3(X_4)$, and $C_G(KX_2) \cong \mathbf{Z}_4/3^{1+2}$. Let β be of order 4 in $N_K(XX_3)$. Then β is faithful on XX_3 so β induces an automorphism on $C_G(XX_3)/XX_3 = D^* \cong U_4(3)$ with $3^{1+2} \leq O_3(C_{D^*}(X_2^*) \cap C_G(\beta))$. Hence by Exercise 16.1.10, $C_G(\langle \beta \rangle XX_3) \cong U_3(3)$

and $N_G(\langle\beta\rangle) \cap C_G(X) \cap N_G(X_3) \le G_2(2)$. Also

$$N_L(X_3) = N_L(K) \cap N_G(X_3) \cap C_G(I) \le N_G(\langle\beta\rangle) \cap C_G(X) \cap N_G(X_3)$$

for suitable choice of β, as $C_{Aut(K)}(I) \cap N_{Aut(K)}(X_3) = \langle\tau\rangle$, where τ is a transposition in $S_6 \le Aut(K)$, since τ acts on some conjugate of β under XX_3. Then as $|G_2(4) : G_2(2)| = |\Gamma|$, we conclude that L is transitive on Γ and (2) holds. Transitivity of $C_G(I)$ on Γ gives (1) and (3).

Lemma 49.12: *Let I be a root A_4-subgroup of G and V of order 7 in $C_G(I)$. Then*

(1) *$C_G(V) \cong \mathbf{Z}_7 \times A_7$ and $N_G(V)/V \cong \mathbf{Z}_3 \times S_7$.*

(2) *$N_G(E(C_G(V)))$ is of index 2 in $PGL_2(7) \times S_7$.*

(3) *Let W be of order 3 in $C_G(E(C_G(V)))$. Then W is fused into $\langle Q, Q^g \rangle$ and $N_G(W) \cong S_3 \times A_9$.*

Proof: By 49.1 we have a subgroup $L_1 \times L_2$ of G with L_1 a root A_5-subgroup of G containing I and $J_2 \cong L_2$. Further by 49.11, L_1 is contained in a root A_6-subgroup K_1 and $U_3(3) \cong C_G(K_1) = K_2 \le C_{L_2}(K_1)$. Next by Exercise 16.5, there exists $h \in L_2$ with $K_2 \cap K_2^h = K \cong L_3(2)$. Then $F = \langle K_1, K_1^h \rangle$, and indeed $C_G(F) = C_G(K_1) \cap C_G(K_1)^h = K_2 \cap K_2^h = K$. Now $S_4 \cong S \le K$ with the involutions in S in z^G, so without loss of generality $z \in O_2(S)$. Then a Sylow 2-subgroup of S is contained in Q, so without loss of generality $E = O_2(S)$, and then by 46.3, $S \le \langle Q, Q^g \rangle$ with $S \cap R = E$. Further for $W \in Syl_3(S)$, by 46.3, $S_3 \cong S_0 = \langle Q, Q^g \rangle \cap N(W)$ is a complement to R in $\langle Q, Q^G \rangle$ and $N_G(W) \cap N_G(R) = S_0 \times K_0$, with $K_0 = C_G(WE) \cong A_8$. Then $S_0 = N_S(W)$ and $C_G(S) = C_G(WE) \cong A_8$. Next $C_G(I) \cong G_2(4)$ by 49.5, and $|G_2(4)|_7 = 7$, so we may take $V \le K$. Then $K = \langle V, S \rangle$ so $F \le C_G(K) = C_G(V) \cap C_G(S)$ and we conclude $F = A_7$ or A_8. But taking $U = O_2(I)$, $N_G(U) \cap C_G(V) = (I \times X)\langle\tau\rangle \times V$, where $\langle\tau\rangle I \cong S_4$ and $X\langle\tau\rangle \cong S_3$. So $F \cong A_7$ and FV is strongly embedded in $C_G(V)$. Then by 7.6, $FV = C_G(V)$. That is, (1) and (2) hold.

Now a 2-central involution d in K_0 is diagonal in the product $U \times U_1$ of root 4-subgroups in K_0, so by 49.5, d is a dodecad involution. Then $W \le C_G(d) = N_G(A) \cap C_G(d)$, where $A = J(T)$ for some $T \in Syl_2(G)$ by 46.13, so if $G = Co_1$ then W is the subgroup of order 3 discussed in 26.4, and by 26.4, $N_G(W) \cong S_3 \times A_9$. In particular by 46.12.4, $C_G(\langle d, W \rangle)$ is determined up to isomorphism as is the fusion of involutions in $C_G(\langle d, W \rangle)$. Similarly $C_G(\langle u, W \rangle) \cong \mathbf{Z}_3 \times (\mathbf{Z}_2/(E_4 \times A_5))$ with fusion of involutions determined, so by the Thompson Order Formula

7.2, $|C_G(W)|$ is determined and thus is $3 \cdot |A_9|$. So $|C_G(W) : WK_0| = 9$ and then as $K_0 \cong A_8$, $C_G(W) \cong \mathbf{Z}_3 \times A_9$.

We term the $U_3(3)$-subgroup of 49.11 a *root $U_3(3)$-subgroup* of G. We term the A_n-subgroups of G, $n = 7, 9$, appearing in 49.12 as *root A_n-subgroups* of G.

Let $\mathcal{C}$ be the closure of the set of all triangles of Δ and all squares $X_0X_1X_2X_3X_0$ with $X_{i+2} \in \Delta_4(X_i)$ for each i.

Lemma 49.13: $\mathcal{C} = \mathcal{C}_4(\Delta)$.

Proof: Let $p = X_0X_1X_2X_3X_0$ be a square in Δ. We first observe that the graph on $\Delta(X_0, X_2)$ obtained by joining X to Y if $X \in \Delta_4(Y)$ is connected. This is because for $K = \langle \Delta(X_0, X_2) \rangle$, $K = \langle N_K(X), N_K(Y) \rangle$. Thus the result follows from 34.7.

Lemma 49.14: Δ *is 4-generated.*

Proof: By 49.10 and 34.5 it suffices to show each 5-gon $p = X_0 \cdots X_5$ is in $\mathcal{C}_4(\Delta)$. If $X_2 \in \Delta_i(X_0)$ for $i = 2$ or 4, we do this by showing $X_0^\perp \cap X_2^\perp \cap X_3^\perp \neq \varnothing$ and appealing to 34.8. Namely if $X_2 \in \Delta_4(X_0)$ this follows from 49.11.3. If $X_2 \in \Delta_2(X_0)$ we will see it follows from 48.20. For by 49.8, $N_G(X_0) \cap N_G(X_2)$ is a complement to X_2 in $N_G(X_2) \cap C_G(z)$ when z is the involution in $\langle X_0, X_2 \rangle$. By 48.20 there is $X \in \Delta(X_2, X_3) \cap H$. Now $X \in X_0^\perp \cap X_2^\perp \cap X_3^\perp$.

Thus we may assume $X_{i+2} \in \Delta_3(X_i)$ for each i. In particular $K = C_G(\langle X_0, X_2 \rangle) \cong J_2$ and there is $X \in \Delta_4(X_0) \cap \Delta_4(X_2) \cap \langle X_0, X_2 \rangle$. Then $M = C_G(\langle X, X_2 \rangle) \cong G_2(4)$ and M is transitive on $\Delta(X_2) - \Delta(X)$ with $C_G(\langle X, W, X_2 \rangle) = U(W) \cong U_3(2)$ for $W \in \Delta(X_2) - \Delta(X)$ by 49.11.

We produce $Y \in \Delta$ such that for $j = 0, 2, 3$, $Y \notin \Delta_3(X_j)$. Then for $0 \leq i \leq 4$, we pick a geodesic p_i from X_i to Y in Δ. Now p is in the closure of the cycles $q_i = p_i \cdot p_{i+1}^{-1} \cdot X_{i+1}X_i$ for each i, so it suffices to show $q_i \in \mathcal{C}_4(\Delta)$. But if $d(X_j, Y) = 1$ for $j = i$ or $i + 1$ then q_i is of length 4 and hence in $\mathcal{C}_4(\Delta)$, while if $d(X_j, Y) = 2$ then at least one $k = j$ or $j + 1$ is 0, 2, or 3, so the cycle q_i of length 5 contains X_k, Y with $Y \notin \Delta_3(X_k)$, and hence $q_i \in \mathcal{C}_4(\Delta)$ by earlier reductions.

Thus it remains to produce Y. If $X_3 \in \Delta(X, X_2)$ let $Y = X$. Thus we may assume $X_3 \notin \Delta(X, X_2)$. In this case we let $V = U(X_3)$ and observe that $K \cap V$ contains an element of order 2 or 7. $|M : K| = 2^5 \cdot 13$ and $|V| = 2^5 \cdot 3^3 \cdot 7$, so $|K \cap V| \geq |V|/|M : K| > 14$. Therefore as maximal subgroups of V of odd order are of order 21 and 27, either $K \cap V$ contains an element of order 2 or 7, or some $Y_0 \in \Delta$. In the latter case let $Y = Y_0$.

If $k \in K \cap V$ is of order 7 then $L = \langle X_3, X_2, X_0 \rangle \leq C_G(k)$ and by

49.12, L is a root A_7-subgroup. So $L^8 = L/O_2(L) \cong A_7$. Similarly if k is an involution in $K \cap V$ then $k \in z^G$ so we may take $k = z$ and $|Q : C_Q(X_i)| = 4$, so $|C_Q(L)| \geq 8$. If $C_Q(L) \cong Q_8$ then L centralizes $Y_0 \in \Delta_H$ with $C_Q(L) = [Y_0, Q]$, and we let $Y = Y_0$. Otherwise we may take $L \leq C_G(E)$, so LR/R is a subgroup of $C_G(E)/R \cong A_8$, so again $L^* \cong A_7$.

Thus in any case $L^* \cong A_7$. Then we may pick $Y \in \Delta \cap L$ with $\langle X_i^*, Y^* \rangle$ not isomorphic to A_5, and hence $Y \notin \Delta_3(X_i)$. Namely representing L^* on $\{1, \ldots, 7\}$ we may take $X_i^* = \langle x_i^* \rangle$, where $x_2^* = (1, 2, 3)$, $x_3^* = (4, 5, 6)$, and $x_0^* = (1, 4, 7)$. Then take $Y^* = \langle (4, 5, 7) \rangle$.

Lemma 49.15: $G \cong Co_1$.

Proof: As usual we construct a uniqueness system for G; then we appeal to Theorem 37.7.

Let W be the subgroup of order 3 in $\langle Q, Q^g \rangle$ discussed in 49.12.3. Then by 49.12.3, $K = N_G(W) \cong S_3 \times A_9$. Let $\bar{G} = Co_1$ and $\bar{W}$ a corresponding subgroup of $\bar{G}$. Then there is an isomorphism $\zeta : K \to \bar{K}$. Pick $X \in \Delta_K = K \cap \Delta$ and let $\bar{X} = X\zeta$, so that $\bar{X} \in \bar{\Delta}_{\bar{K}}$. Then $\mathcal{U} = (G, K, \Delta, \Delta_K)$ and $\bar{\mathcal{U}} = (\bar{G}, \bar{K}, \bar{\Delta}, \bar{\Delta}_{\bar{K}})$ are uniqueness systems.

By Exercise 16.4, there exists an isomorphism $\alpha : G_X \to \bar{G}_{\bar{X}}$. Now $K_X \cong S_3 \times (\mathbf{Z}_2/(\mathbf{Z}_3 \times A_6))$. Let $A(K_X)$ be the subgroup of $Aut(K_X)$ permuting the root 4-subgroups of K_X. Then $A(K_X) = Aut_K(K_X) \cong S_3 \times S_3 \times S_6$. Thus by Exercise 14.3.1, we can choose ζ so that $\zeta = \alpha$ on K_X.

Pick $Y \in \Delta(X) \cap K$ and let $\bar{Y} = Y\zeta$. Then $\bar{Y} = Y\alpha$, $G_{XY}\alpha = \bar{G}_{\bar{X}\bar{Y}}$, and $K(\{X, Y\})\zeta = \bar{K}(\{\bar{X}, \bar{Y}\})$, so α, ζ define a similarity of $\mathcal{U}$ with $\bar{\mathcal{U}}$.

Next G_{XY} is of index 2 in $Aut(E(G_{XY})) = N_G(XY) = L$, where $E(G_{XY}) \cong U_4(3)/E_9$ is quasisimple. Now $C_L(K_{XY}) = C_G(K_{XY}) \leq Z(K_{XY}) = 1$, so

$$C_{Aut(G_{XY})}(K_{XY}) = 1.$$

Therefore by 37.9, α, ζ define an equivalence of $\mathcal{U}$ with $\bar{\mathcal{U}}$.

We now appeal to Theorem 37.7 to complete the proof. By 37.7, it remains to show Δ_K is a base for Δ. Thus by 49.13 and 49.14, it suffices to show that each triangle and each square $p = X_0 \cdots X_4$ with $X_{i+2} \in \Delta_4(X_i)$ for all i in Δ is fused into Δ_K under G. But by 49.9, G is transitive on triangles of Δ and by 49.5 and 49.8, G is transitive on squares satisfying the hypotheses of p. Thus as Δ_K contains triangles and such squares, our proof is complete.

Remarks. The general structure of the the groups J_2 and J_3 was determined by Janko in [J2]. The uniqueness of J_2 as a rank 3 permutation group on the cosets of $U_3(3)$ was proved by M. Hall and Wales in [HW].

Suzuki was the first to investigate Suz; see, for example, [Su]. Soon after, Suz was discovered to be a section of Co_1 and much of the structure of the two groups was determined by Conway and Thompson.

The general structure of Suz and Co_1 was investigated in a systematic way by N. Patterson in his thesis [P], where Patterson also produced uniqueness proofs for Suz and Co_1. To identify the groups, Patterson appealed to a theorem of B. Stellmacher [St] on groups generated by a class Δ of subgroups of order 3 such that for each distinct $X, Y \in \Delta$, $\langle X, Y \rangle$ is isomorphic to E_9, A_4, A_5, or $SL_2(3)$. S. K. Wong also worked on Suz and Co_1 independently and he and Patterson published their work jointly in [PW1] and [PW2].

Exercises

1. Let (V, f) be a 6-dimensional orthogonal space of sign -1 over $F = GF(3)$, $G = \Delta(V, f)$ the group of similarities of the space, and $\bar{G} = P\Delta(V, f)$. Let $L = E(G)$. Prove

   (1) $\bar{L} \cong P\Omega_6^-(3) \cong U_4(3)$, $\bar{G} = Aut(\bar{L})$, and $\bar{G}/\bar{L} \cong D_8$.

   (2) Let z be an involution in G with $[V, z]$ of dimension 4 and sign $+1$. Then $F^*(C_{\bar{L}}(\bar{z})) = \bar{Q} \cong Q_8^2$ and $\bar{L}$ is of type $U_4(3)$.

   (3) Let $K = GF(9)$, $K^\# = \langle \zeta \rangle$, $\zeta^2 = i$, and (V^K, f^K) a 3-dimensional orthogonal space over K with basis $X = \{x_1, x_2, x_3\}$ with $f^K(x_i, x_j) = 0$ for all $i \neq j$, $f^K(x_i) = f^K(x_i, x_i) = 1$ for $i = 1, 2$, and $f^K(x_3) = \zeta$. Let $T = T_F^K$ be the trace from K to F. Take V to be V^K regarded as an F-space. Then $(V, T \circ f^K)$ is 6-dimensional of sign -1, so we may take $f = T \circ f^K$.

   (4) Let $\beta = iI$ be scalar multiplication on V^K via i and regard $\beta \in G$. Then $C_L(\beta) = L^K \langle \tau, -I \rangle$, where $L^K = E(\Delta(V^K, f^K)) \cong \Omega_3(9) \cong L_2(9)$ and τ is the semilinear map $\tau : \sum_i a_i x_i \mapsto \sum_i a_i^3 x_i$ with $x_1\tau = -\zeta(x_1 + x_2)$, $x_2\tau = \zeta(x_2 - x_1)$, and $x_3\tau = x_3$.

   (5) Let A be the set of $\bar{G}$-conjugates of elements $\alpha \in C_{\bar{G}}(\bar{z})$ such that $\alpha^2 \in \langle \bar{z} \rangle$ and $C_{\bar{Q}}(\alpha) \cong D_8$. Then $\bar{G}$ is transitive on A and for each $\alpha \in A$, α is an involution, $\bar{L}$ is transitive on involutions in $\alpha\bar{L}$, and $C_{\bar{L}}(\alpha) \cong M_{10}$.

   (6) $\bar{\beta} \in A$.

   (7) $\langle \beta, C_L(\beta) \rangle = H$ has two orbits S_1 and S_2 on singular points of V, where $S_1 = \{Fv : v \text{ is } K\text{-singular}\}$ is of order 40 and $S_2 = \{Fv : f^K(v) = \pm i\}$ is of order 72. Further L^K has two orbits S_2^ϵ, $\epsilon = \pm 1$, of order 36 on S_2, where $S_2^\epsilon = \{Fv : f^K(v) = \epsilon i\}$.

(8) H has three orbits $\mathcal{L}_j$, $1 \le j \le 3$, on the singular lines of V, where $\mathcal{L}_1 = \{l : l \text{ is a } K\text{-singular point of } V^K\}$ is of order 10, $\mathcal{L}_2 = \{l : |l \cap S_1| = 2, \ |l \cap S_2^\epsilon| = 1, \ \epsilon = \pm 1\}$ is of order 180, and $\mathcal{L}_3 = \{l : |l \cap S_2^\epsilon| = 2, \ \epsilon = \pm 1\}$ is of order 90. Representatives are $K(x_1 + ix_2)$, $F\zeta x_1 + Fi\zeta x_2$, and $F\zeta x_1 + F\zeta x_2$, respectively. That is, $x_1 + ix_2$ is K-singular, $f^K(\zeta x_j) = i$, $f^K(i\zeta x_2) = -i$, $f^K(\zeta x_1 + \epsilon\zeta x_2) = -i$, and $f^K(\zeta x_1 + \epsilon i\zeta x_2) = 0$.

(9) Let $\mathcal{L}$ be the graph on the singular lines of V with l adjacent to k if $l \cap k \ne 0$. Then $\mathcal{L}_3$ is a connected subgraph of $\mathcal{L}$.

(10) Up to conjugation there exists a unique automorphism γ of $\bar{L}$ of order 4 with $\gamma^2 \notin \bar{L}$ such that $O_3(C_{\bar{L}}(X)) \cap C(\gamma)$ contains 3^{1+2}, where X is the center of a Sylow 3-subgroup of $\bar{L}$. Further $C_{\bar{L}}(\gamma) \cong U_3(3)$ and $C_{\bar{G}}(\gamma)/\langle\gamma\rangle \cong G_2(2)$.
(Hint: See Section 2 in [A5] for some help.)

2. Let V be a 4-dimensional unitary space over $GF(9)$, $L = SU(V) \cong SU_4(3)$, $L^* = L/Z(L)$, and H the stabilizer in V of $\{U, U^\perp\}$, where U is a nondegenerate line of V. Prove
   (1) $H = C_L(z^*)$, where z is the involution in L with $U = [V, z]$.
   (2) H has three orbits Γ_i, $1 \le i \le 3$, on the set Γ of singular points of V.
   (3) $\Gamma_1 = (\Gamma \cap U) \cup (\Gamma \cap U^\perp)$ is of order 8.
   (4) $\Gamma_2 = \{\langle u + v\rangle : \langle u\rangle \in \Gamma \cap U, \langle v\rangle \in \Gamma \cap U^\perp\}$ is of order 128 and $W + Wz$ is a singular line for $W \in \Gamma_2$.
   (5) Γ_3 consists of the points $\langle u + v\rangle$, where $\langle u\rangle, \langle v\rangle$ are nonsingular points of $U, U^\perp$, respectively, with $(u, u) = -(v, v)$. Further $|\Gamma_3| = 144$ and $W + Wz$ is a nondegenerate line for $W \in \Gamma_3$.

3. Let $G = U_4(3)$, G_0 the covering group of G, and $\hat{G} = G_0/O^3(Z(G_0))$. Prove
   (1) Let $M \cong A_6/E_{81}$ be a maximal parabolic of G, $\hat{M}$ the preimage of M in $\hat{G}$, and $\hat{Z} = Z(\hat{M})$. Then $\hat{Z} = Z(\hat{G})$ and $\hat{M} \cong M_0/O^3(Z(M_0))$, where M_0 is the covering group of M.
   (2) A Sylow 3-subgroup of the Schur multiplier of G is isomorphic to E_9.
   (3) $O_3(M) = W$ is the dual of the M/W-module $U(L, V)$ of 48.14.
   (4) $B = Aut(\hat{G}) \cong Aut(G)$ and $N_B(\hat{M})$ has two orbits on points of $\hat{Z}$ with representatives $\hat{Z}_1$ and $\hat{Z}_2$, where $N_B(\hat{Z}_1)/W \cong \mathbf{Z}_2 \times S_6$ and $N_B(\hat{Z}_2) \cong \mathbf{Z}_2 \times M_{10}$.
   (5) $\hat{G}/\hat{Z}_2$ is the unique quasisimple group L such that $Z(L) \cong \mathbf{Z}_3$, $L/Z(L) \cong G$, and L admits an involutory automorphism α such that $M\langle\alpha\rangle/O_3(M) \cong M_{10}$.

(Hint: Use 48.14 and its proof; for example, note the role played by M in the first paragraph of the proof of 48.14.)

4. Prove a Sylow 3-subgroup of the Schur multiplier of Suz is of order 3. (Hint: Use Exercise 16.3.2 and 46.5.)

5. Let $G = J_2$. Prove
   (1) G has a subgroup $K \cong U_3(3)$.
   (2) G is rank 3 on $\Omega = G/K$ with parameters $n = 100$, $k = 36$, $l = 63$, $\lambda = 14$, and $\mu = 12$.
   (3) For $\alpha \in \Omega$ with $K = G_\alpha$, $\beta \in \Delta(\alpha)$, and $\gamma \in \Gamma(\alpha)$, $K_\beta \cong L_3(2)$ and $K_\gamma = C_K(t)$ for t an involution in K.
   (Hint: Use 49.11.2 for (1). Then observe that if t is an involution in K then $O_2(C_G(t)) \cap N_G(C_K(t))$ is not contained in K. Conclude $C_K(t) = K_\gamma$ for some $\gamma \in \Omega$. Similarly observe that for $P \in Syl_7(K)$, $N_G(P)$ is not contained in K and $I \cong L_3(2)$ is the unique proper subgroup of K of index at most 36 containing P.)

6. (1) If X is a group with $Z(X) \cong \mathbf{Z}_2$ and $X/Z(X) \cong Q_8$, prove $X \cong \mathbf{Z}_2 \times Q_8$.
   (2) Let $S = Sp_4(3)$. Prove
      (a) S has a subgroup $SL_2(3)wr\mathbf{Z}_2$.
      (b) $S/Z(S) \cong \Omega_6^-(2)$ and involutions of type a_2 in $\Omega_6^-(2)$ lift to involutions of S, while involutions of type c_2 lift to elements of order 4.
   (3) Let $G = \Omega_6^-(2)$, G_0 the covering group of G, and $\hat{G} = G_0/O^2(Z(G_0))$. Prove $\hat{G} \cong Sp_4(3)$, so a Sylow 2-subgroup of the Schur multiplier of G is of order 2.

7. Let $G = \Omega_8^+(2)$, G_0 the covering group of G, and $\hat{G} = G_0/O^2(Z(G_0))$. Let M be the stabilizer of a point in the natural module for G, M_0 the covering group of M, $\tilde{M} = M_0/O^2(Z(M_0))$, and $\hat{M}$ the preimage of M in $\hat{G}$. Prove
   (1) M is the split extension $M = LA$, where $L \cong A_8$ and $A \cong E_{64}$ is the natural module for L. Further $H^1(L, A) \cong \mathbf{Z}_2$.
   (2) $\tilde{M}$ is the split extension $\tilde{M} = \tilde{L}\tilde{A}$, where $\tilde{L} \cong A_8/\mathbf{Z}_2$ is the covering group of A_8 and $\tilde{A} = [O_2(\tilde{M}), \tilde{M}] \cong \mathbf{Z}_2 \times D_8^3$.
   (3) $Z(\hat{G}) = Z(\hat{M}) \cong E_4$ and $\hat{M} = \tilde{M}/\langle x^2y^2\rangle$, where $x \in \tilde{L}$, $y \in \tilde{A}$, and $xZ(\tilde{M}), yZ(\tilde{M})$ are involutions of type c_2 in G.
   (4) $Z(\hat{G}) = \{1, a^2, b^2, c^2\}$, where $aZ(\hat{G})$ and $bZ(\hat{G})$ are involutions of type a_4 in G and $cZ(\hat{G})$ is of type c_2 in G.
   (Hint: See Exercise 2.11 for the discussion of involutions in G. In (2), imitate the proof of 8.17.)

Chapter 17

Subgroups of Prime Order
in Five Sporadic Groups

We have considered five of the sporadic groups in detail: M_{24}, He, J_2, Suz, and Co_1. In particular we have characterized each of these five groups in terms of a hypothesis $\mathcal{H}(w, L)$ for suitable w and L, sometimes with some extra conditions. That is, we have proved that there exists a group G satisfying hypothesis $\mathcal{H}(w, L)$ and that G is unique up to isomorphism. We have also determined the order of G, and we have generated much information about its subgroup structure. In particular we have determined most of the subgroups of G of prime order and their normalizers in G.

This information is summarized in five tables at the end of this chapter. In the case of M_{24} and J_2 all the necessary facts have already been established. Namely Section 21 describes all subgroups of M_{24} of prime order and their normalizers. Similarly Section 47 contains the corresponding information for J_2. However, some work remains to be done for Suz, Co_1, and He. We complete the discussion of the subgroups of prime order for these three groups in this chapter.

50. Subgroups of Suz of prime order

In this section we assume G is Suz. Further we continue the hypotheses and notation of Sections 46 and 48. From 48.7, G has order $2^{13} \cdot 3^7 \cdot 5^2 \cdot 7 \cdot 11 \cdot 13$.

By 48.4.1, G has two classes of involutions. Their centralizers are described in 46.1 and 48.3. The subgroups of G of order 5 and their normalizers are described in 48.8. The normalizer of a subgroup of order 7 is described in 48.6.

According to Table Suz, G should have three classes 3_A, 3_B, and 3_C of subgroups of order 3. The class 3_A is Δ and the normalizer of a member of Δ is described in 46.6. The normalizer of a group in 3_C is described in 48.18. Let $X = \langle x \rangle \in \Delta$ and $Y = \langle y \rangle \in \Delta(X)$; then $\langle xy \rangle \in 3_B$. To verify that $N_G(\langle xy \rangle)$ is as described in Table Suz, using 48.9.2, we only need to show $N_G(\langle xy \rangle) \leq N_G(\{X,Y\})$. If not, $xy = x_1 y_1$ for some $X_1 = \langle x_1 \rangle \in \Delta$ and some $Y_1 = \langle y_1 \rangle \in \Delta(X_1)$ with $\{X,Y\} \neq \{X_1,Y_1\}$. Let $M = C_G(X)$ and $M^* = M/X$, so that $M^* \cong U_4(3)$ and $y^* = (xy)^*$ is a long root element of M^*. Then $xy \in C_M(X_1)$. But by 48.9, $C_M(X_1)^*$ contains no long root elements unless $X_1 \in \Delta(X)$ or $\Delta^2(X)$.

Suppose $X_1 \in \Delta^2(X)$. Then by 48.9, $C_M(X_1)^* = C_M(z)^*$, where z is the involution in $\langle X, X_1 \rangle$. Then z centralizes xy and X, so z centralizes Y. Then $Y \in C_{\Delta(X)}(z) = C_{\Delta(X_1)}(z)$, contradicting $xy \in C(X_1)$.

So $X_1, Y_1 \in \Delta(X,Y)$. But now from 48.9, $x_1 y_1 \notin XY$, contradicting $xy = x_1 y_1$.

So the normalizers of our three classes of subgroups of order 3 are as claimed; it remains to show that each subgroup of order 3 is in one of these three classes. So let B be of order 3. As M contains a Sylow 3-subgroup P of G we may take $B \leq M$. Now $N_G(X)^* \cong \mathbf{Z}_2/U_4(3)$ has three classes of elements of order 3 corresponding to elements with Jordon block structure J_2, J_1^2; J_2^2; and J_3, J_1 on the natural module for $SU_4(3)$. Here J_j^i indicates i Jordon blocks of size j. Elements of the first two types are fused under M^* into $J(P^*) = J(P)^*$ and by 48.5, each subgroup of $J(P)$ of order 3 is in 3_A or 3_B. Thus we may take B^* of type J_3, J_1. But then the three subgroups of order 3 in $BX - X$ are fused in M, so all such subgroups of order 3 must be in 3_C, and our proof that G has just three classes of subgroups of order 3 is complete.

Next we observe

Lemma 50.1: *Let $g \in Co_1$ be of order 13. Then*

(1) $C_{\tilde{\Lambda}}(g) = 0$, *and*

(2) g *centralizes no element of order 11 in* Co_1.

Proof: By 23.3, Co_1 has three orbits on the points of $\tilde{\Lambda}$ with stabilizers Co_2, Co_3, and $M_{24}/E_{2^{11}}$. Hence, using, for example, 22.15, 13 does not divide the order of the stabilizer of a point of $\tilde{\Lambda}$, so (1) holds.

Next the minimal dimension of a nontrivial $GF(2)\langle g \rangle$-module is 12, so $\tilde{\Lambda}$ is the sum of two irreducible $\langle g \rangle$-modules and hence $C_{GL(\tilde{\Lambda})}(g) \cong$ $GL_2(2^{12})$, whose order is not divisible by 11. Thus (2) also holds.

We have seen that the order of centralizers of elements of prime order $p < 11$ in G are not divisible by 11 or 13, and by 50.1 there is no element of order $13 \cdot 11$ in G, so elements of order 11 and 13 are self-centralizing. Therefore the normalizer of a Sylow p-subgroup P of G is a Frobenius group with kernel P for $p = 11$ and 13; further Sylow's Theorem gives us the order of that normalizer, completing the verification of Table *Suz*.

51. Subgroups of Co_1 of prime order

In this section we assume G is Co_1 and continue the hypotheses and notation of Sections 46 and 49. By 49.6, G has order $2^{21} \cdot 3^9 \cdot 5^4 \cdot 7^2 \cdot 11 \cdot 13 \cdot 23$.

By 49.4, G has three classes of involutions and the centralizers of these involutions are described in 46.1, 46.13, and 49.5.

We next consider the subgroups of order 3. First $\Delta = 3_B$ and the normalizer of a member of Δ is described in 46.6. Next 3_C consists of the elements $\langle xy \rangle$, where $X = \langle x \rangle \in \Delta$ and $Y = \langle y \rangle \in \Delta(X)$. We argue as in the previous section that $N_G(\langle xy \rangle) \le N_G(\{X, Y\})$, so $N_G(\langle xy \rangle)$ is as described in 49.8.1. Namely if $M = C_G(X)$, $M^* = M/X$, and $xy = x_1 y_1$ with $X_1 = \langle x_1 \rangle \in \Delta$ and $Y_1 = \langle y_1 \rangle \in \Delta(X_1)$ then $xy \in C_M(X_1)$. Now if $X_1 \in \Delta_i(X)$ for $i = 3$ or 4 then $(xy)^* = y^* \in C_M(X_1)^* \cong J_2$ or $G_2(4)$. But then $Y \le C_M(X_1)$ while $X \cap C_M(X_1) = 1$, contradicting $xy \in C_M(X_1)$. Then we complete the proof as in the previous section.

The subgroups of type 3_D are described in 49.12. Thus it remains to show there exists a subgroup of type 3_A, and that each subgroup of order 3 is in one of these four classes.

Let $P_H \in Syl_3(H)$. Then $\Delta \cap P_H = \{X_1, \ldots, X_4\}$. Let $a = x_1 \cdots x_4$ with $X_i = \langle x_i \rangle$. Then $C_Q(a) = \langle z \rangle$ and $C_{H^*}(a) \cong SU_4(2)$, with a inverted in H. Further if $P_H \le P \in Syl_3(G)$ and $J = J(P)$ then as a is the product of generators of four members of $\Delta \cap J$ and since as a module for $N_G(J)/J \cong M_{12}/Z_2$, J is an image of the module induced from the sign character for $Z_2 \times M_{11}$ on $J \cap \Delta$, a is 3-central in $N_G(J)$, so we can take $a \in Z(P)$. Indeed z is a square in $N_G(J) \cap C_G(a)$, so by Exercise 16.6.3, $C_H(a) \cong Sp_4(3) \times Z_3$.

Let $T \in Syl_2(C_H(a))$; by Exercise 16.6, $T \cong Q_8 wr Z_2$, so $\langle z \rangle = Z(T)$ and hence $T \in Syl_2(C_G(a))$. We claim z is weakly closed in T with respect to $C_G(a)$. Each involution in $T - \langle z \rangle$ is fused into $Z_2(T)$ under $C_H(a)$, while if $z \ne t \in Z_2(T)$ then $T_0 = C_T(t) \cong Q_8 \times Q_8$. Thus if

$h \in C_G(a)$ with $t^h = z$ then $C_T(t)^h = C_{T^h}(z)$ and conjugating in $C_H(a)$, we may assume $h \in N_G(T_0)$. This is impossible as $\langle z \rangle$ is characteristic in T_0.

Therefore z is weakly closed in T. So by Glauberman's Z^*-Theorem [Gl], $C_G(a) = KC_H(a)$, where $K = O(C_G(a))$. Now z inverts $K/\langle a \rangle$ so by a Frattini argument $C_H(a)^\infty \cong Sp_4(3)$ is faithful on $P_1 \in Syl_p(K)$ for each prime divisor p of $|K|$. However, if $p \neq 3$ then $|P_1| \leq p^4$ with equality only when $p = 5$, so $Sp_4(3)$ is not faithful on P_1. That is, $K = O_3(C_G(a))$. Therefore as $P \leq C_G(a)$, $|K| = 3^5$, and then as $C_H(a)$ is faithful on K, $K \cong 3^{1+4}$. Therefore $N_G(\langle a \rangle)$ has the structure described for a member of 3_A and we take $3_A = \langle a \rangle^G$.

It remains to show G has just four classes of subgroups of order 3. First from the structure of $N_G(\langle a \rangle)$, $N_G(J) \cap N_G(\langle a \rangle)$ is of order $2^4 \cdot 3^9$, so $|\langle a \rangle^{N_G(J)}| = 220$. But by 49.7, $|\Delta \cap J| = 12$ and $|3_C \cap J| = 2\binom{12}{2} = 132$, so all 364 subgroups of order 3 of J are accounted for and hence the subgroups of J of order 3 are in 3_A, 3_B, or 3_C.

Next each element of order 3 in $C_G(a)$ is fused into JK under $C_G(a)$. Further each subgroup of order 3 in JK/J is 3-central in $N_G(J)/J$, and hence is fused under $N_G(J)$ into $C_G(X)$. Then from the discussion in Section 50, each subgroup of $C_G(X)$ of order 3 is fused into J or to a subgroup V with V^* of type 3_C in $M^* = C_G(X)/X \cong Suz$. But all subgroups of order 3 in $VX - X$ are conjugate in $C_G(X)$ and from 49.12.3 each is in 3_D. So our treatment of subgroups of order 3 is complete.

Next we consider subgroups of order 5. First $\Omega_8^+(2)$ has three classes of subgroups of order 5 permuted transitively by triality with centralizer $\mathbf{Z}_5 \times A_5$. Thus H also has three classes, which we write as 5_A, 5_B, and 5_C. We choose notation so that $C_Q(B) \cong Q_8 D_8$ for $B \in 5_B$. Then by 49.1, there is a subgroup $(L_1 \times L_2)\langle \alpha \rangle$ with $L_1 \cong A_5$, $L_2 \cong J_2$, α inducing an outer automorphism on L_1 and L_2 with $B \leq L_1$, and $N_G(B) \leq L_1 L_2 \langle \alpha \rangle$. That is, B has the normalizer described in Table Co_1 for a subgroup of type 5_B.

This leaves $\langle z \rangle = C_Q(D)$ for D in 5_A and 5_C, so $C_H(D) \cong \mathbf{Z}_5 \times (A_5/\mathbf{Z}_2)$.

Let $B \leq J \in Syl_5(L_1 L_2)$. Then $J \cong E_{125}$. Now $N_{L_2}(J)$ contains an element of order 3 whose unique fixed point on J is B. Further $J \trianglelefteq P \in Syl_5(G)$ and $J = C_P(B)$. As the same holds for each $B_1 \in 5_B \cap J$, we conclude $N_G(J)$ is transitive on $5_B \cap J$ and $|5_B \cap J| \equiv 10 \mod 15$.

Next by 26.5 we may choose notation so that for $D \in 5_C$, $C_G(D) \cong \mathbf{Z}_5 \times (A_5 wr \mathbf{Z}_2)$. Then we take $J \in Syl_5(C_G(D))$ and a Sylow 2-group of $N_G(J) \cap N_G(D)$ is isomorphic to $\mathbf{Z}_4 \times D_8$ and fixes only the point D

in J. So as in the previous paragraph, we conclude $N_G(J)$ is transitive on $5_C \cap J$ and $|5_C \cap J| \equiv 5 \mod 10$.

Similarly $|5_A \cap J| \geq 6$, so as J has 31 points we conclude $|5_B \cap J| = 10$ and then as $|N_G(J) \cap N_G(B)| = 2^4 \cdot 3 \cdot 5^3$, we have $|N_G(J)| = 2^5 \cdot 3 \cdot 5^4$. Then as $|N_G(J) \cap N_G(D)| = 2^5 \cdot 5^3$ we get $|J \cap 5_C| = 15$. This leaves $J \cap 5_A$ to be an orbit of $N_G(J)$ of order 6.

Now $N_G(J) \cap N_G(D)$ contains an element of order 4 inducing scalar action on J, so $N_G(J)$ induces a 2-transitive group of order 120 on the six points of $J \cap 5_A$, which must then be $PGL_2(5)$. Therefore

Lemma 51.1: *Let* $P \in Syl_5(G)$. *Then*

    (1) $J = J(P) \cong E_{125}$, $N_G(J)$ is the split extension of J by $\mathbf{Z}_4 \times PGL_2(5)$, and J has the structure of a 3-dimensional orthogonal space over $GF(5)$ preserved by $N_G(J)/J$ acting as its group of similarities.

    (2) $J \cap 5_A$ is the set of six singular points of J and $|J \cap 5_B| = 10$ and $|J \cap 5_C| = 15$.

    (3) $N_G(P)$ is the split extension of P by $\mathbf{Z}_4 \times \mathbf{Z}_4$.

Indeed an element of order 5 in $P - J$ has one Jordon block of size 3 on J, so $J = J(P)$. Also $N_G(P)$ has two orbits on subgroups of order 3 in $P - J$ of length 25 and 100.

Next we choose $A \in 5_A \cap J$ to be centralized by z and z to act on P. As $N_G(P) \cap C(A) \cong \mathbf{Z}_4$, $C_H(A) \cong \mathbf{Z}_5 \times SL_2(5)$. Then by the Brauer–Suzuki Theorem (cf. 12.1.1 in [Go]), $C_G(A) = KC_H(A)$, where $K = O(C_G(A))$. Now z inverts K/A, so each element $k \in K$ of prime order $p \neq 5$ is in the center of K. Further $P_K = P \cap K \cong 5^{1+2}$, so P_K is transitive on the five subgroups of order 5 in $P \cap J$ distinct from A. As J has the structure of an orthogonal space this says $J \cap K = A^\perp$ contains subgroups D in 5_C. But $C_G(AD)$ is of order 25 from the structure of $C_G(D)$, so $K \cong 5^{1+2}$ and $N_G(A)$ is transitive on subgroups of order 5 in K distinct from A.

In particular the orbit of $N_G(P)$ on $P - J$ of length 25 is contained in 5_C. We claim the orbit of length 100 is in 5_A or 5_C. This will complete our analysis of subgroups of order 5 by showing each such subgroup is in 5_A, 5_B, or 5_C.

So assume the orbit of length 100 is not as claimed. For each F in this orbit, $N_P(AF)$ is transitive on the five points of $AF - A$, so as J has the structure of an orthogonal space, F is not of type 5_B. Thus F is in some fourth class, and $F^G \cap P = F^{N_G(P)}$. We claim $N_G(F) \leq N_G(P)$. By the Frobenius p-complement Theorem (cf. 39.4 in [FGT]), $C_G(F) = (L \times F)A$, where $L = O_{5'}(C_G(F))$. Notice $C_L(A) = 1$. Hence as $A = [A, t]$ for t an involution inverting AF, Sylow p-subgroups of L are noncyclic for each prime divisor p of $|L|$. This forces $p = 2, 3$. But elements of order 2 and 3 centralize only subgroups of order 5 in 5_I for $I = A, B, C$. For example, each subgroup of order 5 centralizing an element of order 3 also centralizes an involution; we've seen the statement for z, and it is easy to check for the other two classes of involutions.

So indeed $N_G(F) \leq N_G(P)$ and hence F is contained in a unique Sylow 5-subgroup of G. To complete the proof let $T = Z_2(P)F$, so that $T \cong 5^{1+2}$ has a unique subgroup A in 5_A and five subgroups in 5_C (those in $Z_2(P)$ distinct from A), and the remaining twenty-five subgroups of order 5 are conjugate to F under P. In particular as F is in a unique Sylow 5-subgroup of G, J is the unique member of J^G fixed by T. Also the five members $J' \in J^{N_G(A)} - \{J\}$ are the conjugates of J satisfying $N_T(J') = A$.

Next each $D \in T \cap 5_C$ is in 36 members of J^G, so there are $5 \cdot 35 \equiv 50 \mod 125$ members of J^G containing a member of $T \cap 5_C$ but not A. Finally $N_T(J^h) = D \not\leq J^h$ if and only if $N_{J^h}(D)$ is one of the 71 conjugates A^k of A under $N_G(D)$ distinct from A. Then $D \leq O_5(C_G(A^k))$ and there are five such conjugates of J under $C_G((AD)^k)$ fixed by D, so there are $71 \cdot 25 \equiv 25 \mod 125$ such members of J^G. Hence

$$|J^G| \equiv 1 + 5 + 50 + 25 = 81 \mod 125.$$

But $|J^G| = |G : N_G(J)| = 2^{16} \cdot 3^8 \cdot 7^2 \cdot 11 \cdot 13 \cdot 23 \equiv 56 \mod 125$, a contradiction.

This completes our discussion of subgroups of order 5. We next turn to subgroups of order 7. By 49.12 and 26.2 there is $L \leq G$ with $E(L) = L_1 \times L_2$, $L_1 \cong L_2(7)$, $L_2 \cong A_7$, and $L = E(L)\langle\alpha\rangle$ with α an involution inducing an outer automorphism on L_1 and L_2. Further if $P_i \in Syl_7(L_i)$ then $N_G(P_i) \leq L$. Let $P = P_1 P_2$, so that $P \in Syl_7(G)$; observe that $N_L(P)/P \cong \mathbf{Z}_6 \times \mathbf{Z}_3$.

Now $|N_G(P) : P| \equiv |G : P| \equiv 2 \mod 7$, so $|N_G(P) : N_L(P)| \equiv 4 \mod 7$. On the other hand a subgroup Z of order 6 in $N_L(P)$ induces scalar action on P and $N_G(P)/PZ \leq PGL_2(7)$ with $N_L(P)/PZ$ of order 3 and $N_G(P)/PZ$ a $\{2, 3\}$-group, so we conclude $N_G(P)/PZ \cong A_4$. Therefore

Lemma 51.2: *For $P \in Syl_7(G)$, $N_G(P)/P \cong SL_2(3) \times \mathbf{Z}_3$.*

In particular $N_G(P)$ has two orbits of order 4 on points of P, so P_1 and P_2 are representatives of the conjugacy classes of subgroups of G of order 7.

Next let I be a root A_4-subgroup of G and $S \in Syl_{13}(C_G(I))$. From 49.5 and the structure of $G_2(4)$, $C_G(\langle u, S \rangle) = US$ for $1 \neq u \in U = O_2(I)$. Hence by Exercise 16.6 in [FGT], either $N_G(S) \leq N_G(I)$ or $C_G(S) \cong S \times A_5$. The latter is impossible as no element of order 5 centralizes an element of order 13. This takes care of subgroups of order 13.

Let $X \in \Delta$ and $V \in Syl_{11}(C_G(X))$. Then $C_G(XV) = XV$ and V is centralized by a dodecad involution d inverting X. From 46.13, $\langle d \rangle V = C_G(\langle d \rangle V)$. Therefore $C_G(V) = K \langle d \rangle$, where $K = O(C_G(V))$, and d inverts K/V. In particular $X \leq Z(K)$, so $K \leq C_G(XV) = XV$. This completes our discussion of subgroups of order 13.

Finally subgroups of order 23 can be handled using Sylow's Theorem.

52. Subgroups of prime order in *He*

In this section G is He and we adopt the hypotheses and notation of Chapter 14. By 42.13, $|G| = 2^{10} \cdot 3^3 \cdot 5^2 \cdot 7^3 \cdot 17$.

By 40.6, G has two classes of involutions. The centralizers of involutions are described in 39.1, 42.5, and 42.7.

By 42.16 there is X of order 3 in G with $C_G(X) \cong A_7/\mathbf{Z}_3$ quasisimple and $N_G(X)/X \cong S_7$. Thus $X^G = 3_A$. By 42.16.2, the involutions in $C_G(X)$ are in 2_B while an involution $v \in N_G(X)$ with $C_G(\langle v \rangle X) \cong \mathbf{Z}_2 \times S_4$ is in 2_A.

Next by 43.1.5 there is a subgroup $L_1 \times L_2$ with $L_1 \cong S_4$, $L_2 \cong L_3(2)$, and involutions in L_2 in 2_A. Thus if $Y \in Syl_3(L_1)$, then $Y \notin X^G$, so we take $Y^G = 3_B$. Let $M = N_G(Y)$ and $K = Y \langle t \rangle \times L_2$, where t is an involution in L_1 inverting Y. We may take z to be an involution in L_2; then $C_H(Y) \cong D_8$, so $C_M(z) \leq K$ and as z^K is the set of involutions in $C_K(Y)$, $z^M \cap K = z^K$, so z fixes a unique point of M/K. Thus if $M \neq K$, we can apply Exercise 2.10. Now by the construction of 43.1, $t \in 2_B$ and $N_G(\langle t \rangle Y) = N_K(\langle t \rangle Y)$. Let U be maximal in the set $\mathcal{U}$ of Exercise 2.10; as K has more than one class of involutions, $U \neq 1$, so up to conjugation in K, $U = \langle tz \rangle$. But then condition (2c) of Exercise 2.10 is violated. So $M = K$ and $N_M(Y)$ has the desired structure.

Next for $P \in Syl_3(N_G(X))$, $3^{1+2} \cong P$, and $P \in Syl_3(G)$. As $N_G(X)/X \cong S_7$, there are three $N_G(X)$-classes of subgroups of order 3 in P, and X is fused into one of these classes in the centralizer of a 2_B-involution, so G has two classes of subgroups of order 3. This completes our analysis of subgroups of order 3.

Next 42.14 supplies us with a subgroup K of index 2 in $S_5 wr \mathbf{Z}_2$. Let K_1 and K_2 be the components of K and $P \in Syl_5(K)$. Then

$N_K(P)/P \cong \mathbf{Z}_4 * D_8$, so $|N_K(P)| \equiv 1 \mod 5$. But $|N_G(P) : P| \equiv |G : P| \equiv 3 \mod 5$, so $|N_G(P) : N_K(P)| \equiv 3 \mod 5$. Next $N_K(P)$ has a subgroup $Z \cong \mathbf{Z}_4$ inducing scalars on P, so $N_G(P)/PZ \leq PGL_2(5)$. As $N_K(P)/PZ \cong E_4$ and $|N_G(P) : N_K(P)| \equiv 3 \mod 5$, we conclude $N_G(P)/P \cong \mathbf{Z}_4 * SL_2(3)$ or the multiplicative group of $GF(25)$ extended by a field automorphism. The latter case is out as a Sylow 2-group is $Z_4 * D_8$. Thus

Lemma 52.1: *For* $P \in Syl_5(G)$, $N_G(P)/P \cong \mathbf{Z}_4 * SL_2(3)$.

In particular $N_G(P)$ is transitive on the points of P, so G has one class of subgroups of order 5. This completes our analysis of such subgroups.

Recall our subgroup $L_1 \times L_2$; by construction in 43.1, $L_1 = \langle U, U^h \rangle$ for some root 4-subgroup U of G. Then $L_2 \leq C_G(U) \cap C_G(U^h)$. Indeed if $V \in Syl_7(L_2)$ then $N_{L_2}(V) = VX$, where $X \in 3_A$ is faithful on V and $VX = C_G(S)$, where $S_4 \cong S$ with $U = O_2(S)$ and S contains a member of U^h. Therefore $J = \langle S, L_1 \rangle \leq C_G(VX) \leq C_G(X) \cong A_7/\mathbf{Z}_3$. Further, from the structure of $N_G(U)$, $N_G(U) \cap C_G(V) = VS$, so from the subgroup structure of A_7 we conclude $JX/X \cong A_6$ or $L_3(2)$. The former is impossible as the preimage of A_6 in $C_G(X)$ is quasisimple whereas $X \nleq C_G(V)$. Therefore $J \cong L_3(2)$.

Let $M = C_G(V)$ and $M^* = M/V$. As U is weakly closed in a Sylow 2-subgroup of $N_M(U)$, that Sylow group is Sylow in M; notice it is D_8. Similarly if $Y \in Syl_3(S)$ then $Y \in 3_B$ and $YV = C_G(YV)$, so $Y \in Syl_3(M)$. Therefore $|M^*| = 2^3 \cdot 3 \cdot 7^a \cdot 17^b$ with $a = 1, 2$ and $b = 0, 1$. But $N = N_{M^*}(Y) \cong S_3$ so $1 \equiv |M^* : N| \equiv 2^b \mod 3$, and hence $b = 0$. That is, $|M : JV| = 1$ or 7. The latter is impossible as 7^2 does not divide the order of A_7. Therefore $N_G(V) = XV \times J$ and we take $V^G = 7_B$. Notice the involutions centralizing V are in 2_B and the subgroups of order 3 are in 3_B.

Let $Z \in Syl_7(J)$; then $VZ \trianglelefteq P \in Syl_7(G)$ and VZ admits the subgroup XY with orbits of length 1, 1, 3, 3 on the points of VZ. $N_G(VZ)$ is not transitive on the eight points of VZ as $Z(P)$ is not conjugate to V, so $N_G(VZ) \leq N_G(P)$. As V and Z are the fixed points of XY it follows that $Z = Z(P)$. Further there is i of order 3 in XY inducing scalars on VZ, so $C_P(i)$ is a complement to VZ in P and $P \cong 7^{1+2}$. By a Frattini argument, $N_G(VZ) = P(N_G(VZ) \cap N_G(V)) = PXY$. By Sylow's Theorem, $|N_G(P) : P| \equiv |G : P| \equiv 4 \mod 7$, so as $N_G(VZ) = PXY$, we conclude

Lemma 52.2: *Let* $P \in Syl_7(G)$. *Then* $P \cong 7^{1+2}$ *and* $N_G(P)$ *is the split extension of* P *by* $\mathbf{Z}_3 \times S_3$.

In particular $N_G(P)$ has orbits of length 1, 14, 21, 21 on the subgroups of P of order 7.

Now Z is centralized by $X \in 3_A$ and hence by 42.16.2, XZ is inverted by a conjugate v of z. That is, $N_G(P) = PXY\langle v \rangle$. Further subgroups of order 7 centralizing involutions in 2_B are in 7_B, while subgroups of order 7 in H are not inverted in H, so $C_G(Z) = O(C_G(Z))$. As $C_G(XZ) = XZ$, $X \in Syl_3(C_G(Z))$. As elements of order 5 don't centralize elements of order 7, $C_G(Z)$ has order prime to 5. Finally as X doesn't centralize an element of order 17, $C_G(Z)$ has order prime to 17. Therefore $N_G(Z) = N_G(P)$, and we take $7_A = Z^G$.

Let $W = C_P(v)$; then $W\langle v \rangle = C_G(W\langle v \rangle)$, so $C_G(W) = O(C_G(W))\langle v \rangle$ with v inverting $O(C_G(W))/W$. Thus $N_G(W) \leq N_G(ZW) \leq N_G(Z)$, and we set $7_C = W^G$.

It remains to show G has just three classes of subgroups of order 7. Now Z, V, W are representatives for the orbits of $N_G(P)$ on subgroups of P of order 7 of length 1, 14, 21, respectively, so it remains to show the second orbit $D^{N_G(P)}$ of length 21 is in 7_A. Assume not; as D is inverted in $N_G(P)$, $D \notin 7_B$ or 7_C, so $D^{N_G(P)} = D^G \cap P$. Elements of order 2, 3, 5 do not centralize members of D^G, so $N_G(D) \leq N_G(ZD) \leq N_G(Z)$, and hence each member of $DZ^{\#}$ is in a unique Sylow 7-group of G. Therefore $|P^G| \equiv 1 \mod 49$. But $|P^G| = |G : N_G(P)| \equiv 22 \mod 49$, completing our treatment of subgroups of order 7.

The normalizer of a subgroup of order 17 can now be determined using Sylow's Theorem. This completes our discussion of subgroups of He of prime order.

Table M_{24}

$$G = M_{24}; \ |G| = 2^{10} \cdot 3^3 \cdot 5 \cdot 7 \cdot 11 \cdot 23$$

Normalizers of Subgroups of Prime Order

2A	$L_3(2)/D_8^3$
2B	S_5/E_{64}
3A	$S_6/Z_3 = Z_2/\hat{A}_6$
3B	$S_3 \times L_3(2)$
5A	$Z_2/(D_{10} \times A_4)$
7A	$S_3 \times (Z_3/Z_7)$
11A	Z_{10}/Z_{11}
23A	Z_{11}/Z_{23}

Table J_2

$$G = J_2; \; |G| = 2^7 \cdot 3^3 \cdot 5^2 \cdot 7$$

Normalizers of Subgroups of Prime Order

$2A$	A_5/Q_8D_8
$2B$	$E_4 \times A_5$
$3A$	$PGL_2(9)/\mathbf{Z}_3 = \mathbf{Z}_2/\hat{A}_6$
$3B$	$S_3 \times A_4$
$5A$	$D_{10} \times A_5$
$5B$	$D_{10} \times D_{10}$
$7A$	$\mathbf{Z}_6/\mathbf{Z}_7$

Table Co_1

$$G = Co_1; \; |G| = 2^{21} \cdot 3^9 \cdot 5^4 \cdot 7^2 \cdot 11 \cdot 13 \cdot 23$$

Normalizers of Subgroups of Prime Order

$2A$	$\Omega_8^+(2)/Q_8^4$
$2B$	$\mathbf{Z}_2/(E_4 \times G_2(4))$
$2C$	$Aut(M_{12})/E_{2^{11}}$
$3A$	$GSp_4(3)/3^{1+4}$
$3B$	$\mathbf{Z}_2/Suz/\mathbf{Z}_3$
$3C$	$E_4/U_4(3)/E_9$
$3D$	$S_3 \times A_9$
$5A$	$GL_2(5)/5^{1+2}$
$5B$	$\mathbf{Z}_2/(D_{10} \times J_2)$
$5C$	$\mathbf{Z}_2/(D_{10} \times (A_5 wr \mathbf{Z}_2))$
$7A$	$\mathbf{Z}_2/((\mathbf{Z}_3/\mathbf{Z}_7) \times L_3(2))$
$7B$	$\mathbf{Z}_2/((\mathbf{Z}_3/\mathbf{Z}_7) \times A_7)$
$11A$	$(\mathbf{Z}_{10}/\mathbf{Z}_{11}) \times S_3$
$13A$	$\mathbf{Z}_2/((\mathbf{Z}_6/\mathbf{Z}_{13}) \times A_4)$
$23A$	$\mathbf{Z}_{11}/\mathbf{Z}_{23}$

Table *He*

$$G = He; \; |G| = 2^{10} \cdot 3^3 \cdot 5^2 \cdot 7^3 \cdot 17$$

Normalizers of Subgroups of Prime Order

$2A$	$L_3(2)/D_8^3$
$2B$	$\mathbf{Z}_2/L_3(4)/E_4$
$3A$	$S_7/\mathbf{Z}_3 = \mathbf{Z}_2/\hat{A}_7$
$3B$	$S_3 \times L_3(2)$
$5A$	$\mathbf{Z}_2/(D_{10} \times A_5)$
$7A$	$(\mathbf{Z}_3 \times S_3)/7^{1+2}$
$7B$	$L_3(2) \times (\mathbf{Z}_3/\mathbf{Z}_7)$
$7C$	$\mathbf{Z}_3/(\mathbf{Z}_7 \times D_{14})$

Table *Suz*

$$G = Suz; \; |G| = 2^{13} \cdot 3^7 \cdot 5^2 \cdot 7 \cdot 11 \cdot 13$$

Normalizers of Subgroups of Prime Order

$2A$	$\Omega_6^-(2)/Q_8^3$
$2B$	$\mathbf{Z}_2/(E_4 \times L_3(4))$
$3A$	$\mathbf{Z}_2/U_4(3)/\mathbf{Z}_3$
$3B$	$(SL_2(3) * D_8)/3^{2+4}$
$3C$	$Q_8/(E_9 \times A_6)$
$5A$	$\mathbf{Z}_2/(D_{10} \times A_6)$
$5B$	$\mathbf{Z}_2(D_{10} \times A_5)$
$7A$	$\mathbf{Z}_2/((\mathbf{Z}_3/\mathbf{Z}_7) \times A_4)$
$11A$	$\mathbf{Z}_{10}/\mathbf{Z}_{11}$
$13A$	$\mathbf{Z}_6/\mathbf{Z}_{13}$

List of symbols

Bibliography

[A1] M. Aschbacher, Flag structures on Tits geometries, *Geom. Ded.* **14** (1983), 21–32.

[A2] M. Aschbacher, Overgroups of Sylow subgroups of sporadic groups, *Memoirs AMS* **342** (1986), 1–235.

[A3] M. Aschbacher, Chevalley groups of type G_2 as the group of a trilinear form, *J. Alg.* **109** (1987), 193–259.

[A4] M. Aschbacher, The geometry of trilinear forms, *Finite Geometries, Buildings, and Related Topics*, Oxford University Press, Oxford, 1990, pp. 75–84.

[A5] M. Aschbacher, A characterization of some finite subgroups of characteristic 3, *J. Alg.* **76** (1982), 400–441.

[AS1] M. Aschbacher and Y. Segev, Extending morphisms of groups and graphs, *Ann. Math.* **135** (1992), 297–323.

[AS2] M. Aschbacher and Y. Segev, The uniqueness of groups of type J_4, *Invent. Math.* **105** (1991), 589–607.

[AS3] M. Aschbacher and Y. Segev, Locally connected simplicial maps, *Israel J. Math.* **77** (1992), 285–303.

[ASe] M. Aschbacher and G. Seitz, Involutions in Chevalley groups over fields of even order, *Nagoya Math. J.* **63** (1976), 1–91.

[B] R. Bruck, *A Survey of Binary Systems*, Springer-Verlag, New York, 1971.

[BF] R. Brauer and K. Fowler, On groups of even order, *Ann. Math.* **62** (1955), 565–83.

[Bu] F. Buekenout, Diagrams for geometries and groups, *J. Com. Th. Ser. A* **27** (1979), 121–51.

[Ch] C. Chevalley, Sur certains groupes simples, *Tohoku Math. J.* **7** (1955), 14–66.

[CN] J. Conway and S. Norton, Monstrous moonshine, *Bull. London Math. Soc.* **11** (1979), 308–39.

[Co1] J. Conway, Three lectures on exceptional groups, *Finite Simple Groups*, Academic Press, London, 1971, pp. 215–47.

[Co2] J. Conway, A group of order 8,315,553,613,086,720,000, *Bull. London Math. Soc.* **1** (1969), 79–88.

[Co3] J. Conway, A simple construction of the Fischer-Griess monster group, *Invent. Math.* **79** (1985), 513–40.

[Cu] R. Curtis, On the Mathieu group M_{24} and related topics, Thesis: U. Cambridge (1972).

[CW] J. Conway and D. Wales, Construction of the Rudvalis group of order 145,926,144,000, *J. Alg.* **27** (1973), 538–48.

[F] B. Fischer, Finite groups generated by 3-transpositions, *Invent. Math.* **13** (1971), 232–46.

[FGT] M. Aschbacher, *Finite Group Theory*, Cambridge University Press, Cambridge, 1986.

[FLM] I. Frenkel, J. Lepowski, and A. Meurman, *Vertex Operator Algebras and the Monster*, Academic Press, San Diego, 1988.

[G] M. Golay, Notes on digital coding, *Proc. IRE* **37** (1949), 657.

[Gl] G. Glauberman, Central elements of core-free groups, *J. Alg.* **4** (1966), 403–20.

[Go] D. Gorenstein, *Finite Groups*, Harper and Row, New York, 1968.

[Gr1] R. Griess, A construction of F_1 as automorphisms of a 196,883 dimensional algebra, *Proc. Nat. Acad. Sci. USA* **78** (1981), 689–91.

[Gr2] R. Griess, The Friendly Giant, *Invent. Math.* **69** (1982), 1–102.

[GMS] R. Griess, U. Meierfrankenfeld, and Y. Segev, A uniqueness proof for the Monster, *Ann. Math.* **130** (1989), 567–602.

[HW] M. Hall and D. Wales, The simple group of order 604,800, *J. Alg.* **9** (1968), 417–450.

[Ha] K. Harada, On the simple group F of order $2^{14} \cdot 3^6 \cdot 5^6 \cdot 7 \cdot 11 \cdot 19$, *Proceedings of the Conference on Finite Groups*, Academic Press, New York, 1976, pp. 119–95.

[He] D. Held, The simple group related to M_{24}, *J. Alg.* **13** (1969), 253–79.

[Hi] D. Higman, Finite permutation groups of rank 3, *M. Zeit.* **86** (1964), 145–56.

[HiG] G. Higman, On the simple group of D. G. Higman and C. C. Sims, *Illinois J. Math.* **7** (1963), 79–96.

[HM] G. Higman and J. MacKay, On Janko's simple group of order 50,232,969, *Bull. London Math. Soc.* **1** (1969), 89–94.

[HS] D. Higman and C. Sims, A simple group of order 44,352,00, *M. Zeit.* **105** (1968), 110–13.

[J1] Z. Janko, A new finite simple group with abelian 2-Sylow subgroups and its characterization, *J. Alg.* **3** (1966), 147–86.

[J2] Z. Janko, Some new simple groups of finite order, I, *Symposia Math.* **1** (1968), 25–65.

[J4] Z. Janko, A new finite simple group of order 86,775,571,046,077,562,880 which possesses M_{24} and the full cover of M_{22} as subgroups, *J. Alg.* **42** (1976), 564–96.

[JT] Z. Janko and J. Thompson, On a class of finite simple groups of Ree, *J. Alg.* **4** (1966), 274–92.

[Le1] J. Leech, Some sphere packings in higher space, *Canadian J. Math.* **16** (1964), 657–82.

[Le2] J. Leech, Notes on sphere packings, *Canadian J. Math.* **19** (1967), 251–67.

[LS] J. Leon and C. Sims, The existence and uniqueness of a simple group generated by $\{3, 4\}$-transpositions, *Bull. AMS* **83** (1977), 1039–40.

[Ly] R. Lyons, Evidence for a new finite simple group, *J. Alg.* **20** (1972), 540–69.

[M1] E. Mathieu, Mémoire sur le nombre de valeurs que peut acquérir une fonction quand on y permut ses variables de toutes les manières possibles, *J. de Math. Pure et App.* **5** (1860), 9–42.

[M2] E. Mathieu, Mémoire sur l'étude des fonctions de plusieures quantités, sur la manière des formes et sur les substitutions qui les laissent invariables, *J. de Math. Pure et App.* **6** (1861), 241–323.

308   *Bibliography*

[M3] E. Mathieu, Sur la fonction cinq fois transitive des 24 quantités, *J. de Math. Pure et App.* **18** (1873), 25–46.

[Mc] J. McLaughlin, A simple group of order 898,128,000, *Theory of Finite Groups*, Benjamin, New York, 1969, pp. 109–111.

[Mi1] G. Miller, On the supposed five-fold transitive function of 24 elements and 19!/48 values, *Mess. Math.* **27** (1898), 187–90.

[Mi2] G. Miller, Sur plusieurs groupes simples, *Bull. Soc. Math. de France* **28** (1900), 266–7.

[N] S. Norton, The construction of J_4, *Proc. Sym. Pure Math.* **37** (1980), 271–8.

[ON] M. O'Nan, Some evidence for the existence of a new simple group, *Proc. London Math. Soc.* **32** (1976), 421–79.

[P] N. Patterson, On Conway's group ·0 and some subgroups, Thesis: U. Cambridge (1974).

[Pa] N. Paige, A note on the Mathieu groups, *Canadian J. Math.* **9** (1957), 15–18.

[Ph] K. Phan, A characterization of the finite simple group $U_4(3)$, *J. Australian Math. Soc.* **10** (1969), 77–94.

[PW1] N. Patterson and S. Wong, A characterization of the Suzuki sporadic simple group of order 448,345,497,600, *J. Alg.* **39** (1976), 277–86.

[PW2] N. Patterson and S. Wong, The nonexistence of a certain simple group, *J. Alg.* **39** (1976), 138–49.

[Q] D. Quillen, Homotopy properties of the coset of nontrivial p-subgroups, *Adv. in Math.* **28** (1978), 101–28.

[RS] M. Ronan and S. Smith, 2-local geometries for some sporadic groups, *Proc. Sym. Pure Math.* **37** (1980), 283–289.

[Se] Y. Segev, On the uniqueness of the Harada-Norton group, *J. Alg.* **151** (1992), 261–303.

[Si1] C. Sims, On the isomorphism of two groups of order 44,352,000, *Theory of Finite Groups*, Benjamin, New York, 1969, pp. 101–8.

[Si2] C. Sims, The existence and uniqueness of Lyons' group, *Gainesville Conference on Finite Groups*, North-Holland, Amsterdam, 1973, pp. 138–41.

[Sm] S. Smith, Large extraspecial subgroups of width 4 and 6, *J. Alg.* **58** (1979), 251–280.

[Sp] E. Spanier, *Algebraic Topology*, McGraw-Hill, New York, 1966.

[St] B. Stellmacher, Einfache Gruppen, die von einer Konjugiertenklasse von Elementen der Ordnung drei erzeugt werden, *J. Alg.* **30** (1974), 320–54.

[Su] M. Suzuki, A simple group of order 448,345,497,600, *Theory of Finite Groups*, Benjamin, New York, 1969, pp. 113–9.

[T1] J. Tits, A local approach to buildings, *The Geometric Vein*, Springer-Verlag, 1981, pp. 517–47.

[T2] J. Tits, On R. Griess' Friendly giant, *Invent. Math.* **78** (1984), 491–9.

[Th1] J. Thompson, A simple subgroup of $E_8(3)$, *Finite Groups Symposium*, Japan Soc. for Promotion of Science, Tokyo, 1976, pp. 113–6.

[Th2] J. Thompson, The uniqueness of the Fischer-Griess monster, *Bull. London Math. Soc.* **11** (1979), 340–6.

[Tm] F. Timmesfeld, Finite simple groups in which the generalized Fitting group of the centralizer of some involution is extra-special, *Ann. Math.* **107** (1978), 297–369.

[To1] J. Todd, On representations of the Mathieu groups as collineation groups, *J. London Math. Soc.* **34** (1959), 406–16.

[To2] J. Todd, A representation of the Mathieu group M_{24} as a collineation group, *Annali di Math. Pure ed App.* **71** (1966), 199–238.

[Wa1] H. Ward, On Ree's series of simple groups, *Trans. AMS* **121** (1966), 62–89.

[Wa2] H. Ward, Combinatorial polarization, *Discrete Math.* **26** (1979), 185–97.

[Wa3] H. Ward, Multilinear forms and divisors of codeword weights, *Quart. J. Math.* **34** (1983), 115–28.

[W1] E. Witt, Die 5-fach transitiven Gruppen von Mathieu, *Abl. Math. Hamburg* **12** (1938), 256–64.

[W2] E. Witt, Über Steinersche Systeme, *Abl. Math. Hamburg* **12** (1938), 265–74.

Index